# Nanomedicine in Brain Diseases

Xue Xue
Editor

# Nanomedicine in Brain Diseases

## Principles and Application

Springer

*Editor*
Xue Xue
State Key Laboratory of Medicinal
Chemical Biology
Nankai University
Tianjin, Tianjin, China

ISBN 978-981-13-8733-3 ISBN 978-981-13-8731-9 (eBook)
https://doi.org/10.1007/978-981-13-8731-9



This Springer imprint is published by the registered company Springer Nature Singapore Pte Ltd.
The registered company address is: 152 Beach Road, #21-01/04 Gateway East, Singapore 189721, Singapore

# Preface

Nanotechnology had matured significantly during the last two decades as it has transitioned from benchtop science to applied technology. Understanding how the intrinsic properties of nanomaterials translate to complex biological responses is an area of intensive research. The challenge for nanomedicine is to utilize nanotechnology and develop innovative biocompatible nanomaterials that mimic tissue characteristics, cause minimal inflammation and cell loss, and have the abilities to function for a relative long period of time. Nevertheless, the application of nanomaterials to basic and clinical neuroscience is only at an early stage. This book is to overview the current strategies to use nanomaterials for brain disease and to introduce several novel nanomaterials that have excellent potentials in this field. With multiple functionalized modifications, nanomaterials are easy to traverse blood-brain barrier, and they are capable of diagnosing or treating brain diseases. Among these nanomaterials, nanozymes reduce oxidative stress via their intrinsic enzyme-like activation. Magnetic nanomaterials are vital contrast agents for MRI with their excellent magnetic orientation ability. Biomacromolecule-based nanomaterials greatly improve therapeutic efficiency through their self-assembly ability. Carbon-based nanomaterials are capable of forming a variety of special structures due to their strength and toughness, which makes them become feasible carrier for traditional drugs. Polymeric nanomaterials possess various properties that are widely used in diagnosing and treating brain disease. This book provides researchers, graduate students, and clinic practitioners with a cutting-edge and comprehensive summary of research on nanomedicine and their potential applications in clinic. I sincerely thank all the chapter authors for their great contribution to the book.

Tianjin, China Xue Xue

# Acknowledgments

We wish to express our appreciation to the editorial staff members at Springer Nature for their professional and courteous guidance. Uma Maheswari Srinivasan, editorial project manager, offered assistance throughout the entire process of this project, and Lewis Liu also provided help for us. Of course, our greatest gratitude should be given to our colleagues in the past year covering these ten chapters. Without their hard work and enthusiasm, this contributed volume would have not been possible.

# Contents


**1 Introduction: Nanomedicine in the Brain** . . . . . 1
Tian-Qi Li, Li-Wen Huang, and Xue Xue
1.1 Introduction . . . . . 1
1.2 Application of Nanomedicine . . . . . 3
1.2.1 Magnetic-Based Nanomaterials . . . . . 5
1.2.2 Carbon-Based Nanomaterials . . . . . 5
1.2.3 Silicon-Based Nanomaterials . . . . . 6
1.2.4 Lipid-Based Nanomaterials . . . . . 7
1.2.5 Rare-Earth Elements-Based Nanomaterials . . . . . 7
1.2.6 Quantum Dots . . . . . 8
1.2.7 Metallic Organic Framework Compounds . . . . . 8
1.2.8 Others . . . . . 9
1.3 Overview of Neuropsychiatric Disease . . . . . 10
1.3.1 Neurological Disease . . . . . 11
1.3.2 Psychiatric Disease . . . . . 17
1.4 Conclusion and Perspective . . . . . 19
References . . . . . 21

**2 The Strategies of Nanomaterials for Traversing Blood-Brain Barrier** . . . . . 29
Mohd Ahmar Rauf, Fawad Ur Rehman, Meng Zheng, and Bingyang Shi
2.1 Introduction . . . . . 30
2.2 BBB Structure and Passage Mechanism . . . . . 31
2.3 Different Strategies for Targeting of Drugs to the Brain . . . . . 34
2.4 Convection-Enhanced Delivery (CED) . . . . . 34
2.5 Intranasal Delivery to Bypass BBB . . . . . 35
2.6 Employment of NPs for Drug Delivery into the Brain . . . . . 36
2.7 Aspects Affecting the Entry of NPs over the BBB . . . . . 37

2.8 Different Types of Nanoparticles Employed to Cross BBB . . . . 39
2.8.1 Liposomes . . . . . . . . . . . . . . . . . . . . . . . . . . . . . . . 39
2.8.2 Lipid-Based NPs . . . . . . . . . . . . . . . . . . . . . . . . . . . 41
2.8.3 Polymer-Based Nanocarriers Across the BBB . . . . . . . 41
2.9 Employment of Different siRNA-Based Deliveries . . . . . . . . . 43
2.9.1 Exosome-Based Delivery . . . . . . . . . . . . . . . . . . . . . 44
2.9.2 Employment of Cell-Penetrating Peptides (CPPs) . . . . 45
2.9.3 Other Types of NPs . . . . . . . . . . . . . . . . . . . . . . . . 45
2.10 Mechanism of Nanoparticle Route to BBB . . . . . . . . . . . . . . 47
2.11 Neurotoxicity of the Nanoparticles . . . . . . . . . . . . . . . . . . . 49
2.12 A Future Point of View . . . . . . . . . . . . . . . . . . . . . . . . . . 50
2.13 Conclusion . . . . . . . . . . . . . . . . . . . . . . . . . . . . . . . . . . 50
References . . . . . . . . . . . . . . . . . . . . . . . . . . . . . . . . . . . . . 51

**3 The Strategies of Nanomaterials for Drug Delivery and Release** . . . . . . . . . . . . . . . . . . . . . . . . . . . . . . . . . . . 59
Jinjin Wang, Qianqian Huang, and Xing-Jie Liang
3.1 Introduction . . . . . . . . . . . . . . . . . . . . . . . . . . . . . . . . . . 59
3.2 Nanopreparations for Brain Diseases . . . . . . . . . . . . . . . . . . 61
3.2.1 Construction of Nanocarriers . . . . . . . . . . . . . . . . . . 61
3.2.2 Types of Nanopreparations for Brain Disease . . . . . . . 62
3.3 Advances in Drug Delivery and Release Approaches for Brain Diseases . . . . . . . . . . . . . . . . . . . . . . . . . . . . . . 66
3.3.1 Controlled Nano Drug Delivery Systems . . . . . . . . . . 66
3.3.2 Targeted Nano Drug Delivery Systems . . . . . . . . . . . 66
3.3.3 Smart Response Nano Drug Delivery and Release Systems . . . . . . . . . . . . . . . . . . . . . . . . . 72
3.3.4 Intranasal Drug Delivery Systems . . . . . . . . . . . . . . 75
3.4 Conclusion . . . . . . . . . . . . . . . . . . . . . . . . . . . . . . . . . . 76
References . . . . . . . . . . . . . . . . . . . . . . . . . . . . . . . . . . . . . 76

**4 The Strategies of Nanomaterials for Therapy** . . . . . . . . . . . . . . . . 83
Yang Du, Shuying Wang, Fangyuan Li, and Daishun Ling
4.1 Introduction . . . . . . . . . . . . . . . . . . . . . . . . . . . . . . . . . . 83
4.2 Nanomaterials for Brain Tumor Treatment . . . . . . . . . . . . . . 84
4.2.1 Nanomaterial-Based Chemotherapy . . . . . . . . . . . . . . 84
4.2.2 Nanomaterial-Based Gene Therapy . . . . . . . . . . . . . . 85
4.2.3 Nanomaterial-Based Thermotherapy . . . . . . . . . . . . . 87
4.2.4 Nanomaterial-Based Photodynamic Therapy . . . . . . . . 88
4.2.5 Nanomaterial-Based Immunotherapy . . . . . . . . . . . . . 89
4.3 Nanomaterials for Alzheimer's Disease Treatment . . . . . . . . . 89
4.3.1 Nanomaterials for Drug Delivery . . . . . . . . . . . . . . . 90
4.3.2 Nanomaterial-Based Metal Chelation Strategy . . . . . . 91
4.3.3 Nanomaterials for Antioxidant . . . . . . . . . . . . . . . . . 93
4.3.4 Nanomaterial-Based Gene Therapy . . . . . . . . . . . . . . 94

4.3.5 Nanomaterial-Based Immunotherapy . . . . . . . . . . . . . . 95
4.4 Nanomaterials for Parkinson's Disease Treatment . . . . . . . . . . 96
4.4.1 Nanomaterials for Dopamine Replacement . . . . . . . . . 97
4.4.2 Nanomaterials for Dopaminergic Agonist Delivery . . . 97
4.4.3 Nanomaterials for Growth Factor and Peptides Delivery . . . . . . . . . . . . . . . . . . . . . . . . . . . . . . 99
4.4.4 Nanomaterial-Based Gene Therapy . . . . . . . . . . . . . . 100
4.5 Nanomaterials for Stroke Treatment . . . . . . . . . . . . . . . . . . . . 101
4.5.1 Nanomaterials for Thrombolysis . . . . . . . . . . . . . . . . 102
4.5.2 Nanomaterials for Neuroprotection . . . . . . . . . . . . . . . 103
4.5.3 Nanomaterial-Based Gene Therapy . . . . . . . . . . . . . . 105
4.6 Conclusion . . . . . . . . . . . . . . . . . . . . . . . . . . . . . . . . . . . 106
References . . . . . . . . . . . . . . . . . . . . . . . . . . . . . . . . . . . . . . 106

**5 Overcoming the Physiopathologic Barriers: Nanoprobes-Mediated Intracranial Glioma Imaging** . . . . . . . . . . . . 115
Heng Liu, Yu Liu, Fengyuan Man, and Gang Liu
5.1 Introduction . . . . . . . . . . . . . . . . . . . . . . . . . . . . . . . . . . . 116
5.2 Critical Biological Challenges Facing Intracranial Glioma . . . . 117
5.3 Strategies for Bypassing and Crossing BBB . . . . . . . . . . . . . . 118
5.4 Advantages of Nanomaterials for Glioma Imaging . . . . . . . . . . 119
5.5 Applications of Nanoprobes for Intracranial Glioma Imaging . . . . . . . . . . . . . . . . . . . . . . . . . . . . . . . . . . . . 120
5.5.1 Magnetic Resonance Imaging . . . . . . . . . . . . . . . . . . 121
5.5.2 Photoacoustic Imaging . . . . . . . . . . . . . . . . . . . . . . . 125
5.5.3 Fluorescence Imaging . . . . . . . . . . . . . . . . . . . . . . . 127
5.5.4 Multimodal Imaging . . . . . . . . . . . . . . . . . . . . . . . . 129
5.5.5 Intraoperative Glioma Margin Delineation . . . . . . . . . 132
5.6 Challenges and Perspectives . . . . . . . . . . . . . . . . . . . . . . . . . 134
References . . . . . . . . . . . . . . . . . . . . . . . . . . . . . . . . . . . . . . 135

**6 The Advances of Nanozyme in Brain Disease** . . . . . . . . . . . . . . . . 139
Ruofei Zhang, Xiyun Yan, and Kelong Fan
6.1 Introduction . . . . . . . . . . . . . . . . . . . . . . . . . . . . . . . . . . . 139
6.2 ROS-Mediated Oxidative Stress and Its Role in Neurological Diseases . . . . . . . . . . . . . . . . . . . . . . . . . . . 141
6.3 Nanozymes with Antioxidant Activity . . . . . . . . . . . . . . . . . . 144
6.3.1 Nanozymes with SOD-Like Activity . . . . . . . . . . . . . 145
6.3.2 Nanozymes with Catalase-Like Activity . . . . . . . . . . . 150
6.3.3 Nanozymes with Multiple Antioxidant Enzyme-Like Activities . . . . . . . . . . . . . . . . . . . . . . 155
6.4 Nanozymes Used for the Treatment of Neurological Diseases . . . . . . . . . . . . . . . . . . . . . . . . . . . . . . . . . . . . . . 158
6.4.1 Application of Nanozymes in Neuroprotection . . . . . . 158
6.4.2 Application of Nanozymes in Alzheimer's Disease . . . 163

6.4.3 Application of Nanozymes in Cerebral Ischemia . . . . . 168
6.4.4 Application of Nanozymes in Parkinson's Disease . . . 170
6.4.5 Application of Nanozymes in Multiple Sclerosis . . . . . 171
6.5 Challenges and Perspectives . . . . . . . . . . . . . . . . . . . . . . . . . . 171
References . . . . . . . . . . . . . . . . . . . . . . . . . . . . . . . . . . . . . . . . 173

**7 The Advances of Biomacromolecule-based Nanomedicine in Brain Disease** . . . . . . . . . . . . . . . . . . . . . . . . . . . . . . . . . . . . 181
Yuhua Weng and Yuanyu Huang
7.1 Introduction . . . . . . . . . . . . . . . . . . . . . . . . . . . . . . . . . . . . 181
7.2 Key Constraint Factors for Biomacromolecule Delivery into the Brain . . . . . . . . . . . . . . . . . . . . . . . . . . . . . . . . . . . . 182
7.2.1 Blood-Brain Barrier (BBB) . . . . . . . . . . . . . . . . . . . . 182
7.2.2 Blood Cerebrospinal Fluid Barrier (BCFB) and Blood Tumor Barrier (BTB) . . . . . . . . . . . . . . . . 183
7.2.3 Drug Diffusion in the Brain . . . . . . . . . . . . . . . . . . . 183
7.3 Strategies for Biomacromolecular Nanomedicine Delivery into the Brain . . . . . . . . . . . . . . . . . . . . . . . . . . . . . . . . . . . . 184
7.3.1 Direct Injections . . . . . . . . . . . . . . . . . . . . . . . . . . . 184
7.3.2 Convection-Enhanced Delivery (CED) . . . . . . . . . . . 185
7.3.3 Intranasal Administration . . . . . . . . . . . . . . . . . . . . . 185
7.3.4 Oral Administration . . . . . . . . . . . . . . . . . . . . . . . . 186
7.3.5 BBB Disruption . . . . . . . . . . . . . . . . . . . . . . . . . . . 186
7.3.6 Ultrasound and Microbubble with Nanoparticles . . . . . 187
7.3.7 Targeted Brain Delivery . . . . . . . . . . . . . . . . . . . . . . 187
7.4 Preclinical and Clinical Advances of Biomacromolecular Nanomedicines for Brain Disease Treatment . . . . . . . . . . . . . 188
7.4.1 Protein-Based Nanomedicine . . . . . . . . . . . . . . . . . . 188
7.4.2 Enzyme-Based Nanomedicine . . . . . . . . . . . . . . . . . 189
7.4.3 Peptide-Based Nanomedicine . . . . . . . . . . . . . . . . . . 192
7.4.4 Antibody-Based Nanomedicine . . . . . . . . . . . . . . . . . 193
7.4.5 Nucleic Acid-Based Nanomedicine . . . . . . . . . . . . . . 196
7.5 Concerns over Biomacromolecular Nanomedicines . . . . . . . . . 199
7.6 Conclusion and Prospective . . . . . . . . . . . . . . . . . . . . . . . . . . 200
References . . . . . . . . . . . . . . . . . . . . . . . . . . . . . . . . . . . . . . . . 200

**8 Carbon-Based Nanomedicine** . . . . . . . . . . . . . . . . . . . . . . . . . . 209
Peng Zhang, Ming Zhang, and Jia Geng
8.1 Introduction . . . . . . . . . . . . . . . . . . . . . . . . . . . . . . . . . . . . 209
8.2 Application . . . . . . . . . . . . . . . . . . . . . . . . . . . . . . . . . . . . . 210
8.2.1 Drug Carrier . . . . . . . . . . . . . . . . . . . . . . . . . . . . . . 210
8.2.2 Gene Carrier . . . . . . . . . . . . . . . . . . . . . . . . . . . . . . 211
8.2.3 Photodynamic Therapy (PDT) and Photothermal Therapy (PTT) . . . . . . . . . . . . . . . . . . . . . . . . . . . . . . 213
8.2.4 Imaging Agent . . . . . . . . . . . . . . . . . . . . . . . . . . . . . 214

8.3 Pharmacodynamics and Metabolism . . . . . 217
8.3.1 Delivery to the Brain . . . . . 217
8.3.2 Toxicological Characteristics . . . . . 218
8.3.3 Drug Release . . . . . 219
8.4 Diagnosis and Treatment . . . . . 221
8.4.1 Neurodegenerative Disease . . . . . 221
8.4.2 Tumor . . . . . 221
8.4.3 Acute Hyperglycemia . . . . . 223
8.5 Biodistribution of Carbon Nanomaterials . . . . . 224
8.5.1 Brain Distribution . . . . . 224
8.5.2 Body Distribution . . . . . 226
8.6 Summary . . . . . 228
References . . . . . 228

**9 Polymeric Nanomedicine** . . . . . 233
Yu Zhao, Chunxiong Zheng, and Yang Liu
9.1 Introduction: Polymeric Nanomedicine and Polymeric Nanoparticles . . . . . 233
9.2 Various Polymeric Nanoparticles . . . . . 235
9.2.1 Polymeric Micelles . . . . . 235
9.2.2 Core-Shell Nanocarriers . . . . . 237
9.2.3 Nanospheres . . . . . 239
9.2.4 Nanocapsules . . . . . 242
9.2.5 Dendrimers . . . . . 243
9.2.6 Nanogels . . . . . 245
9.3 The Application of Polymeric Nanomedicine in Brain Disease . . . . . 246
9.3.1 Brain Cancers . . . . . 246
9.3.2 Alzheimer's Disease . . . . . 250
9.3.3 Cerebral Amyloid Angiopathy (CAA) . . . . . 255
9.3.4 Parkinson's Disease (PD) . . . . . 256
9.3.5 Stroke . . . . . 259
9.3.6 Multiple Sclerosis . . . . . 262
References . . . . . 262

**10 Magnetic Nanomedicine** . . . . . 269
M. Zubair Iqbal, Gohar Ijaz Dar, Israt Ali, and Aiguo Wu
10.1 Nanomaterials . . . . . 270
10.2 Magnetic Nanoparticles . . . . . 271
10.3 Synthesis Method of MNPs . . . . . 275
10.3.1 Ball Milling Method . . . . . 276
10.3.2 Chemical Methods . . . . . 278
10.3.3 Coprecipitation . . . . . 278
10.3.4 Thermal Decomposition . . . . . 279
10.3.5 Hydrothermal Synthesis . . . . . 280

10.4 Magnetic Nanoparticles as Hyperthermia Agent . . . . . . . . . . . 283
10.4.1 Parameters Affecting the Heating Efficiency . . . . . . . . 284
10.5 Magnetic Nanoparticles for Drug Delivery . . . . . . . . . . . . . . . 289
10.5.1 Drug Delivery Mechanism . . . . . . . . . . . . . . . . . . . . 290
10.5.2 MNPs in Drug Delivery . . . . . . . . . . . . . . . . . . . . . 292
10.6 Magnetic Resonance Imaging (MRI) . . . . . . . . . . . . . . . . . . . 295
10.6.1 MRI Contract Agents . . . . . . . . . . . . . . . . . . . . . . . 296
10.6.2 The Longitudinal Relaxation ($T_1$) Agents . . . . . . . . . . 297
10.6.3 The Transverse Relaxation (T2) Agents . . . . . . . . . . . 299
10.7 Conclusion and Prospective . . . . . . . . . . . . . . . . . . . . . . . . . 300
References . . . . . . . . . . . . . . . . . . . . . . . . . . . . . . . . . . . . . 302

# Chapter 1
# Introduction: Nanomedicine in the Brain

**Tian-Qi Li, Li-Wen Huang, and Xue Xue**

**Abstract** In recent years, as the incidence of brain diseases in the population has gradually increased, the treatment of neuropsychiatric diseases cannot be ignored. There are many pathogenesis of neuropsychiatric diseases, such as neuronal damage, intercellular signaling disorder, and inflammatory reactions. On the other hand, blood-brain barrier (BBB) as the protective layer of the brain, to some extent, hinders the release and delivery of conventional drugs. Nanomedicines, with many excellent physiochemical properties, such as multiple modifications and surface functionalization, have attracted widespread attentions in the scientific community. In this chapter, we provide a broad overview of brain diseases and summarize the applications of nanomedicine in neuropsychiatric disorders, as well as challenges and prospects for future research. We hope that this chapter will enable readers to have a new understanding of brain diseases and nanomedicine and to promote the developments of nanomedicine in brain diseases in clinic.

**Keywords** Nanomaterials · Nanomedicine · Neuropsychiatric disease · Neurological disease · Psychiatric disease

## 1.1 Introduction

Neuropsychiatric disease is a heterogeneous mental disorder including neurological disease and psychiatric disease, which induces serious personal suffering and economic loss. Myriad genetic, immunological, and environmental factors may contribute to individual's susceptibility to neuropsychiatric disease [1]. The pathological characteristics of neuropsychiatric disease contain abnormal protein aggregation, cytokine imbalance, reactive oxygen species overproduction, neuronal degeneration, structural plasticity alternation, etc. [2]. Neurological diseases, including

T.-Q. Li · L.-W. Huang · X. Xue (✉)
State Key Laboratory of Medicinal Chemical Biology, College of Pharmacy, Nankai University, Tianjin, People's Republic of China
e-mail: xuexue@nankai.edu.cn



X. Xue (ed.), *Nanomedicine in Brain Diseases*,
https://doi.org/10.1007/978-981-13-8731-9_1

Alzheimer's disease (AD), Parkinson's disease (PD), ischemia stroke, multiple sclerosis (MS), and glioblastoma, are always accompanied by atrophy of the cortex and hippocampus and lead to the feeling and movement abnormality in the pathogenesis. In the courses of neurological diseases, dementia of AD patients develops gradually along with the loss of neurons. Muscles of PD patients gradually stiffen and vibrate involuntarily together with the increase of α-synuclein in intracellular fibers. One side of limbs of stroke patient gradually loses exercise capacity with the increase of thrombus. Psychiatric disease, including depression, schizophrenia, and autism spectrum disorder, commonly shows a decrease in cortical thickness and surface area and obvious anomaly in mental activities. In the process of psychiatric diseases, the patients' conditions gradually deteriorate accompanying with the decrease of neurotransmitter level until death. Therefore, a large amount of time and money has been put into the research and development of drugs, and many drugs are being marketed for clinical application. However, there are still a number of disadvantages of traditional drugs, which restrict the applications of them. For the traditional treatment of neuropsychiatric disease, various drugs are limited by the poor drug pharmacokinetic and side effects including nonspecific targeting (e.g., proteins, peptides, siRNAs, etc.) [3]. The ability of traditional drugs to interfere with disease progression is still inefficient, and its exercise in the diagnosis of neuropsychiatric disorders is not well developed. In general, drugs enter into patients by oral administration, parenteral administration, intranasal administration, and transdermal administration. Once drugs get into the body, they go through absorption, distribution, metabolism, and excretion stages. However, drugs taken by oral administration are metabolized and inactivated in the gastrointestinal tract, intestinal mucosa, and liver, reducing the effective drug doses entering into the systemic circulation. On the other hand, traditional drugs are isolated from disease area in some ways owing to the protection effect of BBB in preventing pathogen invasion and maintaining brain homeostasis. After drugs enter the bloodstream, they combine with plasma proteins to facilitate the transportation to tissues or organs. In this process, drugs that pass through the BBB become active conditions to bind biomacromolecular targets such as proteins and nucleic acids. Above all, traditional drugs playing therapeutic roles nonspecifically not only decrease the drug dose of target site but inhibit the growth of normal cells for the treatment of brain tumor. In depression, multiple-dose regimen and side effects of drugs make the treatment effect of drugs greatly reduce [4, 5]. The water solubility and drug resistance affect the efficacy of targeted therapeutic agent that further limits the utilization of drugs in clinic. Except these disadvantages, the biocompatibility and instability are also unsatisfactory for the clinical application.

Nanotechnology brings new opportunities by addressing the limitations of neuropsychiatric drugs. Nanotechnology as a multidisciplinary research field, including physics, chemistry, biology, and so on, involves the manipulation of materials at the atomic and molecular level with at least one dimension within 100 nm [6]. Nanotechnology provides various nanomaterials with extraordinary features for diagnostics and therapeutics of neuropsychiatric disease. The high stability, large payload, and flexible design make nanomaterial become a powerful drug delivery carrier to cross the BBB and work effectively [7]. Nanomaterials have higher cellular uptake compare to traditional drugs, which allows nanomaterials to target a wide range of both cellular and intracellular targets. The in vivo pharmacokinetics and efficiently specific target improve the efficacy, safety, sensitivity, and personalization of nanomaterials for molecular diagnostics, in vivo imaging, and neuropsychiatric disease therapy [8]. Among many kinds of nanomaterials, lipid-based nanoparticles, carbon-based nanoparticles, and magnetic nanoparticles are developed to diagnose and treat neuropsychiatric disease. Improved on these basic nanoparticles, the addition of other contents, such as polyethylene glycol (PEG) and poly(lactic-co-glycolic acid) (PLGA) [9], increases the circulation time and decreases the release rate of drugs. As part of traditional medicine lacks the ability to traverse the BBB, multi-modified nanomaterials are considered as promising medicine for brain disease. Moreover, other unique properties of nanomaterials have been developed, such as the superior resolution of imaging accompanied with photothermal effects. The development of nanotechnology offers a platform to combine diagnosis and therapy close. Nanomaterials having dual ability to imaging and delivery are capable to track the theranostics upon successful delivery. The advancements of nanotechnology have promising applications toward the diagnosis and treatment of neuropsychiatric disease. Understanding these functions of nanomedicine in neuropsychiatric disease may lead us to have a comprehensive view of the applications of nanomedicine in the brain. These books about nanomedicine in the brain will provide us new perspective for understanding neuropsychiatric disease and innovating novel nanotherapeutics.

In this chapter, we overview nanomedicine research which has great progress in recent years (Fig. 1.1). The appearance of nanomedicine brings deep insight to the diagnosis and therapies of brain disease; thus, we introduce the characteristics, pathologies, and mechanisms of neuropsychiatric disease in order to understand the design and development of nanomaterials helpfully.

## 1.2 Application of Nanomedicine

Nanomaterial plays an important role in the diagnosis and therapeutic of neuropsychiatric disease. Due to the size effect and multiple functionalization, nanomaterials are able to traverse the BBB, reach the target site to accomplish the delivery, and release of drugs accurately and efficiently. With the advantages of slow controlled release, nanocarrier avoids the problems of free drugs that sometimes are difficult to truly and effectively predict the efficacy in vivo and become a hot topic in the research of new drug delivery systems. With the flexible design, nanocarriers with various sizes, structures, and surface properties avoid the main problems faced by traditional drug carriers, including nonspecific biodistribution, metabolism, and excretion in the body, thus targeting diseased areas specifically [10]. Further, by surface functionalization, nanocarriers are able to control their pharmacokinetics and biodistribution properly.

On the other hand, nanomaterials also function as an agent of magnetic resonance imaging (MRI) for imaging the factors associating with neuroinflammation, thus

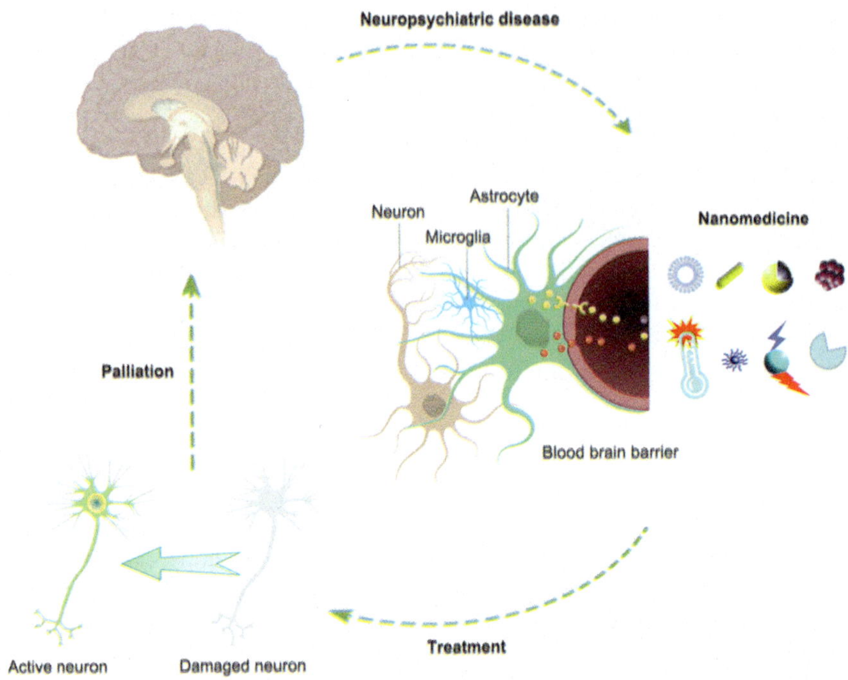


**Fig. 1.1** Function of nanomedicine. Neurons will lose activity after the inflammation happens in the brain, and then, various neuropsychiatric diseases appear. Nanomedicine will perform their functions to protect neurons and improve the symptoms of patients at this moment

diagnosing neuropsychiatric disease. The attractive features of light collection and optical amplification of nanomaterials make them apply in fluorescence imaging and phototherapy widely [11]. Near-infrared quantum dots and magnetic nanoparticles are two kinds of molecular probes that have been widely studied. As an agent of MRI, the size of nanomedicine is an influence factor affecting photothermal therapy. For large particles, long blood circulation is beneficial for enrichment at the target site, while for small particles, strong diffusion ability is beneficial for penetrating into target sites to get a homogeneous distribution. Hence, an applicable nanomedicine should have dimensional variability, which is able to obtain high enrichment in the form of large particles and then split into small particles to obtain high permeability. The targeted nanoprobes with small diameter, stable structure, good water solubility, good biocompatibility, and high affinity are used widely for the diagnosis and therapy of neuropsychiatric disease.

Except for the above applications, nanomaterials are also applied as drugs or enzymes on account of their intrinsic properties, for example, iron nanoparticles and cerium nanoparticles are able to impact redox reaction. In order to give readers a more comprehensive understanding of the application of nanomedicine in neurological diseases, this chapter initially introduces several types of nanomaterials that play an important role in neurological diseases.

### 1.2.1 Magnetic-Based Nanomaterials

The visible absorption bands of most magnetic nanomaterials are able to be activated by NIR, which enables magnetic nanoparticles to be imaging agents and phototherapy agents for the diagnosis and treatment of neuropsychiatric disease [12]. On account of the good tolerance of endogenous substances, iron oxide nanoparticles are used frequently in bioimaging and phototherapy, such as $Fe_3O_4$ and $\gamma Fe_2O_3$ [13]. As a representative MRI developer, the superparamagnetic phenomenon of magnetic nanoparticle occurs when the particle size is less than a certain critical value. In addition to the action of an external magnetic field, magnetic nanoparticles generate a strong local magnetic field that is able to affect hydrogen in the surrounding water molecules. The process of proton relaxation shortens the relaxation time of transverse relaxation (T2), thus exhibits a negative enhancement effect as a T2 developer, which is a potential clinical diagnostic of magnetic nanoparticles. SPIONs, as novel imaging agents providing safe and strong negative contrast enhancement of the target lesion in MRI, are able to track and visualize in vivo monocytes invading the brain. The magnetic moment and the targeted portion are designed by regulating their size, composition, and surface functionalization to further transfer various mechanical and chemical signals with high spatial and temporal precision [14]. Since, magnetocaloric therapy is applied in stimulating action potentials and controlling gene transcription remotely [15]. Antibody-functionalized magnetic nanoparticles catch target bio-molecules sensitively by immunomagnetic reduction, such as ultra-low concentration Aβ and tau protein in human plasma, which makes them break through the poor low-detection limit of enzyme-linked immunosorbent and be a potential method to differentiate PD from pervasive developmental disorder [16]. Drug-embedded acoustic droplets incorporating SPIONs allow both magnetism-assisted targeting and MRI-guided acoustic droplet vaporization, which is a potential theranostic strategy for tumor treatment.

### 1.2.2 Carbon-Based Nanomaterials

Carbon fiber is a one-dimensional carbon nanomaterial with high axial strength and modulus, low density, high specific performance, high temperature resistance, and good fatigue resistance. Due to the dielectric coefficient, the thermal expansion coefficient and the anisotropy, carbon fibers have good electrical conductivity, superior thermal conductivity and excellent electromagnetic shielding performance. Carbon-fiber microelectrodes with minimal tissue response permit longitudinal measurements of neurotransmission in single recording locations during behavior [17]. Graphene is a two-dimensional carbon nanomaterial with a hexagonal honeycomb lattice. On account of the high strength, large specific surface area, and excellent conductivity, graphene is considered to be a revolutionary material in biomedicine and drug delivery. Implantable graphene-based neural electrode interfaces provide reliable and precise validation for detecting functional changes of neuronal activities by measuring the concentration of $H_2O_2$ [18].

Carbon nanotubes (CNTs) are a special one-dimensional material, in which carbon atoms are mainly $sp^2$ hybridized, and the central carbon atoms are connected with three adjacent same carbon atoms to form a hexagonal network [19]. On one hand, CNTs have higher thermal conductivity comparing to the other one-dimensional materials. On the other hand, the manufacturing technique of CNTs is relatively uncomplicated comparing to the other two-dimensional carbon materials. In addition, CNTs have a series of attractive comprehensive properties, such as high mechanical strength, large specific surface area, high conductivity, strong interface effect, low thermal expansion coefficient, good chemical stability, and corrosion resistance. These excellent thermal and mechanical properties indicate that CNTs have the potential as superior heat conduction materials or fillers. CNTs are suitable for the recording of single neuronal activity, thereby reducing inflammation while stimulating neurons effectively [20]. Functional CNTs not only help to immobilize the required biomolecules for sensing applications but also improve their biocompatibility and dispersibility in biomolecular solutions. Functional single-walled carbon nanotubes (SWCNTs) restore normal autophagy by reversing the abnormal activation of mTOR signals and the defects of lysosomal protein hydrolysis, thereby protecting the nerves and treating AD [21]. Aggregated SWCNTs have proven effective in attenuating the behavioral and neurochemical effects of methamphetamine by oxidizing extracellular dopamine that increased in the striatum [22].

### *1.2.3 Silicon-Based Nanomaterials*

Silicon nanoparticle is a commonly used inorganic nanoparticle in MRI and photothermal therapy. As an indirect gap semiconductor, the excited state time of silicon is much longer than direct gap semiconductors such as CdS and CdSe. Spherical silicon nanoparticles ranging from 100 to 200 nm show strong magnetic dipole response in the visible spectrum [23]. Electrochemical etching of porous silicon shows considerable potential in application with high biodegradability, low toxicity, high drug-loading ratio, and inherent photoluminescence characteristics. NIR luminescent nanoparticles made of porous silicon not only possess these features but also are capable to be degraded into components cleared by the kidneys [24]. Hyperpolarized silicon particles have shown long spin-lattice relaxation times at room temperature, which makes them novel promising MRI probes [25]. The multi-gold nanorods crystal-seeded magnetic mesoporous silica nanobeads with multimodal imaging and excellent magnetic guidance are effective for tracking the location of stem cells in the short term and monitoring the long-term fate for the treatment of stroke [26]. Coaxial p-type/i-type/n-type silicon nanowires are sufficient to control neuronal activities, such as cell membrane potential and excitability, in a noninvasive, nongenetic, drug-like manner, and then treat neurological disease [27]. Besides, hyaluronic acid functionalized porous silicon nanoparticle combines with enzyme-responsive hydrogel and pH-responsive polymer is able to overcome the multiple obstacles for efficiently delivering drugs locally [28].

### *1.2.4 Lipid-Based Nanomaterials*

Liposome is a particle containing a solid hydrophobic lipid core, whose size range from 40 to 200 nm. The special structure makes it pass the liver and spleen filtration easily and successful escape from the reticuloendothelial system to the brain [29]. The ability of controlling drug release and improving drug stability and biocompatibility make it a safe nose-to-brain delivery system. After encapsulating neurotransmitters and growth factor, it are able to cure brain disease [30, 31]. The relative low cytotoxicity of liposomes is one of the most promising drug delivery systems [32]. The ability of adjustable size of lipid nanoparticles to enwrap hundreds of polymers at a controlled density allows many liposome products to be successfully transformed into clinical [33]. The rivastigmine liposomes and cell-penetrating peptide-modified liposomes are formulated for improving distribution and reducing the side effect paralleled with enhanced pharmacodynamics [34]. C-Phycocyanin-pertaining liposome and baicalin-loaded liposome have been approved to possess neuroprotective effect to prevent the damages of ischemia-reperfusion [35, 36]. Besides, the modification of PEG also brings high bioavailability for lipid nanoparticles [37, 38]. Liposomes encapsulating bexarotene alter α-helical conformation and inhibit γ-secretase activity by binding to the amyloid precursor protein transmembrane domain, thus achieving the purpose of interfering with Aβ metabolism [39].

### *1.2.5 Rare-Earth Elements-Based Nanomaterials*

Cerium oxide nanoparticle has recently been considered a kind of biomedical agents due to its potent regenerative and antioxidant properties. Cerium oxide nanoparticles are internalized by cells and accumulated at mitochondria in neurons in vivo. The oxidation of $Ce^{4+}$ to $Ce^{3+}$ is able to cause oxygen vacancies and detoxify peroxynitrite and protect neurons by redox reaction [40]. Ceria/polyoxometalates hybrid with both proteolytic and superoxide dismutase activities not only is capable to degrade Aβ aggregates and reduce intracellular reactive oxygen species but also inhibit Aβ-induced BV2 microglial cell activation [41]. The triphenylphosphonium-conjugated ceria nanoparticles possess the ability of mitigating reactive gliosis and damaged mitochondria [42]. Ceria-containing hydroxyapatite has the ability to upregulate oxidative stress markers and recover degenerated neurons in hippocampal and cerebral tissues [43]. The intrinsic properties of cerium oxide nanoparticles make them be internalized by cells and antioxidant easily without any modifications, which results in their excellent biocompatibility comparing with other PEG-modified nanomaterials.

Except for cerium nanoparticles, rare-earth elements are often made into upconversion nanoparticles (UCNPs), which possess single and variable absorption and emission centers. UCNPs with lanthanum ions doped are able to convert two or more low-energy photons into one high-energy photon because they have abundant electron transition in the 4f electron shell [44]. The optical properties of transforming near-infrared excitation into visible and ultraviolet radiation and converting low-energy laser stimulation into high-energy and long-lasting visible emissions of lanthanide-doped UCNPs make them superior in deep-tissue bioimaging [45]. The accuracy and safety of lanthanide-doped UCNPs allow them interact with neural tissue for a long time possibly [46]. For the therapy of brain disease, UCNPs are designed as NIR indicators for photothermal therapy and wavelength converter for photodynamic therapy [47]. On the other hand, molecularly tailored UCNPs are also used as indicators and phototriggers for drug delivery. Combined with near-infrared (NIR) light, NaYF4:Yb/Tm@SiO2 UCNPs cause a rapid increase in dopamine release by noninvasive deep brain stimulation [46]. Similarly, upconversion core-shell nanoparticles coated with mesoporous silica-incorporating dopamine produce a pronounced photothermal effect when they are activated [48].

### *1.2.6 Quantum Dots*

Quantum dots (QDs) represent the first generation of inorganic fluorescent nanoparticles used for fluorescent labeling [49]. There are couple of ways to prepare QDs. The first is depleting two-dimensional electronic gas locally by the nanometer gate electrode. Another is to create a layer of QDs as the semiconductor grows layer by layer [50]. As a semiconductor nanocrystal, QDs show superior fluorescence properties, such as dimensional adjustable emission, narrow symmetrical photoluminescence, wide and strong spectrum of excitation, strong luminescence, and robust optical stability. They are thus good candidates for optical biosensing. For example, QDs modified with transporter could pass through BBB and inhibit HIV-1 replication significantly in the brain. Thus, QDs are capable of integrating high-resolution imaging capability and therapeutic modalities to preserve the brain [51]. Superparamagnetic iron oxide nanoparticles (SPIONs) and quantum dots-loaded cationic polymersomes not only utilize the advantages of quantum dots and SPION but also facilitate a high payload capacity with the unique hollow structure. The new design brings high signal to background ratio and strong signal for long-term monitoring of stem cells via in vivo imaging for the treatment of stroke [52].

### *1.2.7 Metallic Organic Framework Compounds*

Metallic organic framework compounds (MOFs) have potential applications in biomedical diagnosis and drug delivery due to their pore size, central metal ions, and regulatory properties of organic ligands. In particular, enzyme and protein have been conjugated or loaded into their pores to enhance the activity, recyclability, and solvent compatibility by utilizing the three-dimensional structure, high specific surface area, multiple inner porosity, and high crystalline of MOFs [53]. The heterohybridization of MOFs is an effective multi-mode collaborative therapy because it combines the advantages of different materials and endows hybrid materials with new chemical and physiological characteristics. MOF with high biocompability is the point direction according to recent studies; therefore the constructed principle of MOFs is to produce the least toxic combination of inorganic and organic metals. MOFs encapsulating enzyme like an enzymatic nanofactory bring higher efficiency and longer intracellular activity compare to free enzyme [54]. Dual-emission fluorescent MOFs are able to detect pH sensitively to auxiliary diagnose brain cancer [55]. Besides, MOF-based nanoprobe inhibits Aβ aggregation and reduces Aβ-induced cytotoxicity successfully [56].

### *1.2.8 Others*

Polymeric nanoparticles include polymer-organic nanoparticles and polymer-inorganic nanoparticles. The biodegradable and biocompatible polymeric nanoparticles exhibit the superior physicochemical properties, such as enhanced immune escape, controlled drug release, and increased drug payload [57]. In addition, plasmonic nanoparticles have numerous advantages, such as good optical stability, tunable optical absorption and scattering, high electromagnetic field concentration, and high photothermal conversion efficiency [58]. Therefore, polymeric nanoparticle as an optically and electronically active nanomaterial holds promise for both biomedical imaging and drug delivery applications. Implantation of VEGF-releasing PLGA nanoparticles stimulates angiogenesis in the de novo tissue of stroke cavity, thereby promoting the recovery of stroke patients [59]. Poly-γ-glutamic acid nanoparticles and aluminum adjuvant are used as an adjuvant with a single dose of Japanese encephalitis virus-like particles to provide effective protection from Japanese encephalitis virus [60]. PEG-PLGA nanoparticles loaded with haloperidol exhibit good physical stability, great occlusive effect, and low side effects associated with antipsychotics [61]. Combined with magnetic nanomaterials, polymer-labeled magnetic nanoparticles are able to track and control the cellular uptake of nanomaterials with an external magnetic field [62] and then achieve real-time diagnostics and timely treatment. Except for directly providing diagnostic signal, polymeric nanoparticles are also involved in minimally invasive therapeutic technique: photothermal therapy. The addition of other abundant contents promotes polymeric nanoparticles to be stimuli-responsive drug delivery system to have diverse responses under different triggers, including pH, ATP, and ROS [63, 64].

With a host of unusual properties, including drug release modulation, light harvesting and emitting enhancement, and effective expenses with regard to manufacturing processes, nanomaterials with various structures have exhibited

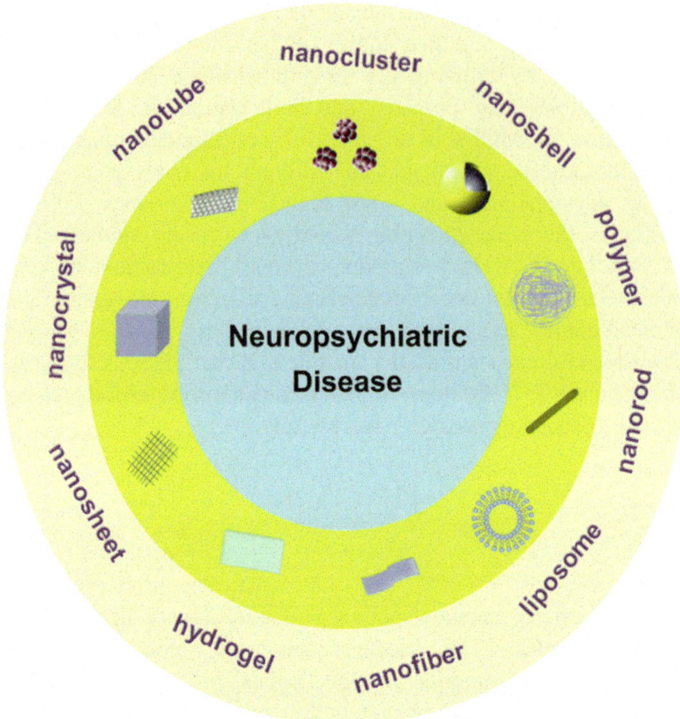


**Fig. 1.2** Various nanomaterials applying in treating neuropsychiatric disease

tremendous potential for diagnostics and therapeutic applications (Fig. 1.2). With the further understanding of nanomaterials, a series of nanomedicines possessing excellent properties are designed and developed to diagnose and treat neuropsychiatric disease. Numerous nanomedicines have so far shown striking therapeutic effects in disease models and become persuasive platform for inhibiting the development of a number of diseases. Despite the advancements described on above, in order to explore other potential applications, the intrinsic superior properties of nanomaterials are still a question worth devoting.

## 1.3 Overview of Neuropsychiatric Disease

The disruption of the early developmental processes of the neural circuit causes neurocognitive and neuropsychiatric dysfunction [65]. With aging, neurons become particularly vulnerable to disruption and then emerge a series of symptoms (Fig. 1.3) [66]. Exploration of indicators reflecting disease status and development of effective drugs in clinic are important for the treatment of neuropsychiatric disease [67]. We will overview the mechanisms of neuropsychiatric disease in the following.

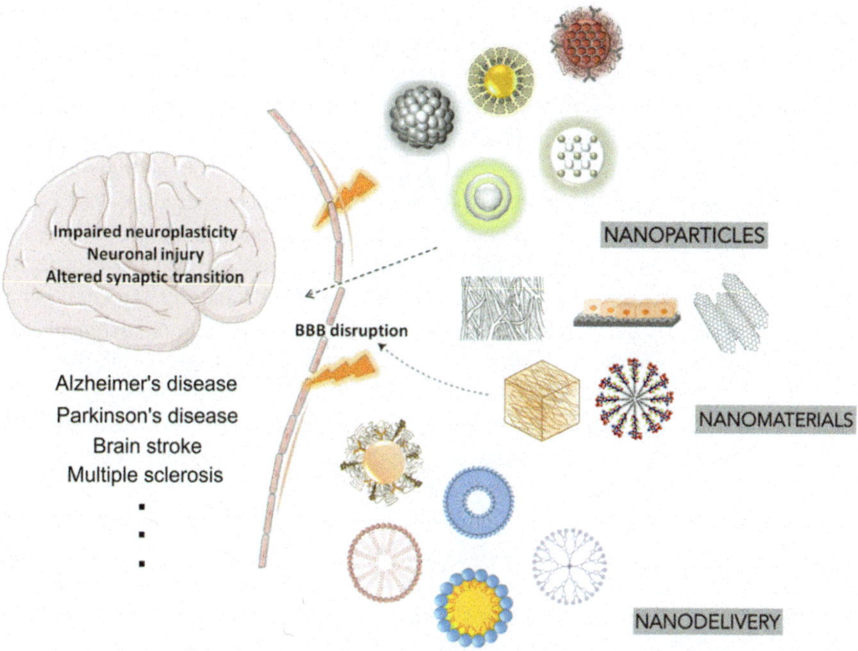


**Fig. 1.3** Schematic illustration of various nanomedicines in neuropsychiatric disease

## *1.3.1 Neurological Disease*

Neurological disease occurs when the central system is broken [66], which leads to the levels of ROS and protein nitration increase. Hence, antioxidants preventing ROS overproduction have been proposed to counteract the development of neurodegenerative diseases. Especially for the glioblastoma with high infiltration, nitroxide radicals-containing redox nanoparticles are able to fight with tumor and prolong the life span of mice with an increased tendency for malignancy [68]. Besides, the electrophysiological approach is another useful tool for nanomedicine to treat neurological disease. The commonly pathological mechanisms of various neurodegenerative diseases include defective protein quality control and degradation pathway, dysfunctional mitochondrial homeostasis, stress granules, and incompatible innate immune responses. Notably, neurodegenerative diseases also show distinct pathologies and symptoms in different brain areas.

### 1.3.1.1 AD

AD is a chronic disease with long preclinical phases and an average clinical time of 8–10 years. Although there is evidence that AD is caused by a complex synergy such as genetic susceptibility, aging, environment, occupation, overexposure to metals, and long-term cofactors, the exact mechanism of AD formation is not fully understood [69]. As an age-related progressive disorder, the nerve damage of AD gradually transits from mild mental retardation to dementia [70]. The prevalent hypothesis regards that two significant pathological features of AD are extracellular plaques formed by extracellular β-amyloid (Aβ) peptide and neurofibrillary tangles formed by intracellular hyperphosphorylated tau protein [71]. The pathogenic forms of Aβ and tau subsequently attenuate synaptic function and trigger a series of events leading to neuronal death [72]. Since the pathogenesis understanding of AD is not thorough, the current therapeutic strategy of AD is to clear Aβ and tau in neurofibrillary tangles in neurons and prevent or interfere with Aβ and tau aggregations to reduce membrane injury, cell apoptosis, intracellular microtubule damage, and reactive oxygen species production [73, 74]. The suitable stage to interfere AD formation is the early stage with only AD biomarkers production, without symptoms of neuronal death and dementia. Nanoparticles modified with Aβ or tau peptide are applied to target Aβ in the plasma or CSF and display abnormal deposition or entanglement of them to boost diagnostic sensitivity and accuracy [75]. For the strategies of AD, nanomedicine inhibiting superoxide dismutase and glutathione peroxidase activity, increasing cortex and hippocampus neuronal densities, and decreasing Aβ and γ-secretase levels have achieved effective therapeutic effects [34]. Solid lipid nanoparticles loaded with resveratrol are able to remodel soluble oligomers, and fibrils form into nontoxic form of Aβ and promote clearance of Aβ [29]. PLGA nanoparticles loaded with curcumin have the ability of reducing the expression of cytokines and chemokines induced by Aβ and decreasing cognitive deficits [76]. Lipid nanoparticles loaded with catechins are applied to increase α-secretase activity and reduce β-secretase activity [77]. Tetrahedral DNA nanostructures have potential of preventing the damage caused by Aβ deposition by activating the ERK1/2 pathway [78]. Recently, chitosan nanoparticles combining with Aβ peptide fragments have dual ability for capturing and clearing Aβ and rescue memory deficits successfully (Fig. 1.4) [79, 80].

#### 1.3.1.2 PD

PD is a long-term chronic disease that alters cortico-basal ganglia-thalamic circuitry. It is reported that PD affects people over 65 years old by about 3% and ranks only second to AD [69]. At early stage, β-oxidation deficiency causes long-chain acylcarnitine decrease [80], and presynaptic protein α-synuclein in intracellular fibers increases gradually [81]. Then, degeneration of dopamine neurons in the midbrain substantia nigra leads to quiescent tremor, bradykinesia, and muscle stiffness [82]. The current program is efforted to evaluate dopamine levels in the brain and stimulate central dopamine receptors. However, there is little evidence to show that dopamine prodrugs and levodopa would help to alleviate dopamine neuron degeneration. In addition, such medical treatment has also been reported to cause kinds of side effects including arrhythmia, gastrointestinal discomfort, extreme emotional changes, etc. The neuroprotective tests to date have not clearly

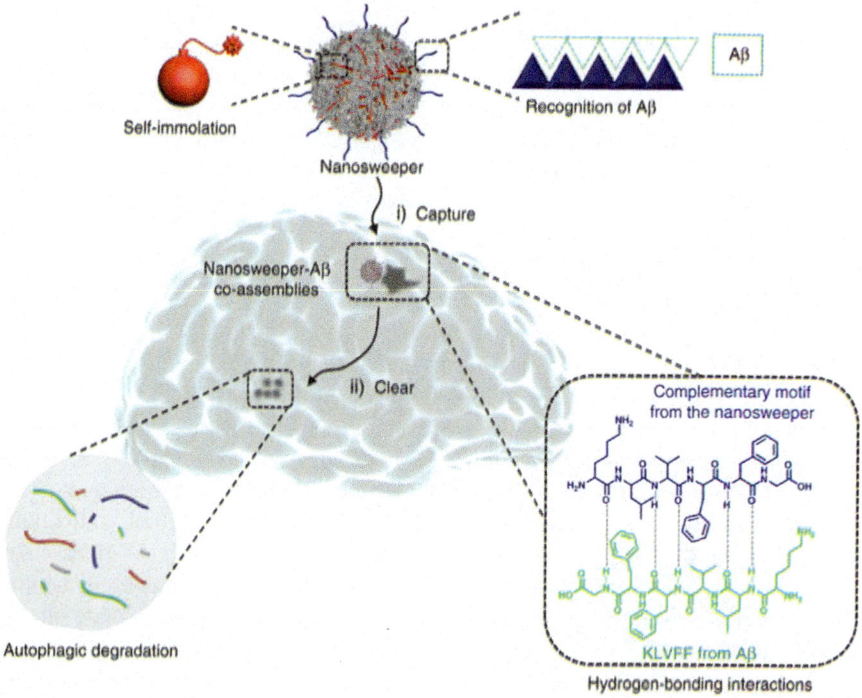


**Fig. 1.4** The schematic illustration of the nanosweeper mechanism of action. The nanosweeper captures Aβ by hydrogen-bonded co-assembly and internalizes a substantial amount into cells carrying Aβ. Then, the nanosweeper activates the cell's autophagic response, resulting in the degradation of Aβ and the nanosweeper itself. Adapted from Luo et al. with permission from the American Association for *Nat Commun*

established a drug that is able to delay or prevent the progression of the disease. The lack of effective biomarkers, the secondary hallucination and delusion effects of drugs, and the resistance of BBB limit the diagnosis and treatment of PD. Therefore, the monitor of α-synuclein and active dopamine neurons and the efficient vectors for carrying therapeutic agents to target sites have been considered as three crucial strategies of nanomedicines for diagnostic improvements. Concurrent strategies of nanomedicine treatments, including targeting dopamine deficiency, caspase-3 activation, α-synuclein accumulation, oxidative stress, mitochondrial dysfunction, inflammation, and growth factor supplementation, present a wide range of possibilities for PD therapy. For dopamine deficiency, ZnO nanowire arrays provide great clinical potential for PD diagnosis by selective dopamine detection [83]. Zn/Al layered organic-inorganic nanocomposite incorporating levodopa prolongs the half-life of levodopa and shows excellent sustained release performance [84]. For caspase-3 activation, functional silicon dioxide nanoparticles combined with electrochemiluminescence present a novel approach for the sensitive determination of caspase-3 activity. Poly-L-lysine nanoparticles encapsulating shRNAs are able to achieve anti-apoptotic and anti-inflammation synergistic therapeutic effects by downregulating the expression and activation of caspase-3 [85]. For mitochondrial dysfunction, ubiquinone-QD is able to simulate an electron-transfer system part of the mitochondrial respiratory chain in the presence of NADH to monitor PD at early stage [86]. Magnetic $Fe_3O_4$ coated with N-isopropylacrylamide derivative is prepared to interfere with the expression of α-synuclein in neurons effectively to avoid synaptic transmission disruption and mitochondrial dysfunction [87].

#### 1.3.1.3 Stroke

As the limitation of neuroregeneration in damaged brain area is difficult to overcome, stroke is the leading cause of long-term disability [88]. With the development of ischemia, accompanying the increasing immune response, angiogenesis and nerve growth are severely restricted, resulting in ischemia cavity in the brain [89]. As to the causes of stroke, about 87% are caused by ischemia, while the rest are caused by hemorrhage [90]. Lack of blood in the brain causes damage to neurons and neural circuits, forming ischemic stroke and behavioral disorders. The level of vascular endothelial growth factor is associated with BBB disruption and brain edema [89]. Therefore, the difficulty to form intracranial blood vessels is the main hurdle in the treatment of stroke. At present, the main repair areas for stroke with tissue repairing technology consist of the adjacent area of the stroke cavity, the area around the infarction, and the area with strong nerve and vascular plasticity after stroke. The conventional treatment for ischemic stroke is to clear the thrombus by mechanical methods and intravenous inject tissue-type plasminogen activator. However, due to safety concerns, for instance, the risk of cerebral hemorrhage after treatment, patients using these two methods are much less than expected. Recently, nanomedicines are applied to eliminate free radical scavengers, reduce free radical scavengers, dissolve fibrin plaques, and provide neuroprotective factors. Retinoic acid, an active metabolite of vitamin A, plays an important role in immune tolerance that consists of regulating the transcription of target genes, participating in the plasticity and regeneration of the nervous system, and regulating the formation of cognition and behavior. Before permanent ischemic injury appears, retinoic acid-loaded polymeric nanoparticles are able to normalize ischemia-altered NO and ROS produced by human endothelial cells and enhance neurogenesis in vivo by promoting human endothelial cell survival and proliferation [91]. In the process of thrombosis, PEG nanoparticle conjugated recombinant tissue plasminogen activator (rtPA) prolongs the half-life of rtPA and decreases infarct volume [92]. After the development of ischemic stroke, ceria nanoparticles as antioxidant prevent free radical proliferating in early stage of acute cerebral ischemia-reperfusion and then avoid the primary injury from ischemic stroke [93]. Solid polymer nanoparticles packaged SOD as nanozyme reduce the volume of infarct areas in rodent models of stroke by over 50% compared with free enzyme [94]. The sensor consists of magnetic disks and

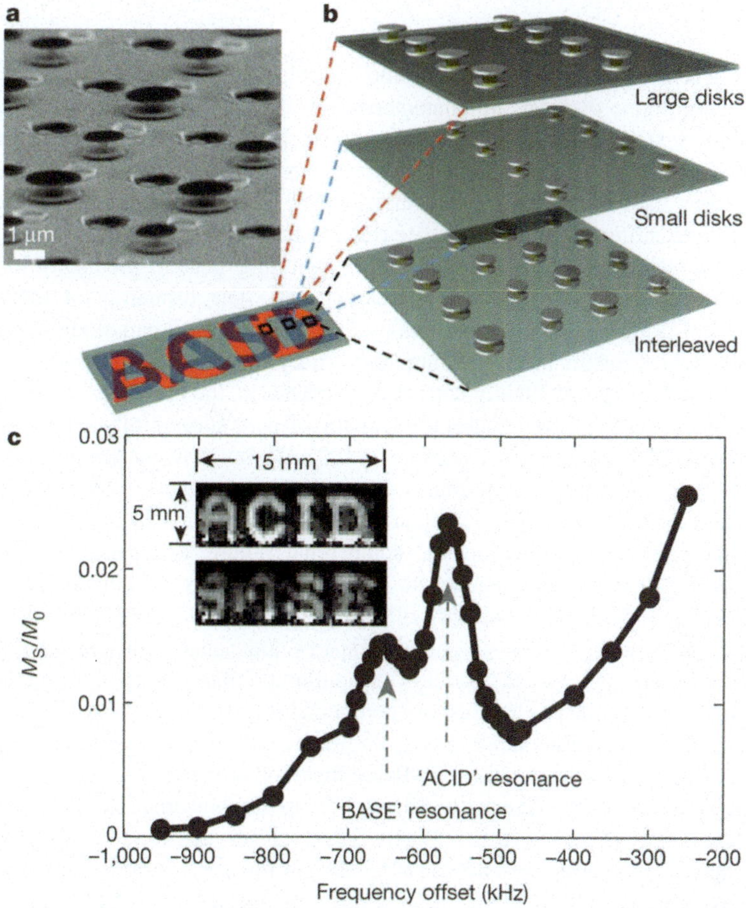


**Fig. 1.5** Inset shows two MRI difference images of same sample region acquired at different offset frequencies, showing selective addressing of different sensor populations. (**a**) SEM showing array of interleaved sensors with alternately larger and smaller disk sizes. (**b**) Schematic of hierarchical, interleaved sensor patterning. The words "ACID" and "BASE" are patterned out of two interleaved arrays of sensors. "ACID" characters are spatially patterned from sensors with larger disks; "BASE" characters are from sensors with smaller disks. Adapted from Zabow et al. with permission from the American Association for *Nature Letter*

swellable hydrogel with high sensitivity responded to pH has the ability of decreasing the severity of ischemic stroke by brain tissue improvement, BBB protection, and neurotoxicity reduction (Fig. 1.5) [95]. Platelet membrane-derived biomimetic nanobubbles not only serve as a sensor to monitor the dynamic development of stroke by real-time contrast-enhanced ultrasound imaging but also have the ability of accurate targeting and microvascular bio-remodeling in the lesion [96].

#### 1.3.1.4 MS

MS is a classic tissue-specific chronic T cell-mediated autoimmune disease [97]. Autoimmune diseases are characterized by loss of tolerance to autoantigens in immune system, which leads to autoimmune cells attacking the body. T lymphocyte plays an important role in the pathogenesis of many autoimmune diseases, which loses their self-tolerance and their ability to distinguish between self and nonself when suffered autoimmune diseases, leading to organ and tissue damage. A variety of factors are able to induce MS, including genetic predisposition and environmental factors (e.g., viral infections) [98]. The formation of multifocal plaques in CNS caused by the disruption of BBB is a typical pathologic feature of MS [99], which results in the inflammatory injury of myelin and axon [100]. The strategies of MS therapy mainly consist of two aspects, the first is to develop novel drugs inducing T cell differentiates to regulatory T cells successfully and the second is to protect BBB from being attacked [101, 102]. However, many current treatments for MS are broad-spectrum immunosuppressive agents that target a variety of immune cells or inflammatory mediators, which induce a wide range of severe side effects, including lymphopenia, acute renal failure, and cancer [103]. It is worth noting that nanotherapeutics to induce tolerance to myelin antigen is an effective way to regulate the activity of effector T cells [104, 105]. Poly (lactic-co-glycolic acid) nanoparticles as antigen carriers are capable of inducing robust tolerance in the MS mouse model with minimal inflammatory cell infiltration, inflammatory cytokine production, and demyelination [101]. For these two diagnostic strategies of MS, CSF examination is limited by the risk of complicated operation accompanying lumbar puncture, and MRI of the brain or spinal cord is limited by the accuracy to differentiate MS and the other CNS disease. Nanomaterial-based sensors array combined with artificial neural networks analyses are used to confirm MS by detecting the level of sulfur dioxide [106]. Except for that, SPIONs combined with MRI have the ability to demonstrate immune cells infiltration in MS lesions accurately, thus detecting MS rapidly and efficiently [107].

#### 1.3.1.5 Glioblastoma

Glioblastoma represents aggressive primary malignant brain tumor with poor prognosis, high recurrence, and heavy mortality rate [108]. The vessels in tumors are abnormal in structure and function, which results in a hostile microenvironment with low oxygen tension and high interstitial fluid pressure [109]. Besides, the activity of mitotic, microvascular, and tumor growth factor receptors also shows abnormal condition in glioblastoma [110, 111]. Traditional surgical resection is the first-line treatment of glioma patients. However, the strong invasiveness of glioma cells makes patients require adjuvant chemotherapy after surgery, which limited the clinical improvement of glioblastoma. The limitations of BBB and the blood-tumor barrier also hinder the efficacy of drugs [112]. Although immunotherapies are developed to treat various tumors, the transformation to brain tumors is still a distinct challenge [111]. Epidermal growth factor receptor (EGFR) is frequently overexpressed in many gliomas and results in the aberrant activities of underlying molecular signaling pathway [113]. EGFR and the mutant EGFRvIII are major focal points in current concepts of glioblastoma therapy; nevertheless, a single target-specific agent of EGFR- or EGFRvIII-specific treatment for glioma patients frequently does not produce the desired results [114]. At present, MRI is a standard and ossified method for diagnosis and monitoring of brain tumors. The photothermal effect of magnetic nanoparticles accompanying with MRI affects the microenvironment brain tumors positively. Except for the photothermal treatment of glioblastoma, nanomedicine mainly aims at increasing cell uptake and autophagy activity to reduce the migration and invasion of glioma cells [115]. The ability of magnetic absorption makes SPIONs overcome the intracellular and extracellular barriers and increase the concentration of drugs in glioma cells locally. The ability of targeting objective area makes nanoparticles deliver drugs to glioma cell selectively to increase the therapeutic index and tumor retention of drugs and avoid drug resistance [116, 117]. Last but not the least, the low intrinsic cytotoxicity and biodegradability of nanomaterials further reduce the side effects of chemotherapy [118].

### *1.3.2 Psychiatric Disease*

Psychiatric disease is associated with genetic variation and often accompanies with the progress of neurological disease, for example, depression is often detected in PD or AD patients [119]. The inducement of psychiatric disease is complicated, including social pressure, personal stress, and psychological stress. The neurogenesis, synaptic plasticity, and dendritic germination of dentate gyrus in the brain reduce gradually accompanying with psychiatric disease. The harm of psychiatric disease is the erosion of normal social function. The social burden of numerous patients accounts for 7.1% of the total disease burden [120]. The main symptoms of psychiatric disease include disturbance of consciousness, movement disorder, language and speech disorders, cognitive dysfunction, and affective disorder. However, the preferred orally dosed pharmacological treatment options available for psychiatric disease are often limited by factors such as low drug aqueous solubility, food effects, high hepatic first-pass metabolism effects, and short drug half-lives. Nanomedicines for psychiatric disease not only have the potential to solve these problems but also enhance oral bioavailability of drugs. On the other hand, nanomaterials are also able to diagnose psychiatric disease by neurotransmitter detection (e.g., dopamine, epinephrine, and norepinephrine).

### 1.3.2.1 Depression

Depression is a widespread and devastating multifactor relating emotional disorder, which poses a serious physical and psychological burden on patients and their families. The related genetic, endocrine, and environmental factors are reported to impact on the progression of depression and lead to individual susceptibility [121]. Research has shown that depression often correlates with inflammatory reaction [122], accompanying with a decrease in norepinephrine activity, serotonin activity, dopamine activity, and dopamine β-hydroxylase activity [123]. Typical antidepressants are mainly for neurotrophic factor regulation, monoamine neurotransmission, and neurologic disorders. Treatment takes weeks or months and even years to relieve depressive symptoms, but severe side effects including acute nausea, sexual dysfunction, and weight gain limit their clinic applications [124]. Hence, it is urgently necessary to conduct kinds of nano-delivery systems for identifying novel therapeutic targets to combine with drugs efficiently and then treat depression effectively by the increasing enzymes and hormones. The immune system is a potential target of nanomedicine. Nanomaterials loading biologics target individual cytokines specifically and effectively in depressive symptoms without off-target effects. For example, curcumin and dexanabinol-loaded solid lipid nanoparticles prevent corticosterone-induced BDNF/NeuN expression reduction to exert neuroprotective effects [125].

### 1.3.2.2 Schizophrenia

Schizophrenia is another psychiatric disease with high morbidity and mortality, which characterized by hallucinations and confusion of thought at present. Schizophrenia patients are inconsistent with the environment from the following aspects: the language, perception, thinking, social activity disorder, and mental activity. The substantial disability associated with this disease, coupled with its early onset and chronicity, imposes a huge burden on patients. Owing to the positive symptoms of schizophrenia may be related to hyperdopaminergic status, existing neuroinhibitors mainly act on dopamine receptor 2, which is highly expressed in the basal ganglia and is effective in reducing positive symptoms in many patients [126]. Therefore, the key to treat schizophrenia is to identify abnormalities in the brain, such as the decrease of apolipoprotein A1 in peripheral tissues and the brain [117]. The turn-on fluorescent dopamine sensing based on in situ synthesized polydopamine nanoparticles is very promising for detection of dopamine-related diseases. Aripiprazole, haloperidol, olanzapine, loxapine, remoxipride, risperidone, and paliperidone have been demonstrated to be effective for the treatment of positive and negative symptoms. However, these antipsychotics are insoluble in water, which makes it very difficult to incorporate them in pharmaceutical products. Nanoparticles encapsulated by antipsychotics have been demonstrated to be a successful production avoiding food influences [61]. PEG-PLGA nanoparticles loaded with haloperidol and solid lipid nanoparticles encapsulated by olanzapine exhibit good physical stability and great occlusive effect and reduce the side effects associated with antipsychotics [127].

#### 1.3.2.3 Other Psychiatric Disease

Except for above applications, there are also important advances of nanomedicine in improving other psychiatric diseases. Electro-responsive hydrogel nanoparticles facilitate the delivery of the antiepileptic drug and overcome drug resistance and disease recurrence after reduction in medication comparing with traditional antiepileptic drugs [128]. Superparamagnetic nanoparticles promote the resection of epileptic foci by MRI, and the combination of interleukin-1β (IL-1β) makes it not only an accurate brain mapping technique but an effective drug delivery system [129, 130]. Mesoporous silica nanoparticles, liposomes, and polymer nanoparticles coated with drugs enhance analgesic effects and increase the overall effectiveness of the drug [131]. "Chemical-pain sensor" based on nanovesicle immobilized on single-walled carbon nanotube is able to detect chemical pain stimuli, which overcome poor sensitivity and selectivity of previous chemical pain stimuli sensors [132]. It is so exciting that using gold nanoparticles to deliver CRISPR DNA into the brain to edit the genes responsible for autism not only controls the release of CRISPR DNA but decreases the off-targeted effects of CRISPR system (Fig. 1.6) [133]. Although the experiment is carried out mainly in mice, the great development makes it possible to replace the micromolecule drug and cure autism spectrum disorder.

## 1.4 Conclusion and Perspective

This chapter introduces the advantages of nanomedicine in neuropsychiatric disorders systematically, including (1) achieving precise targeting and delivery; (2) evaluating efficiently therapeutic effects; and (3) bypassing the limitation of biological barriers. These advantages are particularly important in the treatment of brain diseases. Further elaboration, the occurrence of inflammation and the damage of neuron are two important factors in the development of brain diseases. The diversity, controllability, and easy pass through of the blood-brain barrier of nanomedicine have brought new opportunities for the treatment of brain diseases. Currently, many nanomedicines have been tested in the treatment of brain diseases. The development of nanomedicine makes the materials compatible and promotes nanotechnology close to the clinical applications. Despite the advancements described in this chapter, there are still some challenges to be addressed. The problems of target specificity, instability, and dispersibility in the body of nanomedicine are still difficult to achieve

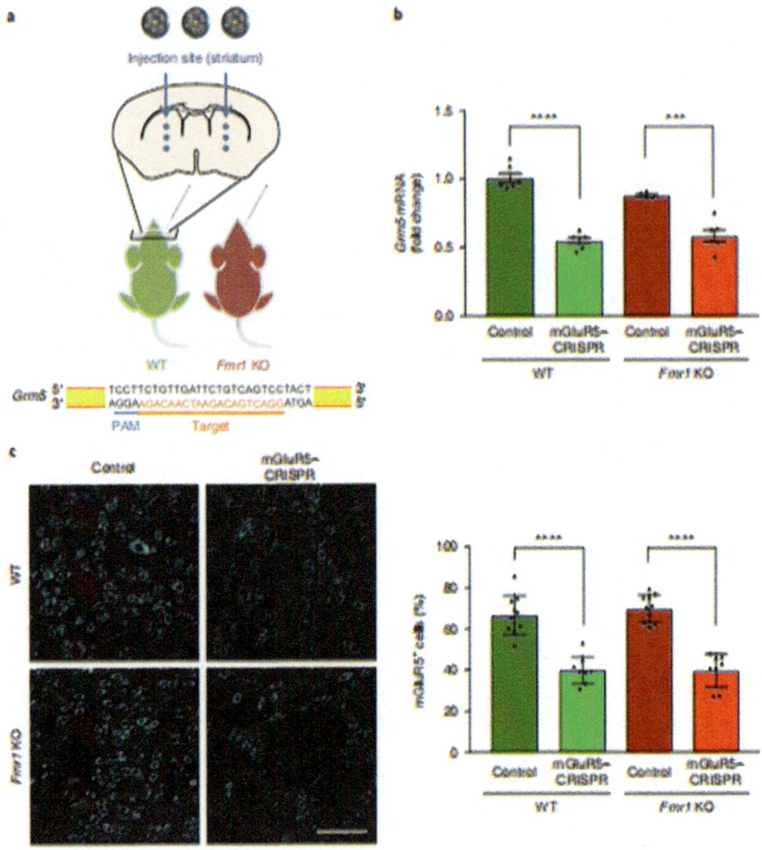


**Fig. 1.6** mGluR5-CRISPR successfully promotes mGluR5 gene editing in the striatum of wild-type and Fmr1 knockout mice. (**a**) Upper: schematic of the injection process for mGluR5-CRISPR into the striatum of wild-type (WT) and Fmr1 knockout (KO) mice. Saline or mGluR5-CRISPR was injected into the striatum (bregma: 0.26 mm, three injection sites per hemisphere are indicated as blue dots, 0.4 mm interval) of WT or Fmr1 KO mice. Lower: schematic of the target sequences of Cas9 RNPs and the protospacer adjacent motif (PAM) for Grm5 knockout. (**b**) RNA was extracted from the saline-injected control side (control) or from the mGluR5-CRISPR-injected side (mGluR5-CRISPR) of WT or Fmr1 KO mice 11 weeks after stereotaxic injections. (**c**) Left: immunostaining of mGluR5 (cyan) 5 weeks after stereotaxic injection of saline (control) or mGluR5-CRISPR into the striatum of WT or Fmr1 KO mice. Scale bar, 100 μm. Right: the number of mGluR5+ cells in WT control, WT mGluR5-CRISPR, Fmr1 KO control, and Fmr1 KO mGluR5-CRISPR groups is counted and normalized to the number of DAPI+ cells. Copyright 2018 The Author(s)



optimal therapeutic effects in practical applications for treating brain diseases. Serious studies are urgently needed in this area for the translation from preclinical to specific clinical applications.

Although many nanomedicines have been studied for the treatment of neurological diseases, there are still many challenges to be taken up for clinical treatments as follows: (1) the balance between toxicity and drug efficacy of nanomedicine; (2) the precise targeting of nanomedicine *in vivo and the effective diffusion at the site of action requiring continued research*; (3) the combined application of nanomaterials between diagnosis and treatment; and (4) the intrinsic activities and accompanying applications of nanomaterials that deserve further exploration. In view of the exponential growth of nanomedicine in neuropsychiatric disease, other innovative tools will emerge in the future to enable early diagnosis and efficient treatment of neuropsychiatric diseases. These outcomes would further broaden the applications of nanomaterials in neuropsychiatric disease. Under the guidance of innovative multifunctional nanomaterials, the diagnosis and treatment of neuropsychiatric disease could be more effectively and accurately carried out.

**Acknowledgment** This work was supported by the National Natural Science Foundation of China (31771031 and 81701829), and the National Key Research and Development Program of China (2018YFA0209800).

## References

1. Misra MK, Damotte V, Hollenbach JA. The immunogenetics of neurological disease. Immunology. 2018;153:399–414.
2. Forrest MP, Parnell E, Penzes P. Dendritic structural plasticity and neuropsychiatric disease. Nat Rev Neurosci. 2018;19(4):215–34.
3. Zhang W, Wang W, Yu DX, Xiao Z, He Z. Application of nanodiagnostics and nanotherapy to CNS diseases. Nanomedicine. 2018;13:2341–71.
4. Singh R, Lillard JW. Nanoparticle-based targeted drug delivery. Exp Mol Pathol. 2009;86:215–23.
5. Ortiz GG, Pacheco-Moises FP, Macias-Islas MA, Flores-Alvarado LJ, Mireles-Ramirez MA, Gonzalez-Renovato ED. Role of the blood-brain barrier in multiple sclerosis. Arch Med Res. 2014;45:687–97.
6. Ramanathan S, Archunan G, Sivakumar M, Tamil Selvan S, Fred AL, Kumar S, Gulyás B, Padmanabhan P. Theranostic applications of nanoparticles in neurodegenerative disorders. Int J Nanomed. 2018;13:5561–76.
7. Zhang XQ, Xu X, Bertrand N, Pridgen E, Swami A, Farokhzad OC. Interactions of nanomaterials and biological systems: Implications to personalized nanomedicine. Adv Drug Deliv Rev. 2012;64:1363–84.
8. Li J, Rao J, Pu K. Recent progress on semiconducting polymer nanoparticles for molecular imaging and cancer phototherapy. Biomaterials. 2018;155:217–35.
9. Son J, Yi G, Yoo J, Park C, Koo H, Choi HS. Light-responsive nanomedicine for biophotonic imaging and targeted therapy. Adv Drug Deliv Rev. 2018;138:133–47.
10. Mura S, Nicolas J, Couvreur P. Stimuli-responsive nanocarriers for drug delivery. Nat mater. 2013;12:991–1003.
11. Qian CG, Chen YL, Feng PJ, Xiao XZ, Dong M, Yu JC, et al. Conjugated polymer nanomaterials for theranostics. Acta Pharmacol Sin. 2017;38:764–81.
12. Liu TM, Conde J, Lipiński T, Bednarkiewicz A, Huang CC. Revisiting the classification of NIR-absorbing/emitting nanomaterials for in vivo bioapplications. NPG Asia Mater. 2016;8:e295.
13. Dobson J. Remote control of cellular behaviour with magnetic nanoparticles. Nat Nanotechnol. 2008;3:139–43.

14. Chen R, Canales A, Anikeeva P. Neural recording and modulation technologies. Nat Rev Mater. 2017;2:16093.
15. Stanley SA, Gagner JE, Damanpour S, Yoshida M, Dordick JS, Friedman JM. Radio-wave heating of iron oxide nanoparticles can regulate plasma glucose in mice. Science. 2012;336:604–8.
16. Yang SY, Chiu MJ, Lin CH, Horng HE, Yang CC, Chieh JJ, et al. Development of an ultra-high sensitive immunoassay with plasma biomarker for differentiating Parkinson disease dementia from Parkinson disease using antibody functionalized magnetic nanoparticles. J Nanobiotechnol. 2016;14:41.
17. Rodeberg NT, Sandberg SG, Johnson JA, Phillips PEM, Wightman RM. Hitchhiker's guide to voltammetry: Acute and chronic electrodes for in vivo fast-scan cyclic voltammetry. ACS Chem Neurosci. 2017;8:221–34.
18. Liu TC, Chuang MC, Chu CY, Huang WC, Lai HY, Wang CT, et al. Implantable graphene-based neural electrode interfaces for electrophysiology and neurochemistry in in vivo hyperacute stroke model. ACS Appl Mater Interfaces. 2016;8:187–96.
19. Hu X, Wei H, Liu J, Zhang J, Chi X, Jiang P, et al. Nanoenvelopes: Wrapping a single-walled carbon nanotube with graphene using an atomic force microscope. Adv Mater. 2018: e1804918. https://doi.org/10.1002/adma.201804918.
20. Vitale F, Summerson SR, Aazhang B, Kemere C, Pasquali M. Neural stimulation and recording with bidirectional, soft carbon nanotube fiber microelectrodes. ACS Nano. 2015;9:4465–74.
21. Xue X, Wang LR, Sato Y, Jiang Y, Berg M, Yang DS, et al. Single-walled carbon nanotubes alleviate autophagic/lysosomal defects in primary glia from a mouse model of Alzheimer's disease. Nano Lett. 2014;14:5110–7.
22. Xue X, Yang JY, He Y, Wang LR, Liu P, Yu LS, et al. Aggregated single-walled carbon nanotubes attenuate the behavioural and neurochemical effects of methamphetamine in mice. Nat Nanotechnol. 2016;11:613–20.
23. Kuznetsov AI, Miroshnichenko AE, Fu YH, Zhang J, Luk'yanchuk B. Magnetic light. Sci Rep. 2012;2:492.
24. Gu L, Hall DJ, Qin Z, Anglin E, Joo J, Mooney DJ, et al. In vivo time-gated fluorescence imaging with biodegradable luminescent porous silicon nanoparticles. Nat Commun. 2013;4:2326.
25. Kwiatkowski G, Jahnig F, Steinhauser J, Wespi P, Ernst M, Kozerke S. Nanometer size silicon particles for hyperpolarized MRI. Sci Rep. 2017;7:7946.
26. Zhao Y, Jiang Y, Lv W, Wang Z, Lv L, Wang B, et al. Dual targeted nanocarrier for brain ischemic stroke treatment. J Control Release. 2016;233:64–71.
27. Parameswaran R, Carvalho-de-Souza JL, Jiang Y, Burke MJ, Zimmerman JF, Koehler K, et al. Photoelectrochemical modulation of neuronal activity with free-standing coaxial silicon nanowires. Nat Nanotechnol. 2018;13:260–6.
28. Li W, Li Y, Liu Z, Kerdsakundee N, Zhang M, Zhang F, et al. Hierarchical structured and programmed vehicles deliver drugs locally to inflamed sites of intestine. Biomaterials. 2018;185:322–32.
29. Loureiro JA, Andrade S, Duarte A, Neves AR, Queiroz JF, Nunes C, et al. Resveratrol and grape extract-loaded solid lipid nanoparticles for the treatment of Alzheimer's disease. Molecules. 2017;22 https://doi.org/10.3390/molecules22020277.
30. Hernando S, Herran E, Figueiro-Silva J, Pedraz JL, Igartua M, Carro E, et al. Intranasal administration of TAT-conjugated lipid nanocarriers loading GDNF for Parkinson's disease. Mol Neurobiol. 2018;55:145–55.
31. Zhao YZ, Li X, Lu CT, Lin M, Chen LJ, Xiang Q, et al. Gelatin nanostructured lipid carriers-mediated intranasal delivery of basic fibroblast growth factor enhances functional recovery in hemiparkinsonian rats. Nanomedicine. 2014;10:755–64.

32. Cacciatore I, Ciulla M, Fornasari E, Marinelli L, Di Stefano A. Solid lipid nanoparticles as a drug delivery system for the treatment of neurodegenerative diseases. Expert Opin Drug Deliv. 2016;13:1121–31.
33. Tokatlian T, Kulp DW, Mutafyan AA, Jones CA, Menis S, Georgeson E, et al. Enhancing humoral responses against HIV envelope trimers via nanoparticle delivery with stabilized synthetic liposomes. Sci Rep. 2018;8:16527.
34. Siddiqi KS, Husen A, Sohrab SS, Yassin MO. Recent status of nanomaterial fabrication and their potential applications in neurological disease management. Nanoscale Res Lett. 2018;13:231.
35. Chung GY, Shim KH, Kim HJ, Min SK, Shin HS. Chitosan-coated C-phycocyanin liposome for extending the neuroprotective time window against ischemic brain stroke. Curr Pharm Des. 2018;24:1859–64.
36. Wei Y, Guo J, Zheng X, Wu J, Zhou Y, Yu Y, et al. Preparation, pharmacokinetics and biodistribution of baicalin-loaded liposomes. Int J Nanomed. 2014;9:3623–30.
37. Kundu P, Das M, Tripathy K, Sahoo SK. Delivery of dual drug loaded lipid based nanoparticles across the blood-brain barrier impart enhanced neuroprotection in a rotenone induced mouse model of Parkinson's disease. ACS Chem Neurosci. 2016;7:1658–70.
38. Lu YM, Huang JY, Wang H, Lou XF, Liao MH, Hong LJ, et al. Targeted therapy of brain ischaemia using Fas ligand antibody conjugated PEG-lipid nanoparticles. Biomaterials. 2014;35:530–7.
39. Kamp F, Scheidt HA, Winkler E, Basset G, Heinel H, Hutchison JM, et al. Bexarotene binds to the amyloid precursor protein transmembrane domain, alters its alpha-helical conformation, and inhibits gamma-secretase nonselectively in liposomes. ACS Chem Neurosci. 2018;9:1702–13.
40. Dowding JM, Song W, Bossy K, Karakoti A, Kumar A, Kim A, et al. Cerium oxide nanoparticles protect against Abeta-induced mitochondrial fragmentation and neuronal cell death. Cell Death Differ. 2014;21:1622–32.
41. Guan Y, Li M, Dong K, Gao N, Ren J, Zheng Y, et al. Ceria/POMs hybrid nanoparticles as a mimicking metallopeptidase for treatment of neurotoxicity of amyloid-beta peptide. Biomaterials. 2016;98:92–102.
42. Kwon HJ, Cha MY, Kim D, Kim DK, Soh M, Shin K, et al. Mitochondria-targeting ceria nanoparticles as antioxidants for Alzheimer's disease. ACS Nano. 2016;10:2860–70.
43. Wahba SM, Darwish AS, Kamal SM. Ceria-containing uncoated and coated hydroxyapatite-based galantamine nanocomposites for formidable treatment of Alzheimer's disease in ovariectomized albino-rat model. Mater Sci Eng C. 2016;65:151–63.
44. Wen S, Zhou J, Zheng K, Bednarkiewicz A, Liu X, Jin D. Advances in highly doped upconversion nanoparticles. Nat Commun. 2018;9:2415.
45. Zhou B, Shi B, Jin D, Liu X. Controlling upconversion nanocrystals for emerging applications. Nat Nanotechnol. 2015;10:924–36.
46. Chen S, Weitemier AZ, Zeng X, He L, Wang X, Tao Y, et al. Near-infrared deep brain stimulation via upconversion nanoparticle-mediated optogenetics. Science. 2018;359:679–84.
47. Feng W, Zhu X, Li F. Recent advances in the optimization and functionalization of upconversion nanomaterials for in vivo bioapplications. NPG Asia Mater. 2013;5:e75.
48. Lv R, Yang P, Chen G, Gai S, Xu J, Prasad PN. Dopamine-mediated photothermal theranostics combined with up-conversion platform under near infrared light. Sci Rep. 2017;7:13562.
49. Jin D, Xi P, Wang B, Zhang L, Enderlein J, van Oijen AM. Nanoparticles for super-resolution microscopy and single-molecule tracking. Nat Methods. 2018;15:415–23.
50. Warburton RJ. Single spins in self-assembled quantum dots. Nat Mater. 2013;12:483–93.
51. Xu G, Mahajan S, Roy I, Yong KT. Theranostic quantum dots for crossing blood-brain barrier in vitro and providing therapy of HIV-associated encephalopathy. Front Pharmacol. 2013;4:140.

52. Wen X, Wang Y, Zhang F, Zhang X, Lu L, Shuai X, et al. In vivo monitoring of neural stem cells after transplantation in acute cerebral infarction with dual-modal MR imaging and optical imaging. Biomaterials. 2014;35:4627–35.
53. He J, Yang H, Zhang Y, Yu J, Miao L, Song Y, et al. Smart nanocomposites of Cu-hemin metal-organic frameworks for electrochemical glucose biosensing. Sci Rep. 2016;6:36637.
54. Lian X, Erazo-Oliveras A, Pellois JP, Zhou HC. High efficiency and long-term intracellular activity of an enzymatic nanofactory based on metal-organic frameworks. Nat Commun. 2017;8:2075.
55. Chen H, Wang J, Shan D, Chen J, Zhang S, Lu X. Dual-emitting fluorescent metal-organic framework nanocomposites as a broad-range pH sensor for fluorescence imaging. Anal Chem. 2018;90:7056–63.
56. Wang J, Fan Y, Tan Y, Zhao X, Zhang Y, Cheng C, et al. Porphyrinic metal-organic framework PCN-224 nanoparticles for near-infrared-induced attenuation of aggregation and neurotoxicity of Alzheimer's amyloid-beta peptide. ACS Appl Mater Interfaces. 2018;10:36615–21.
57. Rhee M, Valencia PM, Rodriguez MI, Langer R, Farokhzad OC, Karnik R. Synthesis of size-tunable polymeric nanoparticles enabled by 3D hydrodynamic flow focusing in single-layer microchannels. Adv Mater. 2011;23:H79–83.
58. Pourcin F, Reynaud C, Carlberg M, Le Rouzo J, Duche D, Simon JJ, et al. Plasmonic nanocomposites based on silver nanocube-polymer blends displaying Nearly Perfect Absorption in the UV region. Langmuir. 2018; https://doi.org/10.1021/acs.langmuir.8b03003.
59. Bible E, Qutachi O, Chau DY, Alexander MR, Shakesheff KM, Modo M. Neo-vascularization of the stroke cavity by implantation of human neural stem cells on VEGF-releasing PLGA microparticles. Biomaterials. 2012;33:7435–46.
60. Okamoto S, Yoshii H, Matsuura M, Kojima A, Ishikawa T, Akagi T, et al. Poly-gamma-glutamic acid nanoparticles and aluminum adjuvant used as an adjuvant with a single dose of Japanese encephalitis virus-like particles provide effective protection from Japanese encephalitis virus. Clin Vaccine Immunol. 2012;19:17–22.
61. Piazza J, Hoare T, Molinaro L, Terpstra K, Bhandari J, Selvaganapathy PR, et al. Haloperidol-loaded intranasally administered lectin functionalized poly(ethylene glycol)-block-poly(D,L)-lactic-co-glycolic acid (PEG-PLGA) nanoparticles for the treatment of schizophrenia. Eur J Pharm Biopharm. 2014;87:30–9.
62. Sun B, Sun MJ, Gu Z, Shen QD, Jiang SJ, Xu Y, et al. Conjugated polymer fluorescence probe for intracellular imaging of magnetic nanoparticles. Macromolecules. 2010;43:10348–54.
63. Mura S, Couvreur P. Nanotheranostics for personalized medicine. Adv Drug Deliv Rev. 2012;64:1394–416.
64. Chen XY, Gambhlr SS, Cheon J. Theranostic nanomedicine. Acc Chem Res. 2011;44:841.
65. Barks A, Hall AM, Tran PV, Georgieff MK. Iron as a model nutrient for understanding the nutritional origins of neuropsychiatric disease. Pediatr Res. 2018; https://doi.org/10.1038/s41390-018-0204-8.
66. Gan L, Cookson MR, Petrucelli L, La Spada AR. Converging pathways in neurodegeneration, from genetics to mechanisms. Nat Neurosci. 2018;21:1300–9.
67. Forrest MP, Parnell E, Penzes P. Dendritic structural plasticity and neuropsychiatric disease. Nat Rev Neurosci. 2018;19:215–34.
68. Lago SG, Tomasik J, van Rees GF, Ramsey JM, Haenisch F, Cooper JD, et al. Exploring the neuropsychiatric spectrum using high-content functional analysis of single-cell signaling networks. Mol Psychiatry. 2018; https://doi.org/10.1038/s41380-018-0123-4.
69. Sadowska-Bartosz I, Bartosz G. Redox nanoparticles: Synthesis, properties and perspectives of use for treatment of neurodegenerative diseases. J Nanobiotechnol. 2018;16:87.
70. Kaushik AC, Bharadwaj S, Kumar S, Wei DQ. Nano-particle mediated inhibition of Parkinson's disease using computational biology approach. Sci Rep. 2018;8:9169.

71. Bateman RJ, Xiong C, Benzinger TL, Fagan AM, Goate A, Fox NC, et al. Clinical and biomarker changes in dominantly inherited Alzheimer's disease. New Engl J Med. 2012;367:795–804.
72. Mann AP, Scodeller P, Hussain S, Braun GB, Molder T, Toome K, et al. Identification of a peptide recognizing cerebrovascular changes in mouse models of Alzheimer's disease. Nat Commun. 2017;8:1403.
73. Eitan E, Hutchison ER, Marosi K, Comotto J, Mustapic M, Nigam SM, et al. Extracellular vesicle-associated abeta mediates trans-neuronal bioenergetic and Ca(2+)-handling deficits in Alzheimer's disease models. NPJ Aging Mech Dis. 2016;2016:2.
74. Zhang M, Mao X, Yu Y, Wang CX, Yang YL, Wang C. Nanomaterials for reducing amyloid cytotoxicity. Adv Mater. 2013;25:3780–801.
75. Shao X, Ma W, Xie X, Li Q, Lin S, Zhang T, et al. Neuroprotective effect of tetrahedral DNA nanostructures in a cell model of Alzheimer's disease. ACS Appl Mater Interfaces. 2018;10:23682–92.
76. Masters CL, Bateman R, Blennow K, Rowe CC, Sperling RA, Cummings JL. Alzheimer's disease. Nat Rev Dis Primers. 2015;1:15056.
77. Tiwari SK, Agarwal S, Seth B, Yadav A, Nair S, Bhatnagar P, et al. Curcumin-loaded nanoparticles potently induce adult neurogenesis and reverse cognitive deficits in Alzheimer's disease model via canonical Wnt/beta-catenin pathway. ACS Nano. 2014;8:76–103.
78. Smith A, Giunta B, Bickford PC, Fountain M, Tan J, Shytle RD. Nanolipidic particles improve the bioavailability and alpha-secretase inducing ability of epigallocatechin-3-gallate (EGCG) for the treatment of Alzheimer's disease. Int J Pharm. 2010;389:207–12.
79. Tang M, Pi J, Long Y, Huang N, Cheng Y, Zheng H. Quantum dots-based sandwich immunoassay for sensitive detection of Alzheimer's disease-related Abeta1-42. Spectrochim Acta Part A, Mol Biomol Spectrosc. 2018;201:82–7.
80. Luo Q, Lin YX, Yang PP, Wang Y, Qi GB, Qiao ZY, et al. A self-destructive nanosweeper that captures and clears amyloid beta-peptides. Nat Commun. 2018;9:1802.
81. Saiki S, Hatano T, Fujimaki M, Ishikawa KI, Mori A, Oji Y, et al. Decreased long-chain acylcarnitines from insufficient beta-oxidation as potential early diagnostic markers for Parkinson's disease. Sci Rep. 2017;7:7328.
82. Ruggeri FS, Mahul-Mellier AL, Kasas S, Lashuel HA, Longo G, Dietler G. Amyloid single-cell cytotoxicity assays by nanomotion detection. Cell Death Dis. 2017;3:17053.
83. Atashrazm F, Hammond D, Perera G, Dobson-Stone C, Mueller N, Pickford R, et al. Reduced glucocerebrosidase activity in monocytes from patients with Parkinson's disease. Sci Rep. 2018;8:15446.
84. Yue HY, Huang S, Chang J, Heo C, Yao F, Adhikari S. ZnO nanowire arrays on 3D hierachical graphene foam: Biomarker detection of Parkinson's disease. ACS Nano. 2014;8:1639–46.
85. Kura AU, Hussein Al Ali SH, Hussein MZ, Fakurazi S, Arulselvan P. Development of a controlled-release anti-parkinsonian nanodelivery system using levodopa as the active agent. Int J Nanomed. 2013;8:1103–10.
86. Liu Y, Guo Y, An S, Kuang Y, He X, Ma H. Targeting caspase-3 as dual therapeutic benefits by RNAi facilitating brain-targeted nanoparticles in a rat model of Parkinson's disease. PLoS One. 2013;8:e62905.
87. Ma W, Qin LX, Liu FT, Gu Z, Wang J, Pan ZG. Ubiquinone-quantum dot bioconjugates for in vitro and intracellular complex I sensing. Sci Rep. 2013;3:1537.
88. Niu S, Zhang LK, Zhang L, Zhuang S, Zhan X, Chen WY. Inhibition by multifunctional magnetic nanoparticles loaded with alpha-synuclein RNAi plasmid in a Parkinson's disease model. Theranostics. 2017;7:344–56.
89. Nih LR, Sideris E, Carmichael ST, Segura T. Injection of microporous annealing particle (MAP) hydrogels in the stroke cavity reduces gliosis and inflammation and promotes NPC migration to the lesion. Adv Mater. 2017;29 https://doi.org/10.1002/adma.201606471.

90. Nih LR, Gojgini S, Carmichael ST, Segura T. Dual-function injectable angiogenic biomaterial for the repair of brain tissue following stroke. Nat Mater. 2018;17:642–51.
91. Shcharbina N, Shcharbin D, Bryszewska M. Nanomaterials in stroke treatment: Perspectives. Stroke. 2013;44:2351–5.
92. Machado-Pereira M, Santos T, Ferreira L, Bernardino L, Ferreira R. Intravenous administration of retinoic acid-loaded polymeric nanoparticles prevents ischemic injury in the immature brain. Neurosci Lett. 2018;673:116–21.
93. Gao Y, Chen X, Liu H. A facile approach for synthesis of nano-CeO2 particles loaded co-polymer matrix and their colossal role for blood-brain barrier permeability in Cerebral Ischemia. J Photochem Photobiol B. 2018;187:184–9.
94. Zhang T, Li CY, Jia JJ, Chi JS, Zhou D, Li JZ. Combination therapy with LXW7 and ceria nanoparticles protects against acute cerebral ischemia/reperfusion injury in rats. Curr Med Sci. 2018;38:144–52.
95. Manickam DS, Brynskikh AM, Kopanic JL, Sorgen PL, Klyachko NL, Batrakova EV. Well-defined cross-linked antioxidant nanozymes for treatment of ischemic brain injury. J Control Release. 2012;162:636–45.
96. Zabow G, Dodd SJ, Koretsky AP. Shape-changing magnetic assemblies as high-sensitivity NMR-readable nanoprobes. Nature. 2015;520:73–7.
97. Li M, Liu Y, Chen J, Liu T, Gu Z, Zhang J. Platelet bio-nanobubbles as microvascular recanalization nanoformulation for acute ischemic stroke lesion theranostics. Theranostics. 2018;8:4870–83.
98. Huq R, Samuel EL, Sikkema WK, Nilewski LG, Lee T, Tanner MR. Preferential uptake of antioxidant carbon nanoparticles by T lymphocytes for immunomodulation. Sci Rep. 2016;6:33808.
99. Poser CM. Multiple sclerosis trait: The premorbid stage of multiple sclerosis. A hypothesis. Acta Neurol Scand. 2004;109:239–43.
100. Tysiak E, Asbach P, Aktas O, Waiczies H, Smyth M, Schnorr J. Beyond blood brain barrier breakdown – In vivo detection of occult neuroinflammatory foci by magnetic nanoparticles in high field MRI. J Neuroinflamm. 2009;6:20.
101. Annunziata P, Cioni C, Masi G, Tassi M, Marotta G, Severi S. Fingolimod reduces circulating tight-junction protein levels and in vitro peripheral blood mononuclear cells migration in multiple sclerosis patients. Sci Rep. 2018;8:15371.
102. Hunter Z, McCarthy DP, Yap WT, Harp CT, Getts DR, Shea LD. A biodegradable nanoparticle platform for the induction of antigen-specific immune tolerance for treatment of autoimmune disease. ACS Nano. 2014;8:2148–60.
103. Tostanoski LH, Chiu YC, Andorko JI, Guo M, Zeng X, Zhang P. Design of polyelectrolyte multilayers to promote immunological tolerance. ACS Nano. 2016; https://doi.org/10.1021/acsnano.6b04001.
104. Bonfield TL, Koloze M, Lennon DP, Zuchowski B, Yang SE, Caplan AI. Human mesenchymal stem cells suppress chronic airway inflammation in the murine ovalbumin asthma model. Am J Physiol Lung Cell Mol Physiol. 2010;299:L760–70.
105. Getts DR, Martin AJ, McCarthy DP, Terry RL, Hunter ZN, Yap WT. Microparticles bearing encephalitogenic peptides induce T-cell tolerance and ameliorate experimental autoimmune encephalomyelitis. Nat Biotechnol. 2012;30:1217–24.
106. Tabansky I, Messina MD, Bangeranye C, Goldstein J, Blitz-Shabbir KM, Machado S. Advancing drug delivery systems for the treatment of multiple sclerosis. Immunol Res. 2015;63:58–69.
107. Broza YY, Har-Shai L, Jeries R, Cancilla JC, Glass-Marmor L, Lejbkowicz I. Exhaled breath markers for nonimaging and noninvasive measures for detection of multiple sclerosis. ACS Chem Neurosci. 2017;8:2402–13.
108. Zhao X, Zhao H, Chen Z, Lan M. Ultrasmall superparamagnetic iron oxide nanoparticles for magnetic resonance imaging contrast agent. J Nanosci Nanotechnol. 2014;14:210–20.

109. Randall EC, Emdal KB, Laramy JK, Kim M, Roos A, Calligaris D. Integrated mapping of pharmacokinetics and pharmacodynamics in a patient-derived xenograft model of glioblastoma. Nat Commun. 2018;9:4904.
110. Jain RK, di Tomaso E, Duda DG, Loeffler JS, Sorensen AG, Batchelor TT. Angiogenesis in brain tumours. Nat Rev Neurosci. 2007;8:610–22.
111. Hekmatara T, Bernreuther C, Khalansky AS, Theisen A, Weissenberger J, Matschke J. Efficient systemic therapy of rat glioblastoma by nanoparticle-bound doxorubicin is due to antiangiogenic effects. Clin Neuropathol. 2009;28:153–64.
112. Touat M, Idbaih A, Sanson M, Ligon KL. Glioblastoma targeted therapy: Updated approaches from recent biological insights. Ann Oncol. 2017;28:1457–72.
113. Wu M, Zhang H, Tie C, Yan C, Deng Z, Wan Q. MR imaging tracking of inflammation-activatable engineered neutrophils for targeted therapy of surgically treated glioma. Nat Commun. 2018;9:4777.
114. Jiang YW, Cheng HY, Kuo CL, Way TD, Lien JC, Chueh FS. Tetrandrine inhibits human brain glioblastoma multiforme GBM 8401 cancer cell migration and invasion in vitro. Environ Toxicol. 2018; https://doi.org/10.1002/tox.22691.
115. Keller S, Schmidt MHH. EGFR and EGFRvIII promote angiogenesis and cell invasion in glioblastoma: Combination therapies for an effective treatment. Int J Mol Sci. 2017;18:1295.
116. Zhang H, Zhu Y, Sun X, He X, Wang M, Wang Z. Curcumin-loaded layered double hydroxide nanoparticles-induced autophagy for reducing glioma cell migration and invasion. J Biomed Nanotechnol. 2016;12:2051–62.
117. Fang K, Liu P, Dong S, Guo Y, Cui X, Zhu X. Magnetofection based on superparamagnetic iron oxide nanoparticle-mediated low lncRNA HOTAIR expression decreases the proliferation and invasion of glioma stem cells. Int J Oncol. 2016;49:509–18.
118. Wadajkar AS, Dancy JG, Roberts NB, Connolly NP, Strickland DK, Winkles J. A decreased non-specific adhesivity, receptor targeted (DART) nanoparticles exhibit improved dispersion, cellular uptake, and tumor retention in invasive gliomas. J Control Release. 2017;267:144–53.
119. Plomin R, Owen MJ, McGuffin P. The genetic basis of complex human behaviors. Science. 1994;264:1733–9.
120. G BDD, Hale Collaborators, Murray CJ, Barber RM, Foreman KJ, Abbasoglu Ozgoren A. Global, regional, and national disability-adjusted life years (DALYs) for 306 diseases and injuries and healthy life expectancy (HALE) for 188 countries, 1990–2013: Quantifying the epidemiological transition. Lancet. 2015;386:2145–91.
121. Ge T, Fan J, Yang W, Cui R, Li B. Leptin in depression: A potential therapeutic target. Cell Death Dis. 2018;9:1096.
122. Lee Y, Subramaniapillai M, Brietzke E, Mansur RB, Ho RC, Yim SJ. Anti-cytokine agents for anhedonia: Targeting inflammation and the immune system to treat dimensional disturbances in depression. Ther Adv Psychopharmacol. 2018;8:337–48.
123. Lieberman A. Depression in Parkinson's disease -- A review. Acta Neurol Scand. 2006;113:1–8.
124. Abdallah CG, Adams TG, Kelmendi B, Esterlis I, Sanacora G, Krystal JH. Ketamine's mechanism of action: A path to rapid-acting antidepressants. Depress Anxiety. 2016;33:689–97.
125. He X, Yang L, Wang M, Zhuang X, Huang R, Zhu R. Targeting the Endocannabinoid/CB1 receptor system for treating major depression through antidepressant activities of curcumin and dexanabinol-loaded solid lipid nanoparticles. Cell Physiol Biochem. 2017;42:2281–94.
126. Raabe FJ, Galinski S, Papiol S, Falkai PG, Schmitt A, Rossner MJ. Studying and modulating schizophrenia-associated dysfunctions of oligodendrocytes with patient-specific cell systems. NPJ Schizophr. 2018;4:23.
127. Joseph E, Reddi S, Rinwa V, Balwani G, Saha R. Design and in vivo evaluation of solid lipid nanoparticulate systems of Olanzapine for acute phase schizophrenia treatment: Investigations on antipsychotic potential and adverse effects. Eur J Pharm Sci. 2017;104:315–25.

128. Ying X, Wang Y, Liang J, Yue J, Xu C, Lu L. Angiopep-conjugated electro-responsive hydrogel nanoparticles: Therapeutic potential for epilepsy. Angew Chem Int Ed Engl. 2014;53:12436–40.
129. Pedram MZ, Shamloo A, Alasty A, Ghafar-Zadeh E. Toward epileptic brain region detection based on magnetic nanoparticle patterning. Sensors (Basel). 2015;15:24409–27.
130. Fu T, Kong Q, Sheng H, Gao L. Value of functionalized superparamagnetic iron oxide nanoparticles in the diagnosis and treatment of acute temporal lobe epilepsy on MRI. Neural Plast. 2016;2016:2412958.
131. Yu YH, Hsu YH, Chou YC, Fan CL, Ueng SW, Kau YC. Sustained relief of pain from osteosynthesis surgery of rib fracture by using biodegradable lidocaine-eluting nanofibrous membranes. Nanomedicine. 2016;12:1785–93.
132. Jin HJ, An JM, Park J, Moon SJ, Hong S. "Chemical-pain sensor" based on nanovesicle-carbon nanotube hybrid structures. Biosens Bioelectron. 2013;49:86–91.
133. Lee B, Lee K, Panda S, Gonzales-Rojas R, Chong A, Bugay V, et al. Nanoparticle delivery of CRISPR into the brain rescues a mouse model of fragile X syndrome from exaggerated repetitive behaviours. Nat Biomed Eng. 2018;2:497–507.
134. Brown NF, Carter TJ, Ottaviani D, Mulholland P. Harnessing the immune system in glioblastoma. Br J Cancer. 2018;119:1171–81.
135. Wang L, Lockstone HE, Guest PC, Levin Y, Palotas A, Pietsch S. Expression profiling of fibroblasts identifies cell cycle abnormalities in schizophrenia. J Proteome Res. 2010;9:521–7.

# Chapter 2
# The Strategies of Nanomaterials for Traversing Blood-Brain Barrier

**Mohd Ahmar Rauf, Fawad Ur Rehman, Meng Zheng, and Bingyang Shi**

**Abstract** Central nervous system (CNS) ailments establish an arrangement of difficult neurotic conditions concerning identification and therapeutics. For the more significant part of these clusters, there is an absence of an early determination, biomarkers to permit legitimate follow-up of infection movement and powerful, effective methodologies to permit a diligent fix.

The engraved prognosis and diagnosis of neurodegenerative disease at advanced stages and in older people are recognized as serious health concern worldwide, especially interminable age-related neurodegenerative malfunctions that are reflected as general well-being issues.

The principal issue related to the management of CNS infections is owed, in any event to some degree, to specific qualities of the brain and spinal cord, when contrasted with peripheral organs. In such manner, the CNS is physically and synthetically secured by the blood-brain barrier (BBB) that is responsible for the maintenance of brain's homeostasis and substantially limits the movement of most therapeutics to the brain parenchyma. Different methodologies for the therapeutics were developed and modified for transposing the BBB and expecting to treat brain disorders, without meddling with the regular functioning of the brain.

In the present chapter, we will try to harness the most recent advances in neurodegenerative diseases amelioration approaches based on distinctive drug delivery systems via nanoscale materials, exosomes, and RNAi (i.e., siRNA, etc.) based frameworks.

**Key Points**

1. The nanobiotechnology assumes an encouraging job while delivering the therapeutics across the blood-brain barrier (BBB) that is considered to be a noteworthy snag to remedial delivery for CNS issue.
2. Attractive attributes of the perfect strategy for remedial delivery over the BBB are depicted in this study. It should allow specific targated medication conveyance without harming the BBB.
3. Polymeric NPs had shown the most positive results among the large number of nanoparticles for the convergence of the BBB.

Md. A. Rauf · F. Ur. Rehman · M. Zheng · B. Shi (✉)
International Joint Centre for Biomedical Innovation, Henan University, Kaifeng, China
e-mail: mohd@henu.edu.cn; fawad@henu.edu.cn; mzheng@henu.edu.cn; bs@henu.edu.cn



X. Xue (ed.), *Nanomedicine in Brain Diseases*,
https://doi.org/10.1007/978-981-13-8731-9_2

4. Diffrent modes of drug delivery vehicles were employed for crossing the BBB, like modifying NPs for improving specific targeting, viz., Trojan steed approach, and employment of peptide-NP conjugates and also two industrial innovations: LipoBridge™ and G-Technology®.
5. Targeting of siRNA employing exosomes.
6. The important part of employing nanotechnology for the treatment of CNS maladies is that in addition to the delivery of therapeutics to the brain, it also allowed its trackings and empowers imaging and therapy of brain tumors.
7. In the coming age, there will be an improvement in the nanotechnology for CNS issues.

**Keywords** Blood-brain barrier (BBB) · Central nervous system (CNS) · Nanoparticles (NPs) · Alzheimer disease (AD) · Parkinson disease (PD) · Polyethylene glycol (PEG) · siRNA

## 2.1 Introduction

The blood-brain barrier (BBB) is a dynamic limit that maintains the central nervous system physiology and prevents undesired substances and pathogens [1]. It is additionally the most important boundary hindering medication transport into the cerebrum using the blood flow. In some pathologies of the brain (e.g., tumors, stroke, and neurodegenerative disorders), the BBB is adjusted and turns out to be progressively porous permitting the section of particles that can incite fiery reactions and neuronal harm [2]. In spite of understanding the functional physiology and anatomic structures, the insight of the cell receptors expressed continuous therapeutic innovations and nanotechnology-based methodologies. Even after a development of good innovations and methods, the treatments of a significant number of the CNS-related ailments remains a challenge [3–5]. This is not because there is an absence of effective therapeutics but rather the failure of numerous drug species to cross the BBB, blood-cerebrospinal fluid barrier (BCSFB), or other particular CNS obstructions to achieve accumulation in the specific regions of the brain [6–8]. Predetermined process in conveying therapeutics to the brain or CNS in understanding the physiology of BBB, its porousness under various neurotic or malady conditions, and its reaction to physical and synthetic stimuli, and the different transport receptors at the BBB. Several approaches have been made to deliver therapeutic agents to the brain across the BBB. However, none of them have yet been successful for clinical translation [9–12]. In the present chapter, we have tried to explain different methodologies employed to achieve the effective delivery of the therapeutics to the brain region by bypassing the BBB through different modes.

Nanoparticles (NPs) are viewed as a standout among the most favorable and flexible mode of drug delivery that reaches to inaccessible regions of the brain and effectively delivers the therapeutics into the damaged regions for a specific delivery.

A few NPs have shown their viability in crossing the BBB when their surface is improved with some ligands or surfactants. Considering this, it is imperative to comprehend BBB adjustments in pathology to exploit these characteristics to grow novel and inventive NPs that have the capability of effectively focusing on harmed regions of the brain.

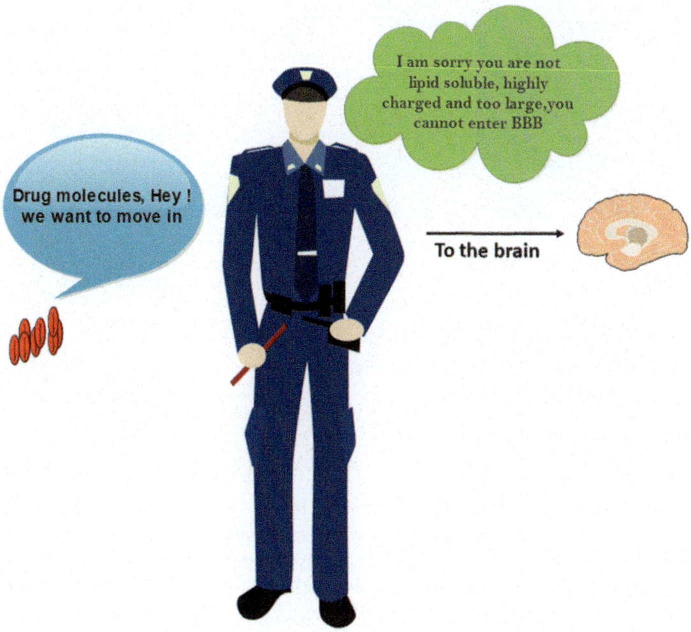


## 2.2 BBB Structure and Passage Mechanism

The vasculature of the central nervous system (CNS) is described by the presence of the BBB, which can be observed as both an anatomical and physiological phenomena. The BBB is considered to be the essential protecting mechanisms of our central nervous system (CNS). It selectively moderates the entry of substances into the brain region and permits the entry of only particular molecules to pass through the capillary endothelial membrane while constraining the movements of noxious agents and thus allowed the precise control of the brain's homeostasis. However, this protection mechanism is also a substantial impediment for the duration of diseased conditions because it dramatically blocks the conveyance of therapeutics to the brain in different hazardous degenerative CNS pathologies like Alzheimer's sickness, Parkinson's illness, cerebrum tumors, and other neurological disorders [13–16].

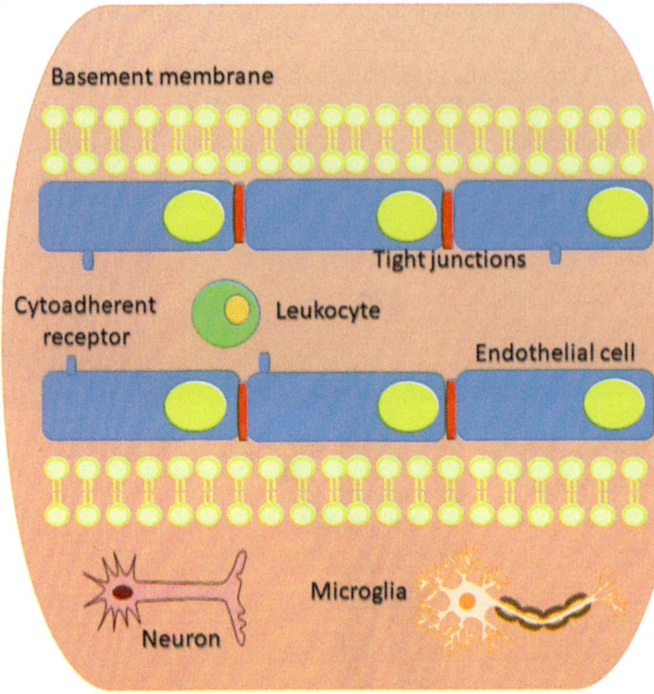


**Fig. 2.1A** Composition and principle adjustments found in neurotic conditions of blood-brain barrier (BBB). The BBB is chiefly made out of vascular endothelial cells, exceedingly associated by adherents and tight intersections (TJs) and a meager layer of pericytes. The basement membrane and a layer of astrocyte endfeet forms encompass the endothelium. Neurons and surveying microglia are additionally imperative go-betweens of BBB integrity in pathophysiological conditions

The BBB is primarily made up of tightly associated endothelial cells of the brain capillaries along with an intermittent layer of pericytes. The BBB cellular engineering and their adjustments in an orthodox manner are portrayed in Figs. 2.1A and 2.1B. The endothelium of the cerebrum has the unique property to keep the BBB integrity, transendothelial cellular transport, and angiogenic capacity to permit revascularization when required [13, 17]. The endothelial region of the brain cells express categorical proteins, to be specific tight junctions (TJs) and adhering junctions (AJs), which ensure the integral nature of BBB [1]. TJs are comprised of transmembrane and cytoplasmic proteins that incorporate claudins, occludin, junction adhesion molecules (JAMs), zonula occludens (ZO), and auxiliary proteins. Although there is a solid framework that keeps endothelial cells firmly associated, still the BBB can permit entry of a specific cell type and molecules to the cerebrum [18, 19].

The module of access between endothelial cells exists as paracellular and is used for particles and solutes that rely upon an inclination of fixation. The entry happening through endothelial cells is named transcellular, and the harmony between paracellular and transcellular transport is definitive to characterize the level of porousness in a solid BBB [20]. The transcellular pathway happens much of the

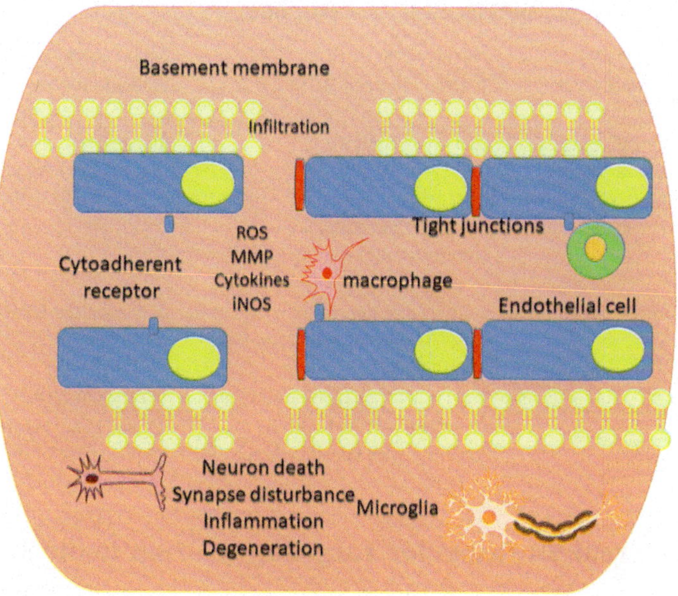


**Fig. 2.1B** In pathological conditions, a few BBB adjustments happened that ended up in expanded penetrability. Large network of metalloproteinase (MMP) action and higher reactive oxygen species (ROS) and nitric oxide levels (got from endothelial cells, by means of endothelial nitric oxide synthase (eNOS), or from microglia/macrophage cells, by means of inducible NOS (iNOS)) alongside arrival of cytokines and chemokines by initiated microglia/macrophages lead to basement membrane disruption, and TJ disturbance (specifically in occludin, zonula occludens (ZO)-1, and claudin 5 integrity) leads to an inflammatory reaction. Finally, these associated abruptions ended up in neurodegeneration

time with aloof dissemination of lipophilic atoms, which happens through specific receptors to transport molecules like CO2 [21–23]. Proteins and peptides that are hydrophilic are transported through receptor molecules, for example, for glucose uptake, there is glucose transporter-1 (GLUT-1) [24, 25].

Different types of transport take place using the arrangement of cell invaginations known as caveolae. These structures frame vesicles around the atom permitting the vehicle in or out of the brain cavity. As mentioned in the above segments, transcytosis components happening at BBB endothelial cells which are in effect at present are investigated as a route for transporting remedial medications into the cerebrum (Figs. 2.1A and 2.1B). In any case, the conveyance of a few of these medications to the brain parenchyma may likewise be diminished by ATP-binding cassette transporters (ABCs), which are dynamic efflux pumps that transport the neurotoxic lipid solvent particles or other pharmaceutic medications into the blood [26, 27]. More profound learning in regard to the systems of entry over this exceedingly specific obstruction will encourage the advancement of innovative procedures for the conveyance of therapeutics.

**Table 2.1** Different nanoparticles employed to cross BBB

| Strategies employed | Advantages | Disadvantages |
|---|---|---|
| Brain permeability enhancer | Transiently open BBB | Specific for few molecules |
| Delivery through permeable BBB under malignancy or faulty conditions | Can cross BBB | Limited excess of specific molecules, random, may lead to erogenous delivery |
| Non-invasive techniques | Bypass BBB through nasal administration, can open BBB and decrease efflux transporters | Toxic and suitable for dosage only |
| Nanoparticles | Active targeting, passive targeting, SiRNA, specific receptors, loading, imaging | Crossing the BBB and limited toxicity |
| Viral vectors | High gene transfection efficiency | Safety concerns |
| Exosomes | Gene delivery and specific targeting with specific receptors | Donor cells, loading issues, requirement of high dosage |

## 2.3 Different Strategies for Targeting of Drugs to the Brain

The diffusion of medications from the blood into the brain cavity depends for the most part upon the capacity of the biologically active molecule to traverse lipid membranes. In this manner, various medications don't have sufficient physicochemical qualities, for example, high lipid dissolvability, small atomic size, and positive charge which are essential to prevail with regard to the crossing of BBB. These limitations lead to the motivation behind why numerous strategies have been developed to cross the BBB, including intrusive and non-obtrusive methods (Table 2.1).

## 2.4 Convection-Enhanced Delivery (CED)

Convection-enhanced delivery (CED) has been considered as an encouraging strategy for delivering the therapeutics over BBB. It depends on the technique of positive pressure and persistent implantation arrangement over hours to days to keep a consistent pressure slope and advance interstitial liquid convection, as a supplement to osmotic impacts. It improves the dispersion of small molecules, macromolecules, macromolecular proteins, and different medications, making it a suitable drug delivery mode to cross the BBB. It has been seen that nanoliposomal irinotecan together with radiation therapy showed synergism with more prominent survival rates than the irinotecan or radiation alone or radiotherapy with vascularly regulated irinotecan. Several clinical trials and studies on model animals using this strategy uncovered that it might need further improvement before it is used securely in people [27–29].

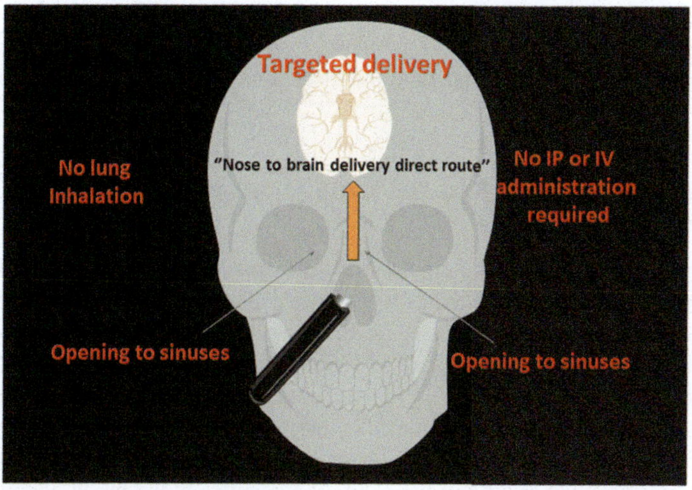


**Fig. 2.2** Delivery of the drug through the nasal cavity

## 2.5 Intranasal Delivery to Bypass BBB

Drug delivery through intranasal route is considered to be a positive one as it is categorized with the quality of usability, rapid absorption, quick beginning, and non-ruinous and non-intrusive method. This method has turned out to be a standout among the most famous strategies in drug delivery. To bypass the BBB, the drugs can be passed to the cerebral spinal liquid or the cerebrum through the olfactory mucosa, along with the connective tissue around the olfactory nerve package or axons of olfactory neurons (Fig. 2.2). A large number of peptides, proteins, and drug molecules can be transported into the cerebrum using the intranasal mode of delivery. Amphiphilic methacrylic copolymer-functionalized poly(epsilon-caprolactone) nanocapsules were proposed by Fonseca et al. that comprise of a much-stick structure to pass on olanzapine through the intranasal association. Studies revealed that these nanocapsules could interface with mucin and nasal mucosa and augment the support of olanzapine (about 40%) on the nasal mucosa after steady washing. The outcome demonstrates that the nanocapsules improved the measure of olanzapine in the rodents' brain (1.5 higher fold comparing to the free form of the drug). Further, olanzapine nanoform did not influence the nasal mucosa uprightness after rehashed portions. The above mentioned information exhibited that the synthesized nanoformulation is a promising framework for nose-to-cerebrum the drugs delivery [30–32].

Intranasal conveyance is restricted, be that as it may by medication portion and physical and concoction properties. The surface zone of the olfactory area of the nasal epithelium is about half in rat and just 5% in human. Along these lines, the intranasal conveyance isn't anticipated that would accomplish therapeutic medication levels in brain regions. Kozlovskaya et al. broke down the accessible quantitative information on intranasal conveyance to uncover the effectiveness of cerebrum tranquilize conveyance and focus on various sorts of nasally managed conveyance systems. Results showed that proficiency of intranasal transport varies drastically between the examinations and does not relate with the medication's physicochemical properties.

## 2.6 Employment of NPs for Drug Delivery into the Brain

An alternative approach is the employment of nanotechnology to deliver therapeutics to the brain. Nanomedicine has lately emerged as a promising field for innovative and practical strategies to cross the BBB and target lethal diseases. The benefit of developing nanoparticles for drug delivery results from their two fundamental properties. Firstly, because of their miniature size, nanoparticles infiltrate into even little vessels and are taken up inside cells, permitting a useful drug accretion at the targeted sites, and the utilization of biodegradable materials for nanoparticle synthesis supports sustained release of the chemotherapeutics at the targeting sites over a more extended period.

A large variety of polymer-based nanoparticles can be synthesized like polyethylenimine (PEI), poly(alkyl cyanoacrylates), polyamidoamine dendrimers (PAMAM), poly(ε-caprolactone) (PCL), polylactic-co-glycolic acid (PLGA), polyesters (polylactic acid) (PLA), or inorganic NPs from inorganic materials, for instance, gold and silicon dioxide (silica), among others (Table 2.1) (Fig. 2.3). These bearers can transport therapeutics to the affected organs either by ensnaring, adsorbing, or covalently connecting to them [33–35]. The NPs having an inorganic origin offer focal points over polymer-based NPs as they have authority over their size and shape and straightforwardness of arranging and surface modifications. Specifically, the metallic NPs are facile to track by conventional microscopy in vogue (e.g., MRI, TEM, or ICP-MS). Even though the metallic NPs displayed positive things, these NPs possess undesirable effects, viz., excretion through the kidneys or their degradation, which may exhibit undesired noxious condition (e.g., carbon nanotubes and fullerenes may provoke lipid peroxidation and oxygen radical development).

The NPs acquired from regular sources like for proteins (gelatin and albumin) or polysaccharides (chitosan and alginate), amino acids (poly(lysine), and poly(aspartic acid) (PASA) have an edge over other NPs, as they provide biological signals to interact with specific receptors/transporters expressed by endothelial cells.

The utilization of NPs to convey medications across BBB may give a preferred critical standpoint over the conventional modalities. As opposed to techniques for constrained drug delivery, NPs can be transported over the BBB via different carriers, additionally alluded to as nanocarriers, without harming BBB [7]. The essential favorable position of NP bearer innovation is that NPs make up for the BBB-constraining attributes of the restorative medication atom. Moreover, this framework may moderate medication discharge in the brain, consequently diminishing peripheral lethality. Certain factors that influence the drug delivery across the BBB are polymer conjugation or surfactant coating. Some NPs have

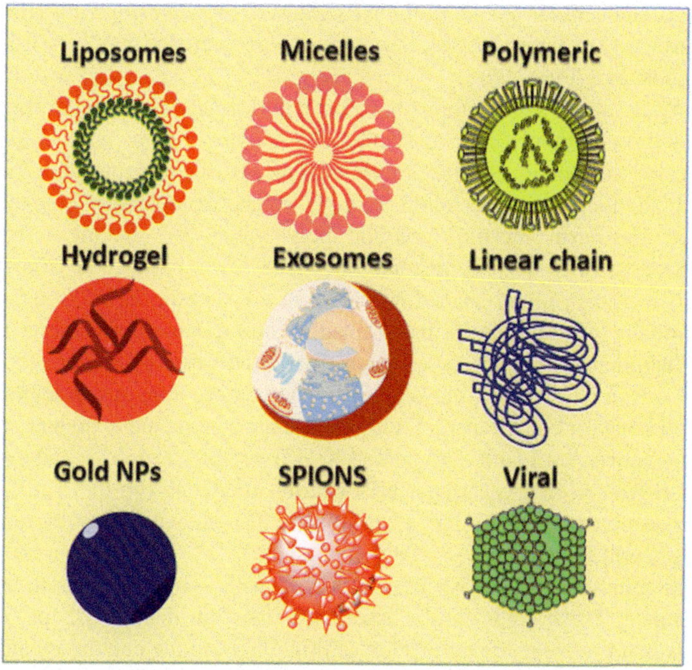


**Fig. 2.3** Different kinds of nanoparticles to cross BBB

been utilized for drug delivery, which were coated with allogeneic components of the body, i.e., cell membranes. However, their clinical applications for BBB amiability remain a challenge. Doxorubicin, an anticancer drug, has been shown to be effectively transported into the brain employing NP approach. A large portion of the methodologies portrayed for the passage of medications through the BBB can be improved by nanotechnology. The main features of nanoscale materials include:

1. NPs empower the drugs to enter the BBB through opening the tight intersections between endothelial cells.
2. The NPs undergo transcyotosis and cross the BBB through endothelial cells layer.
3. Coatings employed for NPs, for example, polysorbates, hinder the transmembrane efflux frameworks (i.e., P-glycoprotein).

## 2.7 Aspects Affecting the Entry of NPs over the BBB

There are a couple of factors that influence the productivity of NP fundamental course, BBB segment, and cell movement. A few examinations exhibited a steady relationship between the penetration of BBB and the size of the NPs and BBB (Fig. 2.3). In a precise manner, the NPs being in size range of 50 to 100 nm showed maximum activity and uptake in tumor models, stroke, AD, or PD models. The movements and cellular uptake of the NPs showed a relationship with NP structure and its surface variability [36–38]. Structure of Nanoparticles may vary from round, cube, to rods, etc. (Fig. 2.3).

Most by far of the examinations have been performed with circular formed NPs since they are commonly easy to synthesize. A number of in vitro investigations have revealed that nanorod surface modified with specific proteins or antibodies has a higher inclination than their round- or cube-shaped nanoparticles counterparts. In particular, transferrin receptor-modified polystyrene NPs having a rod shape ($501 \pm 43.6 \times 123.6 \pm 13.3$ nm) had displayed seven times more efficacy when contrasted with their round-shaped NP under in vivo conditions ($200 \pm 0.01$ nm).

Additionally, one more factor that limits the entry of nanoformulations over BBB is their zeta potential. Studies have shown that the NPs bearing higher zeta potential cause brisk destructiveness to the BBB [39]. Along these lines, a vast segment of the NP depicted that for brain transport, they should have moderate to high negative zeta potential ((between −1 and −15 mV) [40–43] or (between −15 and −45 mV)) [36, 37]. However, in some cases, NPs bearing moderate (up to 15 mV) or high positive zeta potential (more than 15 mV) have also shown the ability to cross the BBB (Fig. 2.3) [37, 38].

A large number of NPs have been modified with various ligands to encourage BBB entrance. These kinds can be assembled into four distinct groups (Fig. 2.3): (i) ligands that intervene the adsorption of proteins from the circulatory system that interface straightforwardly with BBB receptors or transporters [44], (ii) ligands that have coordinate cooperation fundamentally with BBB receptors or transporters, (iii) ligands that enhance charge and hydrophobicity [33, 34], and (iv) ligands that enhance blood flow rate (e.g., PEG) [35, 45].

In the primary case, we can incorporate polysorbate 80 that helps in the adsorption of apolipoprotein E. The surfactant permits the anchoring of apolipoproteins whose connection with lipoprotein receptors empowers the intersection of the BBB. In the second case, we can incorporate a few ligands, like transferrin receptor (transferrin peptide, transferrin protein, or mAb against transferrin) [41, 42], insulin receptor [44, 46], GLUT [8, 47], etc. In the third case, NPs modified with amphiphilic peptide help in the uptake of BBB endothelial cells.

The density of ligand over the NPs depends upon both the NP surface area and the ligand size. Regularly, the ligand proclivity to its receptor is reduced when conjugated to NPs. NP vitality and effectiveness enhanced when a variable amount of ligands are attached. In any case, NP specificity must be balanced for suitable BBB transcytosis. It has been exhibited that when high concentrations of transferrin (100–200 particles of transferrin for every NP) are attached to gold NPs, they remain attached to brain endothelial cells. Whereas when there is a low concentration of transferrin (20–30 molecules of transferrin per NP), they can relate efficiently with the receptor, experience transcytosis, and get moved easily into the parenchymal region of the brain [34, 41, 42].

At the point when NPs enter a physiological situation, there is quick adsorption of proteins from the circulatory system to the NP surface framing a protein covering—the "protein corona" [48, 49]. More than 70 distinct serum proteins have been accounted for to adsorb to the surface of gold NPs [50, 51]. The protein corona may adjust the surface science of the NPs alongside its accumulation state. All the time, it likewise quickens the blood of the NPs through the reticuloendothelial framework generally restricted in the spleen and liver [39, 43], which may diminish the NP dose accessible for brain accumulation.

The most widely recognized approach to address this issue is to utilize particles with the ability to limit surface fouling to keep up execution and well-being of materials. In this sense, antifouling properties can be improved by using PEG-modified NPs. PEGylated NPs present insignificant surface charge prompting lower NP opsonization and lower reticuloendothelial framework uptake [52, 53]. Combining NPs with PEG (5 kDa; somewhere in the range of 0.16 and 0.64 PEG particles per $nm^2$) diminishes protein adsorption and backs off the clearance of the nanomaterials [54, 55]. Additionally, because of its enhanced blood flow rate, PEGylated NPs gather more effectively in the brain [56, 57]. For example, polystyrene NPs (200 nm) covered with PEG (5 kDa; 9 PEG atoms for each 100 $nm^2$) can cross the BBB. Furthermore, PEG-modified PLGA NPs are likewise ready to quickly enter the rodent brain ex vivo, compared to bare NPs.

As a whole, a number of factors impact the delivery of NPs through the BBB at various degrees. The presentation of the NP is a very important factor, and a few viewpoints, for example, the density of ligand and its effect in NPs transport through the BBB, are not all around examined. Up until now, NPs conjugated with ligands ready to interface with BBB receptors at a moderately low thickness (less avidity) have the best execution. However, take note that the best definitions directed intravenously convey up to 5% of the underlying portion viably over the brain. NPs based delivery in brain conveyance enhancement may require frameworks that objective and cross all the more effectively the BBB yet additionally, frameworks that are gradually obvious from the circulatory system. As to the last issue, the surface chemistry and the morphology of the NP have an essential impact on freedom. It has been shown that in animal trials after IV administration, NPs bearing zwitterionic charge have a more extended dissemination time, rather than contrarily and emphatically charged NPs. Also, the small-size NPs are primarily held in the liver and present a fast leeway rate, comparing to the large-size NPs that moved to the spleen and have a slow flow rate [43] (Fig. 2.4).

## 2.8 Different Types of Nanoparticles Employed to Cross BBB

### *2.8.1 Liposomes*

The second class of drug delivery termed as liposomal Trojan horses involves liposomal-based delivery of therapeutics. Liposome properties differ significantly with lipid structure, estimate, surface charge, and the technique for planning.

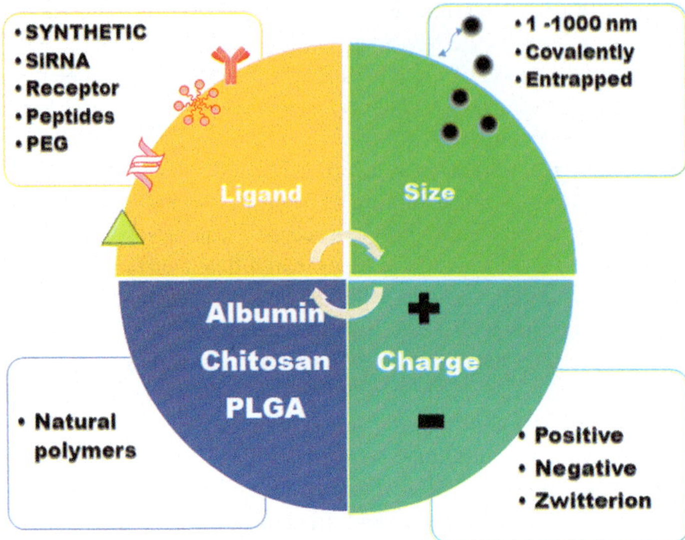


**Fig. 2.4** Primary highlights of the nanoparticles are impacting the systemic delivery and helping in intersecting blood-brain barrier (BBB). At the point when atoms, for example, proteins (egg whites), polysaccharides, chitosan, among others, are utilized for the combination of NPs, these are known as common polymers. Manufactured NPs can be blended from polymers, for example, polylactic-co-glycolic acid (PLGA), polyethylenimine (PEI), and polyesters (polylactic acid) (PLA), or from inorganic materials like gold, silica, silver, and so on. NPs may differ in size (1–1000 nm) and can convey drugs into cells by either capturing, adsorbing, or covalently holding them. NPs can expect diverse shapes (circular, cubic, and rod-like) and charges (negative, zwitterionic, and positive)

Miniature liposomes, nanoliposomes, measure 25–50 nm in breadth. Liposomes are biocompatible, yet generation costs are high. Liposomes can be utilized in systems for effective delivery of therapeutics over the BBB. Functionalization of liposomes with ApoE-determined peptides encourages cell uptake and tranquilizer transport over a model of the BBB [56, 58, 59]. These vehicles have an exceedingly lipophilic character and are hence increasingly effective at infiltrating the BBB. The liposomes might be utilized to convey specialists that can't cross the BBB to their intraparenchymal targets and help keep up a controlled arrival of chemotherapy or quality treatment operators. Liposomal Trojan horse-intervened section over the BBB was first portrayed in 1980, and from that point forward, it has been broadly examined for pharmaceutical conveyance. The liposomal Trojan steed's way to deal with testing the adequacy of conveyance in treating GBM has been connected in numerous creature models, for example, canine, rodent, and nonhuman primate models. Two clinical preliminaries with liposome-stacked doxorubicin for the treatment of strong tumor metastasis and threatening glioma revealed more effective results than in control treatment, even though, in the glioma trials, the middle in general survival with the standard of consideration surpassed that of the liposome-stacked methodology. We additionally note that a great part of the ongoing examination into liposomal vehicles has concentrated on its application in siRNA-based therapy.

### 2.8.2 Lipid-Based NPs

Lipid NPs are spherical structures having a size range in nanometers, comprised of a lipid bilayer and installed with conformationally flawless necessary film proteins. The lipophilic part of their structure helps in traversing the BBB to enter the cerebrum. A few potential medications have been employed with solid lipid NPs (SLNs), for the treatment of CNS ailments [60–62]. SLNs have some limits; however, they have some advantages over traditional delivery modes: unlike classical bilayer liposomes, there is no irregular combination between the drug moieties or with different layers.

Surface chemistry can be effortlessly modified with the multivalent ligands for effective controlled drug delivery. SLNs, around a size of 33–63 nm, and stacked with the chemotherapeutic idebenone, have been seemed to cross an in vitro model of the BBB utilizing a transcellular pathway [63, 64]. Molecular Trojan horses, the antibodies that encourage transcytosis of carrier molecules have been utilized for imaging in a nonhuman primate demonstrate protein transport was dynamic for translocating antibodies over the BBB. An investigation by Farrington et al. [57] included a remedial neutralizer having a recently depicted BBB-transcytosing arm called FC and a bivalent helpful arm intertwined with the human Fc protein. This remedial agent expanded the exchange of 2 locally impermeable particles, neuropeptide Y and neuropeptide dalargin, 30-overlap over the BBB. The FC5 counteracting agent focuses on the epitopes of the BBB-advanced transporter, and this focusing encourages receptor-interceded transcytosis. The low-thickness lipoprotein receptor-related protein-1 was the objective in Phase I clinical preliminary for intermittent dangerous GBM in which GRN1005, a peptide medicate conjugate with paclitaxel, yielded adequate outcomes to warrant further examination in Phase II preliminaries. Different targets likewise have been utilized in atomic Trojan steed approaches, for example, the transferrin, glutathione, and insulin receptors [65, 66].

However, these technologies are not as specific and, in this manner, less successful at upgrading BBB penetrability. The Trojan horse-based strategy was late explored for RNA interference through the transferrin receptor focusing on gliomas, and this approach empowers target-specific treatment.

### 2.8.3 Polymer-Based Nanocarriers Across the BBB

Polymeric nanoparticles are comprised of natural polymers (e.g., chitosan, gelatin, or polysaccharides) or engineered polymers (e.g., polyethyleneimine, polycaprolactone). Some engineered polymers are polyglycolic acid, polylactic acid or their copolymer polylactic-co-glycolic acid, etc. It is favorable that the concoction polymer structure can be effortlessly changed, permitting nanoparticle refinement. At the point when orchestrated as tediously stretched particles, polymers can likewise frame symmetric, circular mixes called dendrimers. The cationic polymers due to their ability to associate with oligonucleotides electrostatically are considered to be most reliable for siRNA delivery. A few classes have been proposed concerning nanoparticle-mediated siRNA delivery, for instance, poly-L-lysine, polyamidoamine dendrimers (PAMAM), polypropylenimine dendrimers, and polyethyleneimine (PEI) and also more explicitly structured polymers like poly(β-amino esters) and cationized cyclodextrins. Additionally, in light of anti-amyloidogenic potential, they have likewise been utilized for the handling of different neurological disorders, for example, AD, PD, and prion infections [67–72].

The polymeric NPs to be employed for crossing should be biocompatible, nontoxic, and non-immunogenic. For example, polysorbate 80 covering on the outside of PBCAs adsorbs ApoB and E, which support their uptake by endothelial cells utilizing receptor-interceded endocytosis and segment over the BBB. Polymeric NPs have been seemed to ensure transporters for CNS-based medication conveyance due to their potential both in various loading drugs, thus protecting them from release and absorption and an appropriation. Polymeric NPs empower focused delivery of drugs at a precise rate and help in the transport over the BBB [68, 71].

A couple of polymer-based frameworks for conveying therapeutics over the BBB using NPs have been examined. For example, leptin sequence, 12–32 (g21) modification on the PLGA NPs, supports their movement over the BBB into the cerebrum following intravenous organization in mice models, demonstrating their drug delivery potential [72–74]. These peptides were altered with the fluorescent labels with the objective that their entrance over the BBB can be tracked with confocal microscopy in rodent models following intravenous infusion. As shown by the eventual outcomes of PLGA NPs that were unfit to cross the BBB at first, they could do like this after the peptide modification. Nanotechnology-based frameworks for medication development over the BBB combine two showcased progressions, LipoBridge™ (Genzyme Pharmaceuticals, MA, USA) and G-Technology® (to BBB, Leiden, the Netherlands) [75, 76].

Another polymer from natural source shrimp shells, chitin, has a couple of therapeutic applications. Adjustment of chitosan with a particular ligand has a proclivity for cell surface receptors and permits its uptake into cells through receptor-mediated endocytosis. Chitosan NPs are known for their ability to support the transportation of different prescriptions, for instance, insulin, antigens, and plasmid DNA. For the dynamic conveyance of a medication crosswise over BBB, chitosan NPs can be adjusted with an assortment of peptides. Their surface can be changed with PEG to upgrade the plasma circulation time through preventing NP uptake by the reticuloendothelial system. Monoclonal antibodies (mAbs) against the Tf receptor were conjugated to NP surface utilizing biotin-streptavidin bonds. The activation of Tf receptors by the NP-mAb complex affects transcytosis and initiates

**Table 2.2** Characteristics of NPs employed under the in vivo condition

| Nanoparticles | Nature of material | Size (in nm) | In vivo application | Pathway involved | Biocompatibility |
|---|---|---|---|---|---|
| Natural polymers | Collagen hydrogels | ~15 | Drug delivery | IV and IA | Non-toxic and biodegradable |
| | Chitosan | | Antigenic carrier | | |
| Lipid-based NPs | Liposomes modified with PEG | ~100 | Drug delivery | | Non-toxic |
| | Cationic lipids | ~100 | Genetic materials | IV | Low toxicity |
| | Solid lipid NPs | ~50–200 | | | |
| Dendrimers | PAMAM G4 | 5–20 | Drug delivery | IV | Non-toxic and biodegradable |
| | | 5–20 | Genetic materials | IP | |
| | PPI glycodendrimers | | | | |
| Synthetic polymers | PLGA | ~100–200 | Drug delivery and genetic materials | IV | |
| | PBCA | ~200 | | | |
| | PEI | ~50–200 | Genetic materials | IP | Non-toxic and biodegradable |
| Quantum dots | CdSe/ZnS | 2–10 | Imaging | Stereotactic | ROS |
| Iron and gold NPs | SPIONS | <10 | Imaging SiRNA loading | IV | Non-toxic |
| | AuNPs | | | | |

the caspase framework in this manner prompting enactment of caspase-3 inhibitor to the brain to limit ischemia-induced caspase-3 development and provide neuroprotection [41, 77] (Table 2.2).

## 2.9 Employment of Different siRNA-Based Deliveries

In the field of cancer therapeutics, nanomedicine has opened new doors because of the likelihood of adjusting pharmacokinetics of existing medications and also taking into consideration the conveyance of drug molecules that are generally not reasonable as therapeutics. These include several small interfering RNAs because of their ability for interference in the genome structure and affect the changes at the cellular level (RNAi) (Fig. 2.5). Since their arrangement explicitness allows the focusing on specific targets, they offer the possibility to a great extent upgrade effective

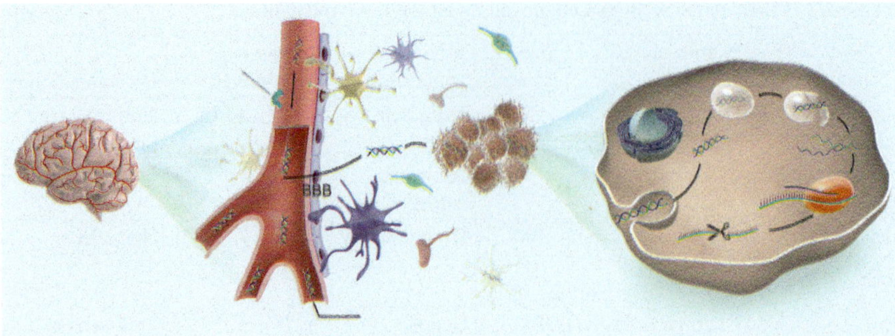


**Fig. 2.5** Delivery of siRNA to treat brain diseases. (Image courtesy: Zheng et al. 2018)

alternatives. siRNAs are an essential part of the RNAi pathway to deal with a few maladies, and some clinical trials are in advancement; however, still, none of these include conveyance to the brain.

Nonetheless, because of their chemical nature, siRNAs are inclined to quick enzymatic (RNAses) and nonenzymatic degradation, and because of their comparably large size and charge, they can't be promptly taken up by cells and, thus, presents significant restrictions as to entering the BBB. All in all, the system for siRNA delivery in the cellular network requires the utilization of potent and methodologies of protection and distribution, including synthetic adjustments and the usage of nanoparticle frameworks. For example, an 29-amino acid peptide (RVG) empowers the trans-vascular delivery of siRNA to the cerebrum. After IV administration into mice, RVG-9R delivered siRNA to the neuronal cells, bringing about particular quality quieting inside the brain. Moreover, intravenous treatment with RVG-9R-bound antiviral siRNA managed vigorous protection against lethal viral encephalitis in mice. Also, some chemical modifications in the siRNA include incorporating adjustments and changes of the backbone phosphates (e.g., phosphorothioate), and nucleotides (e.g., LNA, UNA, 2-OMe) also allowed effective delivery of the siRNA [78–82].

Beyond these factors, another essential query is the selection of relevant targets, therefore requiring a more profound comprehension of the biology of the brain tumor. Further, this may also incorporate the combinational therapies for consolidating siRNA-mediated knockdown with known chemotherapeutic agents for improved efficacy or for investigating twofold knockdown procedures.

### 2.9.1 *Exosome-Based Delivery*

Exosomes are nanovesicles of 40–100 nm size range that have been directed intravenously for the transport of siRNA into the brain region. This system for transport avoids uptake by various tissues outside the brain and any ensuing immune reaction. There is no loss of capability following the reiterated organization of centered exosomes. The exosome-based methodology for siRNA conveyance was given of an impression of being feasible for knockdown of the BACE1 gene in mice models [83, 84].

### *2.9.2 Employment of Cell-Penetrating Peptides (CPPs)*

Cell-penetrating peptides (CPPs) are a general term given for small-size peptides having cell-entering potential; they are commonly under 30 amino acids, made transcendently out of the basic amino acids. CPPs help in the entry of small molecules into the cell, as well as in the macromolecules (multiple times higher than their subatomic weight, e.g., protein, plasmid, nucleic acids, siRNA, nanoparticles, liposome, etc.) [85–87]. In the expansion, CPPs have the upsides of high organic security and minimal effect on normal cells. These highlights of CPPs make them viable transporters for drug delivery and give another incredible asset over other treatment regimes. Inferable from the distinctions in the arrangement, hydrophobicity, and extremity, CPPs can be characterized into three groups: cationic, antibacterial, and chimeric peptides. When the nanoparticles are associated with the CPPs, their efficacy can be significantly progressed. Recent studies on CPP-modified PLGA by Lu et al. revealed that the PLGA NPs modified with CPP may convey insulin into the brain through the nasal course, showing an aggregate brain conveyance productivity of 6%. Chaudhary et al. showed that a surface modification of NPs with CPPs could advance active translocation of drug load into the cell [88–90].

### *2.9.3 Other Types of NPs*

The different variety of nanoparticle frameworks that have been produced for systemic or local delivery for in vivo models comprise inorganic nanoparticles, SNALPS, etc. [12, 91].

SNALPs (stable nucleic acid-lipid nanoparticles) containing a blend of cationic and fusogenic lipids, covered with diffusible polyethylene glycol, are one case of nanoparticles mainly intended for the conveyance of siRNAs. Solid lipid nanoparticles involve a strong lipid center that is stabilized by emulsifiers.

Similarly to that of liposomal nitrogen horse-based delivery, the viral nanoparticles can also be employed for drug delivery, except the technique for BBB circumvention includes a simple diffusion from a convection-upgraded conveyance, intraventricular or intrathecal source. Analyses in creature models have demonstrated that nanoparticles improved in vivo dealing of chemotherapy specialists, for example, camptothecin, TMZ, doxorubicin, irinotecan, and vincristine, and furthermore of quality treatment agents.

Nanoparticle-based methodologies can remarkably serve in focused thermotherapy, for instance, through stereotactically infusing attractive nanoparticles and initiating molecule vibration using MRI. This field speaks in a novel way to deal with the effective delivery of chemotherapy with a guarantee for future research and clinical preliminaries. The nanoparticle-based has now extended to incorporate quality treatment as a load. Current investigation has demonstrated that adenovirus, adeno-related infection, vesicular stomatitis infection, retrovirus, herpes simplex infection, liposomal Trojan steeds, Semliki Forest infection, and manufactured nanoparticles would all be able to be utilized as vehicles for BBB transit. Expansion of quality treatment and variety of diffrent ways to deal and treat neurological malady has incorporated the utilization of viral vectors for GBM treatment. Nonetheless, popular vectors, albeit proficient and adaptable by tropism and replication strategy, are tricky because they may represent a hazard for disastrous immunogenic reactions and may not adequately achieve tumor cells that have dispersed all through the cerebrum parenchyma [92, 93].

The reactions of viral vectors have prompted more noteworthy appropriation of manufactured delivery systems, as depicted above, for example, liposomal Trojan gene delivery and nanoparticle-mediated delivery of genes for quality treatment. The benefits of these engineered delivery systems are that they target tumor cells and can convey expansive gene loads. Also, a significant number of these conveyance techniques evoke almost no insusceptible reaction, making them perfect vehicles for DNA and siRNA delivery [78, 94].

Inorganic nanoparticles are derived from metals, metal oxides/metal sulfides, and silica. Because of their high surface area to volume proportion, they take into consideration proficient siRNA stacking by noncovalent epitome or direct conjugation, and their surface science can be effortlessly altered. Given their exceptional physical and optical properties, various inorganic nanoparticles can be utilized for drug delivery and imaging purposes, consequently making them valuable for indicative and remedial purposes. Issues for the long-haul application may incorporate moderate suspension and nonbiodegradability. On account of siRNA, the extracellular and intracellular (endosomes, lysosomes) security of the payload is of specific significance, before discharging the siRNA in the right intracellular compartment. Directed delivery of nanoparticles may enhance the specificity and efficacy of tumor cells and can be especially critical in glioma treatment as to intersection to the BBB. The in vitro estimation of nanoparticle properties should consequently be taken with caution since they may contrast from the real circumstance in vivo, accordingly. Carbon nanotubes are generally carbon-based nanoparticles for drug delivery comprising of at least one layer of graphite tubes (single or multiwalled tubes). As a result of the enlistment of oxidative pressure and lipid peroxidation, toxicity remains an issue [94–97].

Hybrid nanomaterials are a mix of inorganic and natural nanomaterials, with the end goal that they do not just display the profitable properties of the two materials included, however, can likewise show extra preferences of their own. These hybrid nanomaterials incorporate advancements, for example, nanoscale metal-organic frameworks (MOFs), functionalized nanotubes, and nanogels. Lin and collaborators

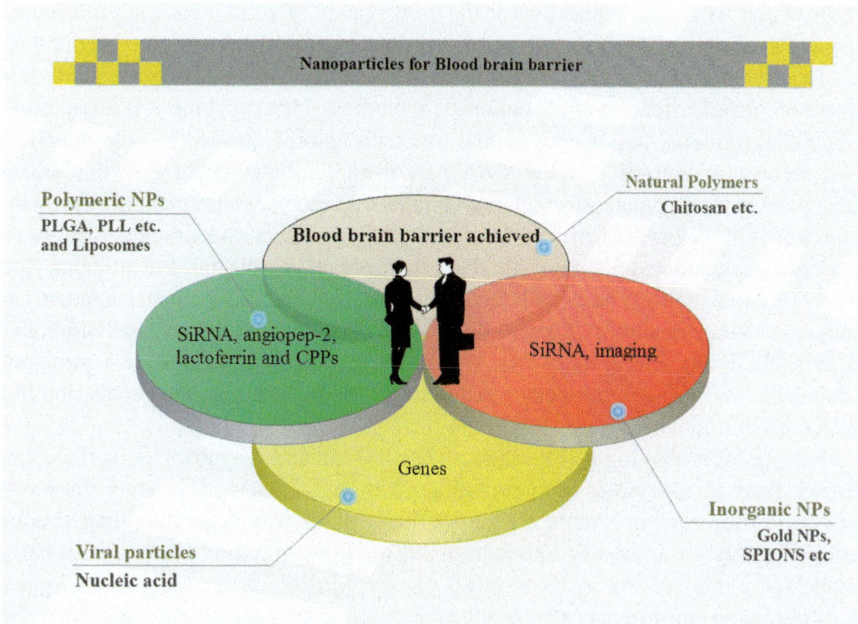


**Fig. 2.6** Flexible nanoparticles to cross the blood-brain barrier (BBB)

utilized a mesoporous silica-based nanoparticle framework and cadmium sulfide nanocrystals as removable caps for controlled release of medications and neurotransmitters. This molecule framework was observed to be biocompatible and was utilized to explore neurochemical communications in astrocytes (Fig. 2.6).

## 2.10 Mechanism of Nanoparticle Route to BBB

The mode of BBB entering framework can be portrayed in dynamic or active and passive transport courses. The passive system involves the transport of substance without the use of energy, for instance, straightforward and simple diffusion. The medications move in tumor cells through the EPR effect [98]. Oppositely, the dynamic transport involves the carrier intervened transport, which all require the hydrolysis of adenosine triphosphate (ATP) [99–101].

Additionally, the medication transported inside the cells varies in different NPs and cellular mechanisms. For example, the small and stereospecific pores in the active targeting framework confine the delivery of extensive subatomic medications. Furthermore, the drug transported inside the cells changes in various NPs and cell instruments. For instance, the little and stereospecific pores in the dynamic focusing on system limit conveyance of broad subnuclear medications. Rather, most carbon dots (CDs) have an ultra-little size (1–10 nm) and adaptable surface functionalities

[59], which would be supportive of the conveyance of substantial drug molecules using the carrier-mediated transport by covalently conjugating with medications.

The targeted delivery of drug has been intensely grown late, particularly for disease drugs for brain tumors, including neuroglioma (astrocytoma, oligodendroglioma, and ependymoma) and meningioma situated outside the BBB and metastatic injuries (for the most part from lung malignant growth, bosom malignant growth, dangerous melanoma, renal carcinoma, colon malignant growth, and different malignancies). A standout among the most encouraging regions of nanoparticle-mediated delivery is the treatment of brain tumors, utilizing the improved penetrability and maintenance (EPR) impact. This impact happens in numerous kinds of tumors that are described by inadequate hyper-vasculature and a deficient lymphatic seepage system. The size of a nanoparticle is a principal trademark that decides the detached focusing on medications and biodistribution inside brain tumors.

Nanoparticles having a size range of 10–100 nm and a hydrophilic surface can escape from phagocytosis of the reticuloendothelial framework in vivo. They can course in the veins for a more drawn out time, in this way expanding their shot of achieving the tumor tissues. The remarkable qualities of tumor tissues additionally empower nanoparticles to be focused on specifically on the target. The quick multiplication of tumor cells requires oxygen and supplements provided by existing and fresh recruit vessels, while the confused development of tumor cells makes a profoundly sporadic age of tumor angiogenesis. This prompts an expanding hole between tumor vascular endothelial cells and lymphatic vessels, called the EPR impact. The quickly developing tumor cells demonstrate a high metabolic state and need unquestionably more oxygen and supplements than the tumor itself can give. In this manner, tumor cells create extra vitality by glycolysis, which can cause fermentation around tumor cells. One of the imperative research headings of nano-sedate conveyance frameworks is to create nanoparticle-stacked medications dependent on the acid-base awkwardness of the smaller-scale condition. Under ordinary nonpartisan pH esteems, nanospheres are steady; however, when they enter the acidic condition of the tumor tissue, the microspheres will quicken the crumbling and in this manner rapidly discharge its inward stacked medications to accomplish a high medication focus in the objective area. Ruan et al. built up a shrinkable nanocarrier, G-AuNPs-DOX-PEG. In this stage, DOX and PEG anchors were fastened to the surface of gold nanoparticles using Au-S bonds. The medication-stacked gold nanoparticles were then covered with gelatin nanoparticles. The arrival of DOX from G-AuNPs-DOX-PEG happened in a pH- and time-subordinate way. At pH 5.0, the arrival of DOX was quicker than at high pH esteem, with the discharge rate of DOX from G-AuNPs-DOX-PEG about 90.9% [85, 102–104].

Receptor-mediated endocytosis is one of the important mechanisms by which endogenous macromolecules traverse the BBB. Few macromolecular polypeptides like insulin and transferrin in the blood cross the BBB specifically by the receptor-mediated transport system. A number of endogenous substances, for example, ligands, consolidate with specific receptors communicated on the BBB cavity surface to shape corpuscles through endocytosis and at that point discharge the ligand by exocytosis and navigate the BBB to enter the brain tissue. Accordingly, different relating ligands of receptors are structured and utilized as a transporter to convey drugs into the brain using the receptor-mediated transport pathway [86, 87, 91].

Similarly one other protein receptor for instance Transferrin receptor (TfR), which is responsible for the transport of iron, is expressed in the brain vessels and enables iron transport into the brain by the TfR-mediated transport pathway. Peptidomimetic mAbs OX26 (to the transferrin receptor) was structured and interfaces with the TfR epitope at a non-characteristic TF restricting site, which evades the opposition of focusing on TF stacked medications with normal ligands; along these lines it can help distinctive medication atoms through the BBB into the brain. Studies have demonstrated that intravenous administration of OX26-BDNF causes a 65–70% decrease in stroke volume in mice models [105–107].

Likewise, low-density lipoprotein receptor-related protein-1 (LRP1) and LRP2 ligands are also utilized for drug delivery. The LDL receptor family is also expressed in BBB and may intervene a progression of substances over BBB through endocytosis, for example, lactoferrin, melanotransferrin, receptor-related protein, and so on. In the meantime, low-thickness lipoproteins additionally show high articulation in glioma cells. Angiopep-2, an explicit ligand of LRP1, can intervene in the framework to infiltrate the BBB and target glioma cells. A large number of studies have shown that the ANG-conjugated NPs displayed biocompatibility for normal cells with better cell uptake in U87 cells [108–111].

## 2.11 Neurotoxicity of the Nanoparticles

A limited knowledge is accessible on the well-being effects of NPs within the size range of 20 nm as they are discharged from the body, and the concern point is the effect of NPs of size 50 nm, but they are nontargeted and come in contact with healthy cells. There are as yet numerous unanswered inquiries regarding the destiny of NPs in the living body. Given the wide going sizes of NPs and the decent variety of materials utilized, some of which might be nontoxic, these impacts will shift incredibly. Modification of NPs with PEG showed potential danger as it can anticipate communication with body tissues and is utilized securely in a few pharmaceutical arrangements of NPs with possible poisonous quality which can counteract connection with body tissues. Some nanomaterials, for example, ceria NPs and water-dissolvable subordinates of buckminsterfullerene C60, are one of a kind class of NP with strong cancer prevention properties [112, 113].

However, currently, there is no downright proclamation on the well-being of NPs (i.e., one can't state that NPs are safe for human trials). Further examinations need to be elaborated before getting any convincing information about their neurotoxicity, albeit no convincing information is accessible.

## 2.12 A Future Point of View

In the context of the noteworthy headway in nanomedicine, both research and clinical applications in various diseases and improvement of protected and convincing techniques for nanobiotechnology-based CNS therapeutics are anticipated in the accompanying 5–10 years. Security and risk issues will be settled, and there will be an improved appreciation of the part of prescription transport through the BBB. Future examinations will move the currently limited data of changes in the BBB that occur in various neurological issues, and these will be contemplated when organizing medications that are to be passed on over the BBB. A couple of clinical primers will be coordinated to set up the practicality of current and furthermore new drugs for neurological disperses, including natural medicines, for instance, quality treatment and siRNAs.

The future accentuation will be on creating drug conveyance frameworks for the mind that can convey sufficient measures of medications in a controlled way in the light of requirements. A portion of these will be incorporated with implantable gadgets, for example, biochips for medication conveyance. Multifunctional NPs will be created as flexible apparatuses; a few NP definitions will serve both symptomatic and helpful capacities. These improvements will add to the movement of customized medication.

## 2.13 Conclusion

The BBB constrains the delivery of medications in the CNS malfunctions. NPs allow the best vision for the delivery of medications and encourage conveyance through the BBB without harming it. NPs proposed the likelihood of consolidating findins with therapeutics. The point of NP details isn't just the uncontrolled entry of medications through the BBB yet controlled and focused on conveyance. A large number of methodologies utilizing nanobiotechnology areas however trial and a couple has been used in human patients, especially in the management of malignant tumors. Due to varieties in the cerebral bloodstream and BBB with age and sickness, any comprehensive strategy for medication conveyance to the cerebrum is probably not going to be appropriate for all cases.

The benefits of employing nanosize-based delivery of regular medications to treat CNS illnesses lies in the unending scope of conceivable outcomes to control surface and auxiliary properties of these nanosystems so as to (i) make them skillful to target infected cells, saving the solid ones; (ii) convey vast heaps of various kinds of particles, giving them assurance against degradation and counteracting early delivery; (iii) furnish drugs with long flow lifetimes; and (iv) permit tranquilizer-controlled and -supported discharge at the ideal target site. Natural organic molecules based nanocarriers and cell-got exosomes advantage from the additional preferred standpoint of having the capacity to convey hydrophobic and hydrophilic atoms in independent compartments inside a similar molecule and to embody or complex nucleic acids with helpful potential.

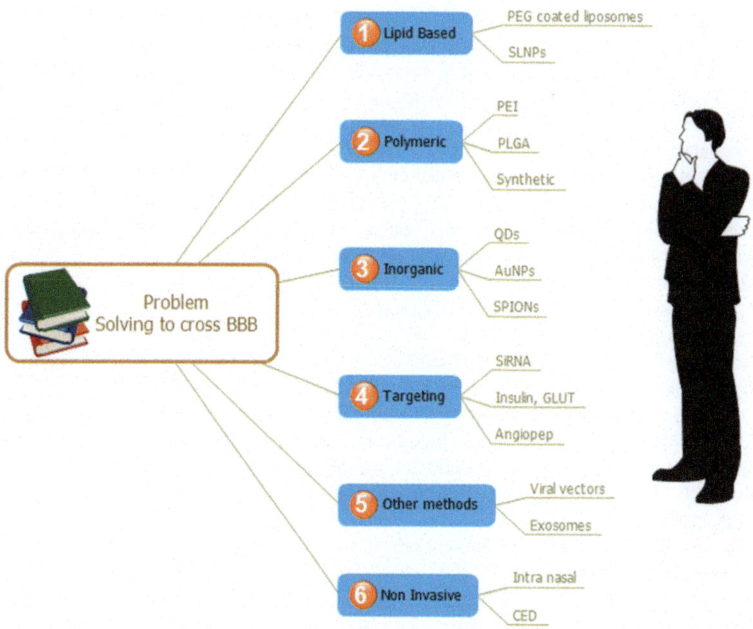


**Fig. 2.7** Achieving BBB

With regard to neurodegeneration disorders, the high lipid substance and specific lipid arrangement of brain cells, just as the prerequisite for lipid go-betweens, make them the most loved focus for bioactive materials containing nanosystems, in which lipid-based nanosystems addressing CNS disease lipids, other than comprising exceptionally adaptable bearer segments, can assume a job as useful gatherings and bioactive atoms. Clinical preliminaries intended to assess the viability of liposomal introductions of the anticancer medications doxorubicin and cytarabine to handle brain tumors underline the capability of lipid-based nanoparticles as medication transporters to CNS. In a rundown, the writing in this part indicates considerable advances in the study of nanosized carriers for drug molecules/siRNAs, supporting the conviction that cutting-edge pharmaceutical will be accessible in a not so distant future, opening new roads toward the clinical trials of a substantial range of CNS maladies. However, there are still a few impediments, and further examinations are presently in advancement (Fig. 2.7).

## References

1. Ballabh P, Braun A, Nedergaard M. The blood–brain barrier: An overview: Structure, regulation, and clinical implications. Neurobiol Dis. 2004;16(1):1–13.
2. Armulik A, Genové G, Mäe M, Nisancioglu MH, Wallgard E, Niaudet C, et al. Pericytes regulate the blood–brain barrier. Nature. 2010;468(7323):557.

3. Schroeder U, Sommerfeld P, Ulrich S, Sabel BA. Nanoparticle technology for delivery of drugs across the blood–brain barrier. J Pharm Sci. 1998;87(11):1305–7.
4. Garcia-Garcia E, Andrieux K, Gil S, Couvreur P. Colloidal carriers and blood–brain barrier (BBB) translocation: A way to deliver drugs to the brain? Int J Pharm. 2005;298(2):274–92.
5. Wohlfart S, Gelperina S, Kreuter J. Transport of drugs across the blood–brain barrier by nanoparticles. J Control Release. 2012;161(2):264–73.
6. Abbott NJ. Astrocyte–endothelial interactions and blood–brain barrier permeability. J Anat. 2002;200(6):629–38.
7. Lockman PR, Mumper RJ, Khan MA, Allen DD. Nanoparticle technology for drug delivery across the blood-brain barrier. Drug Dev Ind Pharm. 2002;28(1):1–13.
8. Kreuter J, Shamenkov D, Petrov V, Ramge P, Cychutek K, Koch-Brandt C, et al. Apolipoprotein-mediated transport of nanoparticle-bound drugs across the blood-brain barrier. J Drug Target. 2002;10(4):317–25.
9. Scherrmann J-M. Drug delivery to brain via the blood–brain barrier. Vasc Pharmacol. 2002;38 (6):349–54.
10. Greig NH. Optimizing drug delivery to brain tumors. Cancer Treat Rev. 1987;14(1):1–28.
11. Groothuis DR. The blood-brain and blood-tumor barriers: A review of strategies for increasing drug delivery. Neuro-Oncology. 2000;2(1):45–59.
12. Vyas TK, Shahiwala A, Marathe S, Misra A. Intranasal drug delivery for brain targeting. Curr Drug Deliv. 2005;2(2):165–75.
13. Hawkins BT, Davis TP. The blood-brain barrier/neurovascular unit in health and disease. Pharmacol Rev. 2005;57(2):173–85.
14. Pan W, Banks WA, Fasold MB, Bluth J, Kastin AJ. Transport of brain-derived neurotrophic factor across the blood–brain barrier. Neuropharmacology. 1998;37(12):1553–61.
15. Bell RD, Zlokovic BV. Neurovascular mechanisms and blood–brain barrier disorder in Alzheimer's disease. Acta Neuropathol. 2009;118(1):103–13.
16. Zlokovic BV. The blood-brain barrier in health and chronic neurodegenerative disorders. Neuron. 2008;57(2):178–201.
17. Frank RT, Aboody KS, Najbauer J. Strategies for enhancing antibody delivery to the brain. Biochim Biophys Acta (BBA)-Rev Cancer. 2011;1816(2):191–8.
18. Abbott NJ, Patabendige AAK, Dolman DEM, Yusof SR, Begley DJ. Structure and function of the blood–brain barrier. Neurobiol Dis. 2010;37(1):13–25.
19. Lossinsky AS, Shivers RR. Structural pathways for macromolecular and cellular transport across the blood-brain barrier during inflammatory conditions. Histol Histopathol Cell Mol Biol. 2004;19(2):535–64.
20. Stamatovic SM, Keep RF, Andjelkovic AV. Brain endothelial cell-cell junctions: How to "open" the blood brain barrier. Curr Neuropharmacol. 2008;6(3):179–92.
21. Tsuji A. Small molecular drug transfer across the blood-brain barrier via carrier-mediated transport systems. NeuroRx. 2005;2(1):54–62.
22. Tamai I, Tsuji A. Transporter-mediated permeation of drugs across the blood–brain barrier. J Pharm Sci. 2000;89(11):1371–88.
23. Bradbury MWB. The blood-brain barrier. Transport across the cerebral endothelium. Circ Res. 1985;57(2):213–22.
24. Oldendorf WH. Brain uptake of radiolabeled amino acids, amines, and hexoses after arterial injection. Am J Physiol Content. 1971;221(6):1629–39.
25. Bradbury MW. The blood-brain barrier. Exp Physiol. 1993;78(4):453–72.
26. Schinkel AH, Jonker JW. Mammalian drug efflux transporters of the ATP binding cassette (ABC) family: An overview. Adv Drug Deliv Rev. 2012;64:138–53.
27. Dean M, Hamon Y, Chimini G. The human ATP-binding cassette (ABC) transporter superfamily. J Lipid Res. 2001;42(7):1007–17.
28. Yi X. Protein modification by pluronic block copolymer for brain delivery. Nebraska: University of Nebraska Medical Center; 2010.

29. Sampson JH, Raghavan R, Provenzale JM, Croteau D, Reardon DA, Coleman RE, et al. Induction of hyperintense signal on T2-weighted MR images correlates with infusion distribution from intracerebral convection-enhanced delivery of a tumor-targeted cytotoxin. Neuroradiology. 2006; https://doi.org/10.2214/AJR.06.0428.
30. Ross TM, Martinez PM, Renner JC, Thorne RG, Hanson LR, Frey Ii WH. Intranasal administration of interferon beta bypasses the blood–brain barrier to target the central nervous system and cervical lymph nodes: A non-invasive treatment strategy for multiple sclerosis. J Neuroimmunol. 2004;151(1–2):66–77.
31. Hanson LR, Frey WH. Intranasal delivery bypasses the blood-brain barrier to target therapeutic agents to the central nervous system and treat neurodegenerative disease. BMC Neurosci. 2008;9(3):S5.
32. Zhu J, Jiang Y, Xu G, Liu X. Intranasal administration: A potential solution for cross-BBB delivering neurotrophic factors. Histol Histopathol. 2012;27(5):537–48.
33. Liu L, Guo K, Lu J, Venkatraman SS, Luo D, Ng KC, et al. Biologically active core/shell nanoparticles self-assembled from cholesterol-terminated PEG–TAT for drug delivery across the blood–brain barrier. Biomaterials. 2008;29(10):1509–17.
34. Yang SC, Lu LF, Cai Y, Zhu JB, Liang BW, Yang CZ. Body distribution in mice of intravenously injected camptothecin solid lipid nanoparticles and targeting effect on brain. J Control Release. 1999;59(3):299–307.
35. Georgieva JV, Kalicharan D, Couraud P-O, Romero IA, Weksler B, Hoekstra D, et al. Surface characteristics of nanoparticles determine their intracellular fate in and processing by human blood–brain barrier endothelial cells in vitro. Mol Ther. 2011;19(2):318–25.
36. Lockman PR, Koziara JM, Mumper RJ, Allen DD. Nanoparticle surface charges alter blood–brain barrier integrity and permeability. J Drug Target. 2004;12(9–10):635–41.
37. Jallouli Y, Paillard A, Chang J, Sevin E, Betbeder D. Influence of surface charge and inner composition of porous nanoparticles to cross blood–brain barrier in vitro. Int J Pharm. 2007;344(1–2):103–9.
38. Fenart L, Casanova A, Dehouck B, Duhem C, Slupek S, Cecchelli R, et al. Evaluation of effect of charge and lipid coating on ability of 60-nm nanoparticles to cross an in vitro model of the blood-brain barrier. J Pharmacol Exp Ther. 1999;291(3):1017–22.
39. Lynch I, Cedervall T, Lundqvist M, Cabaleiro-Lago C, Linse S, Dawson KA. The nanoparticle–protein complex as a biological entity; a complex fluids and surface science challenge for the 21st century. Adv Colloid Interf Sci. 2007;134:167–74.
40. Tang J, Xiong L, Wang S, Wang J, Liu L, Li J, et al. Influence of silver nanoparticles on neurons and blood-brain barrier via subcutaneous injection in rats. Appl Surf Sci. 2008;255 (2):502–4.
41. Ulbrich K, Hekmatara T, Herbert E, Kreuter J. Transferrin-and transferrin-receptor-antibody-modified nanoparticles enable drug delivery across the blood–brain barrier (BBB). Eur J Pharm Biopharm. 2009;71(2):251–6.
42. Chang J, Jallouli Y, Kroubi M, Yuan X, Feng W, Kang C, et al. Characterization of endocytosis of transferrin-coated PLGA nanoparticles by the blood–brain barrier. Int J Pharm. 2009;379(2):285–92.
43. Alkilany AM, Nagaria PK, Hexel CR, Shaw TJ, Murphy CJ, Wyatt MD. Cellular uptake and cytotoxicity of gold nanorods: Molecular origin of cytotoxicity and surface effects. Small. 2009;5(6):701–8.
44. Ulbrich K, Knobloch T, Kreuter J. Targeting the insulin receptor: Nanoparticles for drug delivery across the blood–brain barrier (BBB). J Drug Target. 2011;19(2):125–32.
45. Wang H, Zhao Y, Wu Y, Hu Y, Nan K, Nie G, et al. Enhanced anti-tumor efficacy by co-delivery of doxorubicin and paclitaxel with amphiphilic methoxy PEG-PLGA copolymer nanoparticles. Biomaterials. 2011;32(32):8281–90.
46. Kreuter J. Nanoparticles and microparticles for drug and vaccine delivery. J Anat. 1996;189 (Pt 3):503.

47. Anraku Y, Kuwahara H, Fukusato Y, Mizoguchi A, Ishii T, Nitta K, et al. Glycaemic control boosts glucosylated nanocarrier crossing the BBB into the brain. Nat Commun. 2017;8 (1):1001.
48. Lundqvist M, Stigler J, Elia G, Lynch I, Cedervall T, Dawson KA. Nanoparticle size and surface properties determine the protein corona with possible implications for biological impacts. Proc Natl Acad Sci. 2008;105(38):14265–70.
49. Lesniak A, Fenaroli F, Monopoli MP, Åberg C, Dawson KA, Salvati A. Effects of the presence or absence of a protein corona on silica nanoparticle uptake and impact on cells. ACS Nano. 2012;6(7):5845–57.
50. Walkey CD, Olsen JB, Song F, Liu R, Guo H, Olsen DWH, et al. Protein corona fingerprinting predicts the cellular interaction of gold and silver nanoparticles. ACS Nano. 2014;8 (3):2439–55.
51. Cedervall T, Lynch I, Lindman S, Berggård T, Thulin E, Nilsson H, et al. Understanding the nanoparticle–protein corona using methods to quantify exchange rates and affinities of proteins for nanoparticles. Proc Natl Acad Sci. 2007;104(7):2050–5.
52. Gref R, Lück M, Quellec P, Marchand M, Dellacherie E, Harnisch S, et al. 'Stealth'corona-core nanoparticles surface modified by polyethylene glycol (PEG): Influences of the corona (PEG chain length and surface density) and of the core composition on phagocytic uptake and plasma protein adsorption. Coll Surf B Biointerfaces. 2000;18(3–4):301–13.
53. Niidome T, Yamagata M, Okamoto Y, Akiyama Y, Takahashi H, Kawano T, et al. PEG-modified gold nanorods with a stealth character for in vivo applications. J Control Release. 2006;114(3):343–7.
54. Ventola CL. The nanomedicine revolution: Part 2: Current and future clinical applications. Pharm Ther. 2012;37(10):582.
55. Tavano R, Segat D, Reddi E, Kos J, Rojnik M, Kocbek P, et al. Procoagulant properties of bare and highly PEGylated vinyl-modified silica nanoparticles. Nanomedicine. 2010;5(6):881–96.
56. Bana L, Minniti S, Salvati E, Sesana S, Zambelli V, Cagnotto A, et al. Liposomes bi-functionalized with phosphatidic acid and an ApoE-derived peptide affect Aβ aggregation features and cross the blood–brain-barrier: Implications for therapy of Alzheimer disease. Nanomed Nanotechnol, Biol Med. 2014;10(7):1583–90.
57. Farrington GK, Caram-Salas N, Haqqani AS, Brunette E, Eldredge J, Pepinsky B, et al. A novel platform for engineering blood-brain barrier-crossing bispecific biologics. FASEB J. 2014;28(11):4764–78.
58. Tavazoie SF, Pencheva N. Treatment and diagnosis of melanoma. 2018; US Patent App. 15/881,231.
59. Re F, Cambianica I, Zona C, Sesana S, Gregori M, Rigolio R, et al. Functionalization of liposomes with ApoE-derived peptides at different density affects cellular uptake and drug transport across a blood-brain barrier model. Nanomed Nanotechnol Biol Med. 2011;7 (5):551–9.
60. Gastaldi L, Battaglia L, Peira E, Chirio D, Muntoni E, Solazzi I, et al. Solid lipid nanoparticles as vehicles of drugs to the brain: Current state of the art. Eur J Pharm Biopharm. 2014;87 (3):433–44.
61. Lu C-T, Zhao Y-Z, Wong HL, Cai J, Peng L, Tian X-Q. Current approaches to enhance CNS delivery of drugs across the brain barriers. Int J Nanomed. 2014;9:2241.
62. Pourgholi F, Farhad J-N, Kafil HS, Yousefi M. Nanoparticles: Novel vehicles in treatment of glioblastoma. Biomed Pharmacother. 2016;77:98–107.
63. Montenegro L, Ottimo S, Puglisi G, Castelli F, Sarpietro MG. Idebenone loaded solid lipid nanoparticles interact with biomembrane models: Calorimetric evidence. Mol Pharm. 2012;9 (9):2534–41.
64. Sun Y, Liu L, Xue Y, Wang P. Effects of insulin combined with idebenone on blood–brain barrier permeability in diabetic rats. J Neurosci Res. 2015;93(4):666–77.
65. Kurzrock R, Gabrail N, Chandhasin C, Moulder S, Smith C, Brenner A, et al. Safety, pharmacokinetics, and activity of GRN1005, a novel conjugate of angiopep-2, a peptide

facilitating brain penetration, and paclitaxel, in patients with advanced solid tumors. Mol Cancer Ther. 2012;11(2):308–16.
66. Drappatz J, Brenner A, Wong ET, Eichler A, Schiff D, Groves MD, et al. Phase I study of GRN1005 in recurrent malignant glioma. Clin Cancer Res. 2013; https://doi.org/10.1158/1078-0432.CCR-12-2481.
67. Patil Y, Panyam J. Polymeric nanoparticles for siRNA delivery and gene silencing. Int J Pharm. 2009;367(1–2):195–203.
68. Dahlman JE, Barnes C, Khan OF, Thiriot A, Jhunjunwala S, Shaw TE, et al. In vivo endothelial siRNA delivery using polymeric nanoparticles with low molecular weight. Nat Nanotechnol. 2014;9(8):648.
69. Cun D, Foged C, Yang M, Frøkjær S, Nielsen HM. Preparation and characterization of poly (DL-lactide-co-glycolide) nanoparticles for siRNA delivery. Int J Pharm. 2010;390(1):70–5.
70. Klajnert B, Cortijo-Arellano M, Bryszewska M, Cladera J. Influence of heparin and dendrimers on the aggregation of two amyloid peptides related to Alzheimer's and prion diseases. Biochem Biophys Res Commun. 2006;339(2):577–82.
71. Luo D, Haverstick K, Belcheva N, Han E, Saltzman WM. Poly (ethylene glycol)-conjugated PAMAM dendrimer for biocompatible, high-efficiency DNA delivery. Macromolecules. 2002;35(9):3456–62.
72. Davis ME. The first targeted delivery of siRNA in humans via a self-assembling, cyclodextrin polymer-based nanoparticle: From concept to clinic. Mol Pharm. 2009;6(3):659–68.
73. Lalatsa A, Schatzlein AG, Uchegbu IF. Strategies to deliver peptide drugs to the brain. Mol Pharm. 2014;11(4):1081–93.
74. Cosco D, Di Marzio L, Marianecci C, Trapasso E, Paolino D, Celia C, et al. Colloidal supramolecular aggregates for therapeutic application in neuromedicine. Curr Med Chem. 2014;21(36):4132–53.
75. Lü J-M, Wang X, Marin-Muller C, Wang H, Lin PH, Yao Q, et al. Current advances in research and clinical applications of PLGA-based nanotechnology. Expert Rev Mol Diagn. 2009;9(4):325–41.
76. Cai Q, Wang L, Deng G, Liu J, Chen Q, Chen Z. Systemic delivery to central nervous system by engineered PLGA nanoparticles. Am J Transl Res. 2016;8(2):749.
77. Monsalve Y, Tosi G, Ruozi B, Belletti D, Vilella A, Zoli M, et al. PEG-g-chitosan nanoparticles functionalized with the monoclonal antibody OX26 for brain drug targeting. Nanomedicine. 2015;10(11):1735–50.
78. Whitehead KA, Langer R, Anderson DG. Knocking down barriers: Advances in siRNA delivery. Nat Rev Drug Discov. 2009;8(2):129.
79. Dominska M, Dykxhoorn DM. Breaking down the barriers: siRNA delivery and endosome escape. J Cell Sci. 2010;123(8):1183–9.
80. Son S, Jang J, Youn H, Lee S, Lee D, Lee Y-S, et al. A brain-targeted rabies virus glycoprotein-disulfide linked PEI nanocarrier for delivery of neurogenic microRNA. Biomaterials. 2011;32(21):4968–75.
81. Park T-E, Singh B, Li H, Lee J-Y, Kang S-K, Choi Y-J, et al. Enhanced BBB permeability of osmotically active poly (mannitol-co-PEI) modified with rabies virus glycoprotein via selective stimulation of caveolar endocytosis for RNAi therapeutics in Alzheimer's disease. Biomaterials. 2015;38:61–71.
82. Gooding M, Malhotra M, McCarthy DJ, Godinho BMDC, Cryan JF, Darcy R, et al. Synthesis and characterization of rabies virus glycoprotein-tagged amphiphilic cyclodextrins for siRNA delivery in human glioblastoma cells: In vitro analysis. Eur J Pharm Sci. 2015;71:80–92.
83. Lakhal S, Wood MJA. Exosome nanotechnology: An emerging paradigm shift in drug delivery. BioEssays. 2011;33(10):737–41.
84. Alvarez-Erviti L, Seow Y, Yin H, Betts C, Lakhal S, Wood MJA. Delivery of siRNA to the mouse brain by systemic injection of targeted exosomes. Nat Biotechnol. 2011;29(4):341.
85. Kumar SA, Peter Y-A, Nadeau JL. Facile biosynthesis, separation and conjugation of gold nanoparticles to doxorubicin. Nanotechnology. 2008;19(49):495101.

86. Decuzzi P, Ferrari M. The role of specific and non-specific interactions in receptor-mediated endocytosis of nanoparticles. Biomaterials. 2007;28(18):2915–22.
87. Gupta AK, Berry C, Gupta M, Curtis A. Receptor-mediated targeting of magnetic nanoparticles using insulin as a surface ligand to prevent endocytosis. IEEE Trans Nanobioscience. 2003;2(4):255–61.
88. Heitz F, Morris MC, Divita G. Twenty years of cell-penetrating peptides: From molecular mechanisms to therapeutics. Br J Pharmacol. 2009;157(2):195–206.
89. Stalmans S, Bracke N, Wynendaele E, Gevaert B, Peremans K, Burvenich C, et al. Cell-penetrating peptides selectively cross the blood-brain barrier in vivo. PLoS One. 2015;10(10): e0139652.
90. Li T, Bourgeois J-P, Celli S, Le Sourd A-M, Mecheri S, Weksler B, et al. Cell-penetrating anti-GFAP VHH and corresponding fluorescent fusion protein VHH-GFP spontaneously cross the blood-brain barrier and specifically recognize astrocytes: Application to brain imaging. FASEB J. 2012;26(10):3969–79.
91. Tortorella S, Karagiannis TC. Transferrin receptor-mediated endocytosis: A useful target for cancer therapy. J Membr Biol. 2014;247(4):291–307.
92. Shriver LP, Koudelka KJ, Manchester M. Viral nanoparticles associate with regions of inflammation and blood brain barrier disruption during CNS infection. J Neuroimmunol. 2009;211(1–2):66–72.
93. Kuo Y-C, Su F-L. Transport of stavudine, delavirdine, and saquinavir across the blood–brain barrier by polybutylcyanoacrylate, methylmethacrylate-sulfopropylmethacrylate, and solid lipid nanoparticles. Int J Pharm. 2007;340(1–2):143–52.
94. Elbakry A, Zaky A, Liebl R, Rachel R, Goepferich A, Breunig M. Layer-by-layer assembled gold nanoparticles for siRNA delivery. Nano Lett. 2009;9(5):2059–64.
95. Na HB, Song IC, Hyeon T. Inorganic nanoparticles for MRI contrast agents. Adv Mater. 2009;21(21):2133–48.
96. Sokolova V, Epple M. Inorganic nanoparticles as carriers of nucleic acids into cells. Angew Chem Int Ed. 2008;47(8):1382–95.
97. Kim ST, Chompoosor A, Yeh Y, Agasti SS, Solfiell DJ, Rotello VM. Dendronized gold nanoparticles for siRNA delivery. Small. 2012;8(21):3253–6.
98. Maeda H, Wu J, Sawa T, Matsumura Y, Hori K. Tumor vascular permeability and the EPR effect in macromolecular therapeutics: A review. J Control Release. 2000;65(1–2):271–84.
99. Maeda H, Nakamura H, Fang J. The EPR effect for macromolecular drug delivery to solid tumors: Improvement of tumor uptake, lowering of systemic toxicity, and distinct tumor imaging in vivo. Adv Drug Deliv Rev. 2013;65(1):71–9.
100. Lennernäs H, Palm K, Fagerholm U, Artursson P. Comparison between active and passive drug transport in human intestinal epithelial (Caco-2) cells in vitro and human jejunum in vivo. Int J Pharm. 1996;127(1):103–7.
101. Baldwin SA. Mammalian passive glucose transporters: Members of an ubiquitous family of active and passive transport proteins. Biochim Biophys Acta (BBA)-Rev Biomembr. 1993;1154(1):17–49.
102. Cho W-S, Cho M, Jeong J, Choi M, Cho H-Y, Han BS, et al. Acute toxicity and pharmacokinetics of 13 nm-sized PEG-coated gold nanoparticles. Toxicol Appl Pharmacol. 2009;236 (1):16–24.
103. Cho W-S, Kim S, Han BS, Son WC, Jeong J. Comparison of gene expression profiles in mice liver following intravenous injection of 4 and 100 nm-sized PEG-coated gold nanoparticles. Toxicol Lett. 2009;191(1):96–102.
104. Ruan S, Yuan M, Zhang L, Hu G, Chen J, Cun X, et al. Tumor microenvironment sensitive doxorubicin delivery and release to glioma using angiopep-2 decorated gold nanoparticles. Biomaterials. 2015;37:425–35.
105. Aktaş Y, Yemisci M, Andrieux K, Gürsoy RN, Alonso MJ, Fernandez-Megia E, et al. Development and brain delivery of chitosan– PEG nanoparticles functionalized with the monoclonal antibody OX26. Bioconjug Chem. 2005;16(6):1503–11.

106. Pardridge WM, Kang Y-S, Buciak JL. Transport of human recombinant brain-derived neurotrophic factor (BDNF) through the rat blood– brain barrier in vivo using vector-mediated peptide drug delivery. Pharm Res. 1994;11(5):738–46.
107. Zhang T-T, Li W, Meng G, Wang P, Liao W. Strategies for transporting nanoparticles across the blood–brain barrier. Biomater Sci. 2016;4(2):219–29.
108. Demeule M, Currie J, Bertrand Y, Ché C, Nguyen T, Régina A, et al. Involvement of the low-density lipoprotein receptor-related protein in the transcytosis of the brain delivery vector Angiopep-2. J Neurochem. 2008;106(4):1534–44.
109. Wolf BB, Lopes MB, VandenBerg SR, Gonias SL. Characterization and immunohistochemical localization of alpha 2-macroglobulin receptor (low-density lipoprotein receptor-related protein) in human brain. Am J Pathol. 1992;141(1):37.
110. Ren J, Shen S, Wang D, Xi Z, Guo L, Pang Z, et al. The targeted delivery of anticancer drugs to brain glioma by PEGylated oxidized multi-walled carbon nanotubes modified with angiopep-2. Biomaterials. 2012;33(11):3324–33.
111. Chen G-J, Su Y-Z, Hsu C, Lo Y-L, Huang S-J, Ke J-H, et al. Angiopep-pluronic F127-conjugated superparamagnetic iron oxide nanoparticles as nanotheranostic agents for BBB targeting. J Mater Chem B. 2014;2(34):5666–75.
112. Hu Y-L, Gao J-Q. Potential neurotoxicity of nanoparticles. Int J Pharm. 2010;394 (1–2):115–21.
113. Karmakar A, Zhang Q, Zhang Y. Neurotoxicity of nanoscale materials. J Food Drug Anal. 2014;22(1):147–60.

# Chapter 3
# The Strategies of Nanomaterials for Drug Delivery and Release

**Jinjin Wang, Qianqian Huang, and Xing-Jie Liang**

**Abstract** Various strategies have sprung to explore appropriate solutions for delivering therapeutic agents into the brain. Recently, breakthroughs of nanomedicines provide an excellent window for brain targeting and thus can be effectively used to treat a wide variety of brain diseases. This chapter comprehensively depicts various types of nanoformulations and novel drug delivery and release approaches of nanotechnology-based drug delivery systems in brain disease. Moreover, the shortages of advanced nanotechnology-based drug delivery system are also discussed in the last part to broaden the understanding of the design of nanomedicines for brain drug delivery and release.

**Keywords** Brain diseases · Nanotechnology-based drug delivery system · Response release · Target

## 3.1 Introduction

Brain diseases are a serious hazard to human health. Millions of people in the world suffer from complicated kinds of brain diseases, including brain tumors, cerebrovascular disease, bacterial infection of the brain, and Parkinson's disease (PD) [1]. The effect of pharmaceuticals is unsatisfactory due to its poor efficiency of directly delivering active pharmaceutical ingredient to target organs, tissues, cells, and organelles. The lack of specificity at treatment sites and accumulation of drug in other tissues not only impose more therapeutic doses but also spread therapeutic agents randomly in brains tissues, causing undesirable side effects [2]. When delivering drug through different administration, there remain some challenges (Fig. 3.1).

J. Wang · Q. Huang · X.-J. Liang (✉)
CAS Center for Excellence in Nanoscience, CAS Key Laboratory for Biomedical Effects of Nanomaterials and Nanosafety, Chinese Academy of Sciences, National Center for Nanoscience and Technology of China, Beijing, China

University of Chinese Academy of Sciences, Beijing, China
e-mail: liangxj@nanoctr.cn



X. Xue (ed.), *Nanomedicine in Brain Diseases*,
https://doi.org/10.1007/978-981-13-8731-9_3

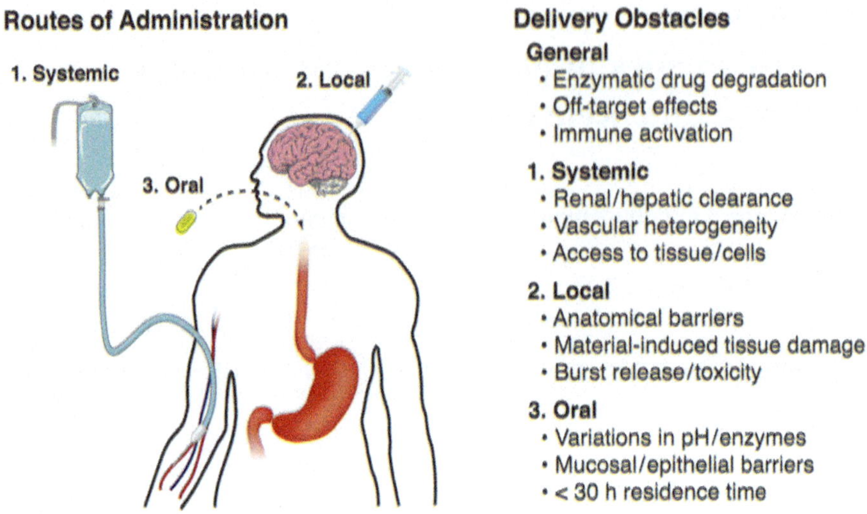


**Fig. 3.1** The obstacles that systemic, local, and oral administration may meet. (Adapted permission from [108]. Copyright 2016 American Chemical Society)

By controlling drug delivery systems, the behavior of drugs in vivo can be regulated in temporal, spatial, and dose, thereby increasing their utilization and therapeutic effect and reducing costs and adverse reactions. Unfavorable factors, such as renal excretion, enzymatic degradation, short circulation lifetime, insufficient tissue penetration, and different pathological conditions, may dilute greatly therapeutic index. Therefore, the source underlying challenge of drug delivery and release for brain-associated diseases after administration mainly is how to control the behavior of drug delivery and release in vivo aiming to those unfavorable factors.

Inspiringly, nanotechnology brings great hope to treatment of brain diseases. In 1959, *Richard Feynman* first discussed the idea of nanotechnology, and later he put forward the concept of medical applications of nanomaterials [3]. Owing to its special size, nanomaterials offer a series of particular properties including high drug loading and physical and chemical stability. Nanomaterials have applications in various aspects from diagnostics and monitoring to treatments [4]. Compared with conventional preparations, nanotechnology-based preparations have advantages in enhancing the solubility of poorly soluble drugs, reducing the side effects of drugs, thereby improving bioavailability. It can provide a new window of improving drug therapeutic index and realizing drug-controlled release. According to their unique properties including small size and customizable surface and the molecular pathogenesis of brain diseases that has been elucidated gradually at the same time, nanotechnology-based delivery systems have emerged as promising methods for potential drug delivery systems to various parts of body (Fig. 3.2). Nanodrug delivery systems can address brain drug delivery problems by improving therapeutic index of drugs.

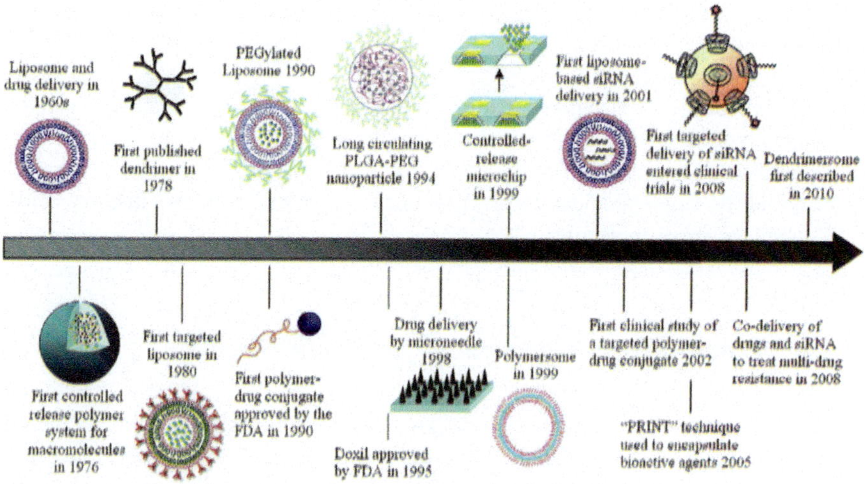


**Fig. 3.2** The history of the development of nanodrug delivery systems. (Adapted permission from [109]. Copyright 2010 American Chemical Society)

To date, various types of nanoformulations have been extensively developed to support different types of brain drug delivery and release. It has been widely exploited that different types of nanoformulations support different kinds of brain drug delivery system like liposomes, micelles, nanoparticles, and nanoemulsions. Hence, the purpose of this article is to demonstrate various types of nanotechnology-based drug delivery systems and highlight advances in drug delivery and release methods for brain diseases. Lastly, we discuss the shortage of advanced nanotechnology-based drug delivery systems in the future.

## 3.2 Nanopreparations for Brain Diseases

### *3.2.1 Construction of Nanocarriers*

Nanocarriers can act like macromolecules in certain circumstances in virtue of its small size and easily tailored feature. Moreover, they could carry a wide range of drugs to control drug release. These characteristics make nanocarriers become an attractive tool for transporting drugs into the brain.

Ideal properties of nanocarriers for drug delivery to the brain can be summarized as follows [5–7]:

(a) Safe, nontoxic, non-immunogenic, noninflammatory, can be degraded or eliminated, high biocompatibility
(b) Stability in vivo/vitro
(c) Controlled circulation time

(d) Particle size should be <200 nm and should have a narrow particle size distribution.
(e) Can target special site
(f) Packing drugs in an appropriate manner, with a high drug loading capacity, capable of achieving an effective concentration for diagnosis or therapy
(g) Controlled drug release
(h) Should be applicable to carry antibodies, peptides, proteins, nucleic acids, sugars, or small molecules

### 3.2.2 *Types of Nanopreparations for Brain Disease*

Recently, a spurt of progress has witnessed in nanotechnology-based medicine for delivery and release of therapeutic for diseases [8]. Multiple nanopreparations, such as micelles, liposomes, dendrimers, solid lipid nanoparticles, nanogels, and nanoemulsions, have been designed to target brain or brain diseased tissues/cells via appropriate administration routes, and most of them are made of natural or synthetic polymeric materials. Besides, great efforts have been utilized in the fabrication and modification of multifunctional, target-specific nanopreparations with good stability. There are some different types of nanopreparations used to treat brain disease listed in Table 3.1.

**Table 3.1** Different types of nanoformations used to treat brain diseases

| Nanoformations | Materials | Drugs | Administration | References |
|---|---|---|---|---|
| Nanoparticle | Chitosan, Pluronic | B-galactosidase | Intravenous | [9] |
| | Chitosan, PEG | siRNA | | [10] |
| | PLGA, chitosan | Huperzine A | Intranasal | [11] |
| | P85, PBCA | Phenytoin | Intravenous | [12] |
| | PEI | MicroRNA | Intravenous | [13] |
| Solid nanoparticle | DOPE | Coumarin-6 | Intravenous | [14] |
| Magnetic nanoparticle | PEG, $Fe_3O_4$ | DOX | Intravenous | [15] |
| | $Fe_3O_4$, $Gd_2O_3$ | Pt | Intravenous | [16] |
| | BSA, $Fe_3O_4$ | | Intravenous | [17] |
| Dendrimer | PAMAM | DNA | Intravenous | [18] |
| Micelle | PEG, PCL | siRNA | Intranasal | [19] |
| | Stearic acid, chitosan | Doxorubicin | Intravenous | [20] |
| Liposome | Mal-PEG2000-DSPE | Dopamine derivative | Intravenous | [21] |
| Nanoemulsion | Capmul MCM, Tween 80 | QTP | Intranasal | [22] |
| Nanogel | Deacetylated gellan gum | Resveratrol | Intranasal | [23] |
| Nanocrystal | Tween80, TPGS | Baicalin | Intravenous | [24] |

#### 3.2.2.1 Nanoparticles

Nanoparticles are a type of nanodrug delivery systems which have compact structures formed by encapsulating drugs in a skeleton of carrier material or modifying drugs on the surface of a carrier material by covalent linkage or absorption. The materials of carrier are generally composed of polymers, lipid, or a mixed system. Chitosan, poly(lactic acid) (PLA), poly(lactic-co-glycolic acid) (PLGA), poly(butyl -cyanoacrylate) (PBCA), and poly(ethylene imine) (PEI) are often used materials in nanomedicine.

Chitosan is obtained by deacetylation of chitin which is widely found in nature. Its structural unit is glucosamine monomer. As a drug carrier material, chitosan has high biocompatibility, good stability, biodegradability, low toxicity, and low immunogenicity and can be easily prepared [25]. It has been widely used in in vivo/vitro delivery of oral drugs, gene drugs, protein drugs, and vaccines. Adopting a certain brain-targeted modification strategy, chitosan could be used for brain drug delivery and release.

PLA is a polymer with biodegradable function mainly obtained by polymerization of lactic acid monomers. PLA also has excellent properties such as high biocompatibility, good stability, and low toxicity. Chen et al. [26] developed noninvasive drug delivery systems for brain disease. In their study, PEG-co-PCL nanoparticles modified by lactoferrin have an ability to enhance delivering peptide into the brain following intranasal administration.

PLGA is a kind of pharmaceutical excipient. It is a random polymerization of lactic acid and glycolic acid. It is similar to PLA, with good biocompatibility, safety, and non-toxicity. As a mature carrier, PLGA has many applications in drug delivery. For example, in order to deliver iA β 5 into the brain for AD treatment, Joana A.L. et al. [27] designed functionalized PLGA nanoparticles loading iAβ5, whose surface was modified by two types of specific antibodies, anti-transferrin receptor monoclonal antibody (OX26) against transferrin receptors and anti-Aβ-(DE2B4) recognizing the Aβ peptide. The functionalized PLGA NPs have good properties such as homogeneous distribution (PDI$\leq$0.1) and spherical shape with diameters ranging approximately about 150–170 nm.

PBCA is a potential drug carrier with rapid degradation in vivo and has good clinical applications. Fang Z et al. [12] developed new and efficient poly(butyl -cyanoacrylate) nanoparticles to increase drug concentrations in the brain for overcoming drug-resistant epilepsy. Besides, cholesterol-polyethylene glycol (PEG)-modified poly(N-butyl) cyanoacrylate nanoparticles were designed by Hu X et al. for brain drug delivery [28].

PEI, a cationic polymer that combines with gene compression, is a highly efficient gene drug delivery system with a proton sponge effect and strong cytotoxicity. Chertok B et al. [29] developed PEI-modified iron oxide nanoparticles (GPEI) for brain tumor drug delivery. Besides, with high cell penetration ability and low cell toxicity, GPEI can be applied in magnetic targeting for targeting brain tumor in the near future.

Solid lipid nanoparticles (SLNs) are nanodrug delivery systems prepared by using a solid or synthetic lipid such as lecithin, triacylglycerol, or the like as a carrier to encapsulate or clamp the drug into lipid nucleus, which has the advantage of controlling drug release and protecting the drug from degradation or leakage. It can also be applied to the study of drug delivery and release in the brain through certain brain-targeting strategies [30].

#### 3.2.2.2 Dendrimer

The dendrimer is a polymer with a highly branched structure and a monodisperse spherical molecule formed by continuously adding repeating units to a central core. Dendrimers can target the brain through structural modifications, and the drug can be trapped in space present in dendrimer to protect the drug from the external environment. The most commonly used material at present is polyamidoamine (PAMAM). Dendrimers have the ability to carry a specific DNA/gene or drug to cells. Avinash Gothwal et al. [31] successfully designed PAMAM dendrimers with lactoferrin in order to deliver RIV to the brain. Furthermore, serine-arginine-leucine (SRL)-modified PAMAM dendrimers are a great brain-targeted gene delivery system to target DNA to the brain [18]. The G4PAMAM dendrimers and the SRL peptide are linked by a bifunctional polyethylene glycol to form the carrier PAMAM-PEG-SRL. In addition, the biodistribution of nanoparticles was investigated after administration. In vitro studies showed that SRL-modified nanoparticles have better transfection efficiency and lower toxicity than unmodified nanoparticles.

#### 3.2.2.3 Micelle

The micelle is a drug carrier having a core-shell structure formed by self-assembly of amphiphilic molecules. The hydrophobic core of micelle can entrap hydrophobic or insoluble drugs, and hydrophilic fragments of micelles help to reduce the RCES (reticuloendothelial system). Therefore, micelle can also be frequently used in the drug delivery and release in brain disease. Various types of nanomicelles recently have been developed, and many kinds of research indicated that the polymer micelles delivery system can improve the delivery efficiency of drugs in the body. In order to realize delivering siRNA to the brain, Kanazawa T et al. [19] designed nanomicelles modified with a cell-penetrating peptide via nasal route administration. In their study, the result showed that CPP-modified nanomicelles-loaded dextran (model siRNA) via intranasal administration can effectively accumulate in the brain. Xie et al. [20] designed micelles containing doxorubicin composed of stearic acid and chitosan for the delivery of brain drugs. The results showed that doxorubicin could reach the brain 15 min after administration, while the free drug did not appear in the brain; in addition to the brain, the micelles were mainly distributed in the lungs, liver, and spleen, thereby reducing the cardiac toxicity of its own. In addition, Shao et al. [32] used polymer micelles for CNS infection treatment. In their job,

PE-PEG were used as a carrier material to encapsulate the antifungal drug amphotericin B, and angiopeptin-2 was used as a targeting molecule to modify and construct brain-targeting systems. The studies demonstrated that angiopeptin-2 could effectively increase the distribution of polymer micelles in the brain of mice and significantly reduce the number of bacteria in the brain of model mice.

#### 3.2.2.4 Liposome

Liposomes can load different forms of drugs with a hydrophilic core and lipophilic bilayer structure. As a classic pharmaceutical carrier, liposomes are closed vesicles of phospholipids dispersed in water and encapsulating a part of the aqueous phase. Its particle size is at least about 20 nm. Usually, it contains one or more phospholipid membranes, which can be used as drug carriers for hydrophobic, hydrophilic, and amphiphilic drugs. Liposomes are highly lipophilic that can enter into the brain parenchyma by passive transport, membrane fusion with cerebral vascular endothelial cells, or endocytic pathways. At present, liposomes have received extensive research as carriers for drug delivery in the brain. For example, Tanifum et al. [33] designed PEG-modified liposomes, the surface of which modified targeting molecules recognizing amyloid-precipitating in AD. After administration in AD transgenic model mice, the modified liposome that can cross the blood-brain barrier and also recognize and bind amyloid plaque realized the diagnosis of AD in vivo. RVG-modified liposomes also represent a very promising method for brain-targeted treatment of PD. In order to further improve the effectiveness of performance of a dopamine derivative N-3,4-bis(pivaloyloxy)-dopamine (BPD), Qu M et al. developed RVG-modified liposomes [21]. In vivo and in vitro distribution studies have shown modified liposomes are selectively distributed in the brain, striatum, and substantia nigra. With improving therapeutic effect in the PD, the modified liposomes showed significantly clinical value.

#### 3.2.2.5 Others

The research of nanopreparations including nanogels, nanoemulsions, and nanocrystals also developed rapidly. Nanoemulsions improve drug bioavailability in the brain by extending its blood circulate times. The size of the nanoemulsion is between 20 and 200 nm. The oil-in-water formulation consists of an oil phase that disperses the droplets and also contains a surfactant that stabilizes it and a co-surfactant. Dordevic SM et al. [34] prepared nanoemulsions and successfully enhanced the delivery efficiency of the antipsychotic drug risperidone. Nanogels are polymers of a cross-linked network that can carry biomolecular drugs and small-molecule drugs. Hap J. et al. [23] developed ionic-sensitive in situ gels to improve brain treatment efficiency by enhancing nasal mucosal permeability and increase nasal residence time.

## 3.3 Advances in Drug Delivery and Release Approaches for Brain Diseases

### *3.3.1 Controlled Nano Drug Delivery Systems*

Drug delivery systems are often designed to improve the pharmacological and therapeutic properties of drug molecules. The superior delivery system requires slow and sustained drug release into the CNS for sustainable treatment with low side effects. Several studies have shown advances that can effectively control the process of drug release aiming to brain disease, including sustained controlled drug release and multistaged controlled release.

There are so many promising drugs that have unsatisfied therapeutic concentrations due to its poor water solubility and rapid serum clearance. In order to solve this problem, Chen et al. designed small-sized mPEG-PLGA nanoparticles that can sustain control drug release. It intended to strengthen brain uptake through sustained release of anti-Parkinsonism natural product schisantherin A for treating PD, which will improve oral bioavailability of Biopharmaceutics Classification System (BCS) Class II compound [35]. Elena Sánchez-López et al. developed nanospheres loading dexibuprofen for AD prevention. Compared with free drug, drug-loaded nanospheres present sustained and slower release in vitro and avoid damage from degradable factor in vivo, thereby improving the efficiency of treatment [36]. Shi et al. designed nanoparticles that can trace drug delivery and multistaged controlled release drug in neurons to solve problem of delivery and release of sulfydryl-containing drugs. This multifunctional nanoparticles coated fluorescent nanogel and loaded gold and captopril. Au nanoparticles were combined with silica nanoparticles through electrostatic interaction and the chemical interaction between Au and amino group, and it combined with drugs through strong binding interaction. The result showed that there are two different stages of drug release: most drugs release with degradation of coating gels in the first stage, and the remaining drugs release via competitive adsorption in the second stage [37].

### *3.3.2 Targeted Nano Drug Delivery Systems*

#### 3.3.2.1 Passive Targeting

Passive targeting is beneficial accumulation of drugs in diseased tissues due to different physiological conditions. A promising area of nanoparticle-mediated brain tumor therapy is the use of enhanced permeability and retention (EPR) effect. In many types of tumors, the EPR effect occurs with a defective vascular system and incomplete lymphatic drainage system [38]. This unique tumor vasculature along with macromolecules-prolonged half-life period makes nanoparticles accumulate in tumor and inflammatory tissues, which promotes more selective extravasation of macromolecular drugs in tumor sites. Nanomedicines which can selectively and efficiently target the tumor tissues have been extensively developed [39].

Feng et al. [40] designed a simple method via cross-linking gold nanospheres with dithiol-polyethylene glycol (HS-PEG-SH) to prepare self-assembled gold nanoclusters to treat brain tumor. The nano-components take advantage of EPR effects of brain tumor, successfully increasing nanoparticle retention in tumor tissues by passive targeting strategy. Further, functionalized nanospheres with epidermal growth factor peptides greatly enhance tumor targeting efficiency. Moreover, the study shows that in the tumor microenvironment, the release of anticancer drugs is controllable due to high-permeability, acidic, and redox conditions, presenting an effective and feasible brain-targeted delivery and release system.

In addition to the osmotic effect of the tumor itself, the additional use of a hypertonic solution can also increase brain drug accumulation. The BBB can be disrupted by infusing a hypertonic solution. The osmotic opening is one of the most general ways to deliver the drug into the brain. It is usually the hyperosmotic agent infused into the carotid artery. The vascular endothelial cells are dehydrated during this process of inducing a high osmotic pressure by hyperosmotic materials, thereby causing shrinkage and destruction of the tight junctions. Osmotic disruption allows large molecules along with the infusion of drugs to enter the CNS, which is helpful in the treatment of brain disease.

However, harmful substances may also enter the brain during this process which may harm the physiological functions of the brain. And the also brain will be damaged due to the excessive concentration of hypertonic solution [41]. To avoid those problems, convection-enhanced delivery (CED) has been exploited as a hopeful strategy to deliver therapeutic drugs. CED enhances the diffusion ability of macromolecules and drugs by positive pressure and continuous infusion of solutions [42]. In the mouse model of intracranial glioblastoma, the combination of systemic administrating liposome loading irinotecan and radiation therapy has a higher antitumor activity and longer survival time than using each of them alone. Either systemic administration or CED administration of irinotecan liposomes has no significant influence on the therapeutic effect.[43].

#### 3.3.2.2 Active Targeting

Active targeting refers to the use of biological specificity, including special antigen, antibody binding, or ligand and receptor binding, to achieve the targeting of drug delivery and release. The drug delivery system that utilizes active targeting molecular modification can mimic the process that receptors or transporters mediate the transport of the corresponding endogenous ligands or substrates, commonly referred to as a "Trojan horse" strategy.

#### 3.3.2.2.1 Receptors-Related Endocytosis

NPs modified with specific ligands or antibodies will be more likely to target the lesion sites than unmodified NPs. Additionally, the BBB membrane is negatively charged, so it exhibits high affinity with positively charged substances due to electrostatic interaction, which also triggers active cellular internalization. The receptors overexpressed on the BBB mainly include the transferrin (Tf) receptor, insulin receptor, acetylcholine receptor, low-density lipoprotein receptor, and diphtheria toxin receptor [44–46]. The contact of either natural ligands or artificial antibodies on NPs with those specific receptors overexpressed on the BBB surface/diseased brain cells could induce endocytosis of NPs into an intracellular transport vesicle; the modified NPs cross BBB/diseased brain cells and release the ligands to play a biological role [47].

The transferrin receptor containing two approximately 90 kDa subunits is highly expressed on brain capillary endothelial cells. Brain iron balance is maintained by a transferrin-mediated transport pathway, which is widely used for brain-targeted delivery [48]. Mahajan et al. [49] used fluorescent quantum dots modified with Tf as a drug carrier and further encased anti-HIV-1 drug amprenavir to successfully construct multifunctional NPs for brain imaging and disease treatment. In fact, the endogenous transferrin may bind with the brain-targeting Tf receptors. So direct use of Tf as a targeted modification is often interfered by endogenous ligands, greatly affecting delivery efficiency. To avoid this problem, NPs with specific antibody are good alternatives to endogenous transferrin. Murine OX26, a common MAb, has been shown to induce receptors-related endocytosis. As reported by Bao et al. [50], PLGA nanoparticles modified with OX26 antibody produced a very efficient analgesic effect on animal models of chronic nerve damage. Anti-Tf receptor antibodies like RI7217 and single-chain fragment also reported as commendable NP modifications to improve brain drug delivery via receptors-related endocytosis [51, 52]. Lactoferrin (Lf) receptor has a certain level of expression on the BBB. Research showed that the brain uptake of Lf-modified PEG-PLGA nanoparticle is three times higher than that of unmodified NPs [53], indicating that receptor Lf is the potential targeting ligand for brain drug delivery.

In addition to the two receptors mentioned above, several other receptors, like low-density lipoprotein receptor [54], insulin receptor [55], diphtheria toxin receptor [56], and nicotinic acetylcholine receptor [57], have been developed for brain-targeting drug delivery. Currently, the method of receptor-mediated drug-targeted delivery systems becomes more hopeful in the field of brain drug delivery. A diphtheria toxin mutant covalently bound to transferrin (Tf-CRM107) was used to treat glioblastoma, whose performance was excellent with low toxicity and tumor response in patients with recurrent high-grade brain tumors in clinical trials [58].

Moreover, some receptors are overexpressed on diseased cells, so directly targeted brain tumor cells or other diseased brain cells can be regarded as a trying way for activating brain drug delivery. When cancer cells invade neighboring healthy cerebral tissues in the brain tumor, tumor neovasculature forms and the permeability is altered. As a tumor grows further, BBTB integrity is compromised along with the formation of inter-endothelial gaps between CECs [59]. Therefore, NPs can directly target brain tumor, especially malignant brain tumor, through blood circulation due to the compromised BBB. Owing to its target integrin receptor αvβ3 that is overexpressed on brain tumor cells, RGD peptide (Arg-Gly-Asp) is widely used in brain tumor drug delivery [60, 61]. Wang F et al. designed RGD peptide-modified functionalized gold nanorods (RDG) for the delivery of small hairpin (sh) RNA. The RGD-modified gold nanorods triggered rapid cytoplasm shRNAs release in brain tumors and have better tumor gene silencing efficiency in U87 tumor-bearing BALB/c mice than non-RGD-grafted particles. [62]. Gao et al. [63] developed a DTX-incorporated albumin-lipid nanoparticles (DNPs) using bovine serum albumin (BSA) in combination with DTX and egg yolk lecithin (EYL). The DNPs could directly target brain tumors and prolong the median survival time of glioma-bearing mice, exhibiting good tolerance and antitumor effect. In some brain disorders, other diseased brain cells can also be potential targets for direct brain drug delivery. Astrocytes play a crucial role in the brain [64]. Astrocyte proliferation exists in pathological damages of the brain, accompanied by the changes of cellular components and functional activities [65]. Astrocytes are potential drug delivery targets for neurological disorder treatment. One study reported that targeting ephedrine-A5 on astrocytes potentiated the recovery process in stroke [66]. Amyloid β (Aβ) is deposited in the brain of AD patients, which caused associated neurofibrillary pathological changes in the brain; therefore, NPs can directly target Aβ through blood circulation to achieve targeted therapeutic effects [67]. Agyare et al. demonstrated that cyclophosphamide-loaded theranostic NPs modified with anti-amyloid antibody, IgG4.1, could target cerebrovascular amyloid and inhibit the secretion of cytokines triggered by the Aβ40 exposure [68].

#### 3.3.2.2.2 Transporters-Mediated Active Targeting

Most of the nutrients that our brain requires are provided by blood. There are several systems on the BBB for nutrients transportation, including amino acid transporters, glutathione transporters, and choline transporters. Those transporters are often used as brain-targeted drug carriers. Rip J et al. [69] demonstrated that glutathione-grafted liposomes could achieve four times higher accumulation in brain microdialysis study in rats after intravenous injection. In in vitro uptake study, the more fluorescent tracer was found in brain endothelial cells homogenates incubated with GSH-PEG liposomes than unmodified liposomes by using a fluorescent probe. Glutathione transporter-based brain delivery systems are extremely attractive because of bio-safety, ease of large-scale production, and controllable drug release of liposomes. Several drugs have been delivered into the brain using liposomes modified with glutathione transporters [70]. Nevertheless, transporters-mediated active targeting system must have a molecular structure mimicking the endogenous nutrients, and because it may interfere with the transport of nutrients, this approach is generally less favored.

#### 3.3.2.2.3 Cell-Penetrating Peptides-Mediated Active Targeting

CPPs are a class of short polypeptides, typically about 5–30 amino acids, which are positively charged and rich in basic amino acid residues like arginine and lysine [71]. CPPs, also known as protein transduction domains, can deliver not only small molecules but also macromolecules such as proteins, NPs, siRNA, and nucleic acids into cells [72]. As positively charged peptides, CPPs show high affinity with the BBB. In addition, CPPs have remarkable advantages in high biosafety and low cytotoxicity, making them one of the most promising and effective tools for active targeting brain drug delivery.

Wang et al. [73] used Tat peptide and tumor targeting molecule RGD peptide together to modify NPs for diagnosis and treatment of gliomas across the BBB. After in vivo administration, targeted NPs can effectively inhibit the growth of glioma. Other penetrating polypeptides, such as polyarginine, low molecular weight protamine, and octaarginine, were also used in the study of brain-targeted drug delivery systems [74–76].

#### 3.3.2.2.4 Adsorptive-Mediated Active Targeting

Under physiological conditions, the positively charged carriers achieve transshipment by endocytosis and polar transport after binding to the surface of the negatively charged BCEC membranes, so the basic principle of adsorptive-mediated active targeting is electrostatic interaction.

Cationized bovine serum albumin (CBSA), obtained from bovine serum albumin through cationization, can commonly be used for brain-targeted drug delivery via AMT [77]. CBSA could significantly improve cellular uptake by BCECs. Lu et al. used CBSA-based drug carrier to modify the surface of PEG for cationic albumin NPs preparing. The modified NPs inhibited the growth of gliomas dramatically [78]. Other adsorptive-mediated particles, such as cationized immunoglobulin G and mono antibodies, also significantly increase BBB penetration [79].

However, most biofilms are negatively charged, during in vivo circulation, and cationic NPs may also bind to cell membranes of other tissues, causing non-specific uptake. This, undoubtedly, poses a challenge to limit the concentration of drugs in nontarget organs and achieve a desired therapeutic concentration in the brain.

### 3.3.2.3 Magnetic Targeting

The magnetic targeting method mainly uses the external magnetic field to attract magnetic materials in NP-based drug delivery systems. When exposing the magnetic NPs to the alternating magnetic field, NPs will release heat during magnetic hysteresis vanishment and disturb the brain locally, namely, magnetic field-induced hyperthermia [80]. Based on this method, Seyed et al. designed a novel brain-targeting drug delivery system. They founded that the thermal energy generated by magnetic nanoparticles (MNPs) in rat brain capillaries through magnetic heating can increase the BBB permeability. When hyperthermia was applied, the BBB displayed a substantial but reversible opening [81].

This targeting method is usually combined with focused ultrasound (FUS) to further improve the brain accumulation of magic NPs [82], increasing the brain deposition of drug-load MNPs synergistically. FUS techniques concentrate acoustic energy at a focal point deep in the body. This allows to induce local biological effects noninvasively, leading to selective BBB disruption without surgical intervention [83]. FUS has been investigated since 1940s [84, 85]. Over the past 10 years, the viable and popular FUS has been developed to disturb the BBB noninvasively and transiently in a targeted manner. A study showed that a noninvasive targeted neuroregulation via FUS controlled drug release from nanoemulsions. These biocompatible nanoparticles are activated by sonication which are easily got by current clinical transcranial FUS systems [86]. Studies have shown promise both in vitro and in vivo for the combination of magnetic targeting and FUS to improve treatment delivery to brain tumors. *Fan* et al. presented a multifunctional circulating microbubbles-loaded therapeutic (doxorubicin) and conjugated to superparamagnetic iron oxide (SPIO) nanoparticles, which are proved to simultaneously open the BBB and deliver drugs under FUS exposure. These NPs act as dual magnetic resonance imaging and ultrasound contrast agents, allowing magnetic targeting for enhanced drug delivery. They are stable and provide significant superparamagnetic properties for both imaging and enhanced drug delivery, which could be an excellent theranostic tool for future image-guided brain drug delivery [87].

#### 3.3.2.4 Dual Targeting

The brain is more sensitive to toxic agents compared with other organs of our body. In order to achieve an optimal therapeutic effect, the NP-based brain-targeting delivery systems should not only penetrate the BBB or bypass the BBB effectively but also selectively target pathological region to minimize the distribution of normal cells. To address this issue, dual-targeting brain drug delivery systems were developed. NPs were decorated with multiple targeting motifs, displaying better BBB penetrating and higher drug accumulation. These strategies hold promise for the treatment of brain diseases with improved therapeutic efficacy and reduced side effects.

According to different target sites, dual targeting can generally be categorized into three applications: targeting BBB and diseased brain cells using different ligands [88], dual targeting with one ligand [89], and targeting two cell types in brain disease [90]. Research reported dual-functional gold nanoparticles (AuNP-A&C-R) could improve drug accumulation in glioblastoma. AuNP-A&C-R (Fig. 3.3), which is composed of two functional particles, could target the integrin $\alpha v\beta 3$ receptor actively, which mediates NP transcytosis, and then those NPs targeted the receptors on glioblastoma cells. Further investigation of the AuNP-A&C-R

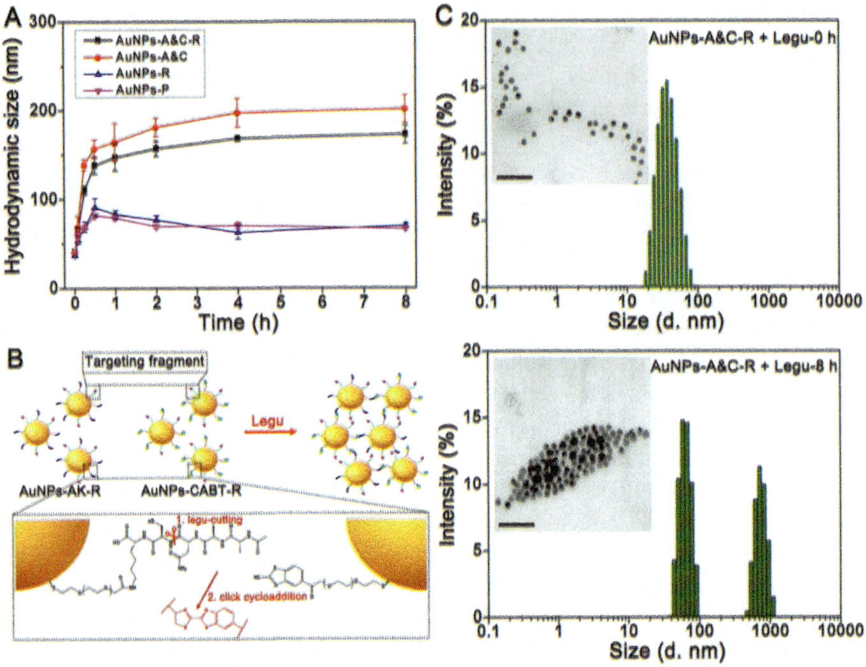


**Fig. 3.3** Characterization of the dual-functional gold nanoparticles. (**a**) The hydrodynamic size of different groups after incubation with legumain for different time intervals. (**b**) Scheme of mechanism of the aggregation process of the dual-functional gold nanoparticles. (**c**) The images of DLS and TEM showed the characterization of particle size and intensity distribution of dual-functional gold nanoparticles after incubation with legumain at 0 and 8 h. (Adapted permission from [91]. Copyright 2017 American Chemical Society)

loaded with doxorubicin revealed a smart therapeutic effect to C6 glioblastoma-bearing mice. These dual-targeting NPs are promising for brain tumor therapy [91]. A novel approach by fusing an aptamer that binds to cancer cells to an aptamer-targeting transferrin receptor has been proved for treatment of brain diseases. This bifunctional aptamer can overcome BBB and specifically target brain. The fusion of the bifunctional sequences not only enhances the binding affinity of the two aptamers but also maintains specificity [92]. Normally, dual-targeting delivery systems are superior to single-targeted strategies, but for better clinical translation, these systems still need to be further optimized [93].

### 3.3.3 *Smart Response Nano Drug Delivery and Release Systems*

It is clear that the drug needs to be transported into the location of the response so that the drug can be released after the effective response to characteristic environment. In most cases, the current types of responsive drug release drugs are mainly divided into directly release and gradually release. Based on different pathological environments to brain diseases and the advantages of nanodrug delivery systems, there are currently a large number of environmentally responsive drug delivery systems that respond to physiological and external stimuli, which can be achieved by manufacturing a spatiotemporal release system, a multimodal system, or a dual/multiple stimuli response delivery system loaded with one or more bioactive components.

#### 3.3.3.1 PH-Sensitive Nanopreparations

The clinical treatment effect of epirubicin (Epi) is largely limited by its poor penetration through BBB/BBTB. Inspiringly, based on the research of Epi-loaded micelles progressed into human clinical trials, S. Quader et al. reported RGD-modified pH-sensitive polymeric micelles against glioblastoma multiforme by improving the efficiency of drug delivery and release. Epi and their analogues were conjugated to a poly(aspartic acid) segment of PEG-b-poly(aspartic acid) copolymers. The results showed that the penetration depth of drug release from cRGD-Epi/m was improved in the U87MG 3D spheroid compared to Epi-loaded micelles and free Epi [94]. In addition, Gao et al. designed acid-responsive loaded DOX nanodrug delivery systems to avoid trapping drug within BBB endothelium via escaping from endo-/lysosomes. This drug delivery system specially recognized transferrin receptor on BBB endothelium cells and entered into cells via transcytosis. After acid-responsive cleavage of Tf, detached nanodrug delivery system releases into the cytoplasm and then enters into the brain parenchyma via exocytosis and then targeted brain tumor [95].

#### 3.3.3.2 ROS-Sensitive Nanopreparations

Oxidative stress sometimes triggers in neurodegenerative diseases and nerve damage. Yang et al. designed ROS-responsive PLA-coated mesoporous silica nanoparticles loading resveratrol that can eliminate excess reactive oxygen species (ROS). PLA prevent the resveratrol from burst release on the journey of drug delivery and enhanced the process of releasing resveratrol under high oxidative stress [96]. In order to improve the self-repairing function of neurons in AD through a proliferation of neural stem cells, Zhang et al. designed positively charged ROS-sensitive prodrug with tracking function (Fig. 3.4). Combined with the characteristics of poly(carboxybetaine) (PCB) that can maintain the stability of NPs and escape from endosomal/lysosomal with ROS-response bond (-Se-Se-) that act as a linker, they synthesized positively charged multi-amphiphile PCB-Se-Se-Sim that were loaded with SPIONs and let-absorbed 7b antisense oligonucleotide. Once the self-assembled NPs entered into cells, it entered into endosomal/lysosomal through internalization and escaped from endosomal/lysosomal through protonation of PCB, thereby releasing the let-absorbed 7b antisense oligonucleotide to the cytoplasm.

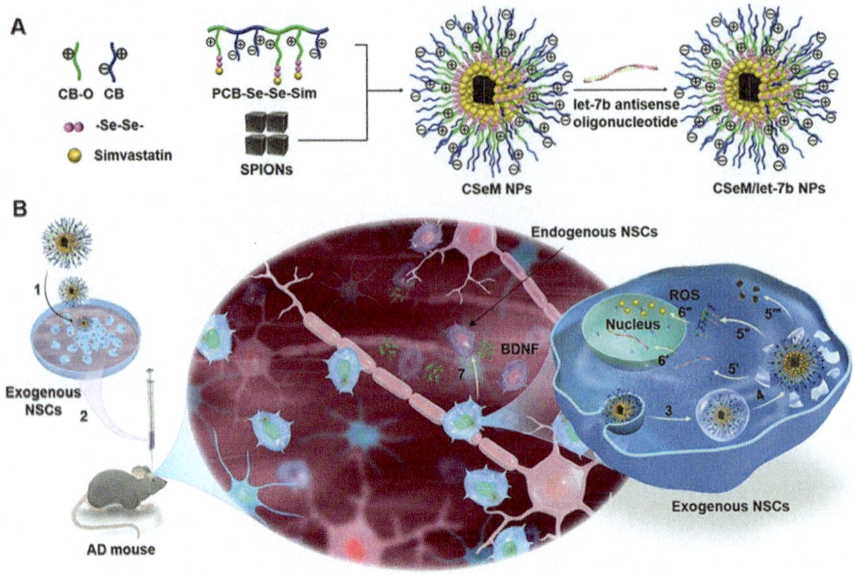


**Fig. 3.4** The strategies of ROS-sensitive prodrug. (**a**) The construct of the CSeM/let-7b NPs. (**b**) The behavior of drug release in vivo. (Adapted permission from [97]. Copyright 2018 American Chemical Society)

With the cleavage of ROS-response bond (-Se-Se-), simvastatin was released into the nucleus. MRI was used to track the migration of transplant sites and exogenous NSCs in real time through SPIONs [97].

#### 3.3.3.3 Temperature-Sensitive Nanopreparations

Polymers containing a low critical solution temperature (LCST) undergo a dramatic phase transition at temperatures approaching LCST, while LCST can be regulated by the ratio of hydrophilic and hydrophobic components in the polymer. When the LCST is between room temperature and body temperature, the polymer is sensitive to physiological temperatures. Compared with traditional nasal or oral preparations, nasal thermosensitive gels can largely be used to settle the problem of high first-pass metabolism and low oral bioavailability, which have good drug release characteristics and therapeutic potential in brain disease. *Tamer M. Sakr* et al. developed thermosensitive gels loading rivastigmine tartrate (RV) for treatment of AD through enhancement of acetylcholine in the brain. In their study, the PluronicF127 (PF127) that acts as a thermogelling agent and the characters of sol-gel temperature, gel strength, and adhesion work were optimized by different prescriptions. A significant proportion of RV can be transported directly to the brain after RV in situ gel nasal delivery [98]. Sridhar V et al. designed Poloxamer 407-chitosan thermosensitive gels loading anti-Parkinson's agent selegiline hydrochloride (SL), which significantly increased levels of dopamine and decreased levels of monoamine oxidase B in the brain after treatment with SNT gels [99].

#### 3.3.3.4 Others

Specific enzymes in different pathological environments, chemical environment, and external stimuli factors such as light, ultrasound, and other specific responses can be used for stimuli-responsive release drugs. Due to the important role of enzymes in different biological processes, disease-related enzymes can also be a targeted target of medicine. Infrared (NIR) light-responsive nanodrug delivery systems provide noninvasive spatially and temporally controlled drug release process without affecting healthy tissue. For instance, Airan R et al. designed focused ultrasound-gated drug release nanoemulsion [86]. Besides, Qu et al. designed smart near-infrared-responsive nanodrug delivery systems to control drug release to eradicating amyloid aggregates through the changeable intensity of near-infrared light. This NIR cage drug delivery system not only removed $Cu^{2+}$ but also clean excess ROS. When the system was irradiated with a low-intensity NIR laser, metal chelating agent was first released to remove free metal ions that can reduce the efficacy of drugs. Subsequently, the trapped drugs were released as the intensity of NIR light increases [100].

### *3.3.4 Intranasal Drug Delivery Systems*

Intranasal delivery allows direct access to a wide range of therapeutics to the brain from the nasal cavity. It offers a noninvasive treatment with enhanced pharmacological effects for topical brain drug delivery safely and effectively; thus, it is regarded as a promising strategy for brain-targeting drug delivery.

Intranasal transportation directly delivers the drugs to the brain without systemic absorption. It has the advantages of low side effects and easy to use [101]; thus, intranasal drug delivery has become a way in brain drug prescription. After administration, drugs can transport across nasal mucosa either transcellularly or paracellularly [102]. Due to the unique connection between the olfactory and trigeminal nerves [103], drugs and NPs can be directly transported to the cerebral spinal fluid or brain with minimal systemic exposure and metabolism. Therefore, direct intranasal delivery offers tremendous potential to bypass peripheral clearance, reduce systemic toxicity, and lower the required dose.

Intranasal delivery of olanzapine is a promising mucoadhesive for intranasal delivery. Scientific research shows that one amphiphilic nanocapsule could interact with nasal mucosa. After continuous wash, the accumulation of drug on the nasal mucosa increased by about 40%. The drug in the rat brains is 1.5-fold higher, and the mucosa integrity wasn't affected after drug administration [104]. Other nose-to-brain drug delivery systems of therapeutic NPs also show amazing management of brain diseases [105]. NP-based nose-to-brain drug system provides a practical, noninvasive approach for delivering therapeutics to the brain and potentially enhance the efficacy of neurotherapeutics. Intranasal delivery is limited, however, due to the capability to deliver concentrated drugs only in a limited volume. In most brain region, the drug dose of an intranasal delivery is restricted. Some quantitative studies of drug delivery efficiency by nasal route even show significant variation between different studies [106, 107]. Therefore, to deeply understand the mechanism of intranasal delivery, more attention should be paid on reliable quantification methodologies.

## 3.4 Conclusion

Nanotechnology-based drug delivery strategies offer great potential in allowing various types of drugs to overcome many existing barriers on the journey and achieve therapeutic effects in the brain. Although the results published are numerous, there are rarely commercialized products. The nanopreparations for brain diseases still face several significant challenges. To date, the first challenge is that the efficiency of drug delivery is still too low to meet the requirements of clinical trials. Second, it is not clear that the mechanism of brain disease and specific biomarkers is associated with brain diseases. Third, the drug behavior in vivo/vitro of nanopreparations does not achieve complete precision and control. In response to external stimuli, such as thermoresponsive, the stimulating factors can be manually controlled. Conversely, nanodrug delivery systems sensitive to internal stimuli are affected by inconsistent biological parameters between different patients. Finally, disease models may be unreliable, and good animal models of brain disease are still lacking. Thus, we should have a better understanding of other challenges as mentioned above to deploy a new and smart drug delivery nanopreparations for brain diseases.

**Acknowledgment** This work was supported by the Natural Science Foundation key project (31630027 and 31430031) and NSFC-DFG project (31761133013).

**Compliance and Ethics** The author(s) declare that they have no conflict of interest.

## References

1. Gao H, Pang Z, Jiang X. Targeted delivery of nano-therapeutics for major disorders of the central nervous system. Pharm Res. 2013;30(10):2485–98.
2. Pardridge WM. Drug targeting to the brain. Pharm Res. 2007;24(9):1733–44.
3. Goswami L, Kothiyal P. Current and future trend in nanotechnology: a review. Dialogue. 1981.
4. Sarkar A, Fatima I, Jamal QMS, Sayeed U, Khan MKA, et al. Nanoparticles as a carrier system for drug delivery across blood brain barrier. Curr Drug Metab. 2017;18(2):129–37.

5. Chen Y, Liu L. Modern methods for delivery of drugs across the blood-brain barrier. Adv Drug Deliv Rev. 2012;64(7):640–65.
6. Dikpati A, Madgulkar AR, Kshirsagar SJ, Bhalekar MR, Chahal AS. Targeted drug delivery to CNS using nanoparticles. J Adv Pharm Sci. 2012;2:179–91.
7. Bhaskar S, Tian F, Stoeger T, Kreyling W, de la Fuente JM, et al. Multifunctional Nanocarriers for diagnostics, drug delivery and targeted treatment across blood-brain barrier: perspectives on tracking and neuroimaging. Part Fibre Toxicol. 2010;7(1):3.
8. Koo YE, Reddy GR, Bhojani M, Schneider R, Philbert MA, et al. Brain cancer diagnosis and therapy with nanoplatforms. Adv Drug Deliv Rev. 2006;58(14):1556–77.
9. Kim JY, Choi WI, Kim YH, Tae G. Brain-targeted delivery of protein using chitosan- and RVG peptide-conjugated, pluronic-based nano-carrier. Biomaterials. 2013;34(4):1170–8.
10. Malhotra M, Tomaro-Duchesneau C, Prakash S. Synthesis of TAT peptide-tagged PEGylated chitosan nanoparticles for siRNA delivery targeting neurodegenerative diseases. Biomaterials. 2013;34(4):1270–80.
11. Meng Q, Wang A, Hua H, Jiang Y, Wang Y, et al. Intranasal delivery of Huperzine A to the brain using lactoferrin-conjugated N-trimethylated chitosan surface-modified PLGA nanoparticles for treatment of Alzheimer's disease. Int J Nanomedicine. 2018;13:705–18.
12. Fang Z, Chen S, Qin J, Chen B, Ni G, et al. Pluronic P85-coated poly(butylcyanoacrylate) nanoparticles overcome phenytoin resistance in P-glycoprotein overexpressing rats with lithium-pilocarpine-induced chronic temporal lobe epilepsy. Biomaterials. 2016;97:110–21.
13. Hwang DW, Son S, Jang J, Youn H, Lee S, et al. A brain-targeted rabies virus glycoprotein-disulfide linked PEI nanocarrier for delivery of neurogenic microRNA. Biomaterials. 2011;32(21):4968–75.
14. Song H, Wei M, Zhang N, Li H, Tan X, et al. Enhanced permeability of blood-brain barrier and targeting function of brain via borneol-modified chemically solid lipid nanoparticle. Int J Nanomedicine. 2018;13:1869–79.
15. Lee K, David AE, Zhang J, Shin MC, Yang VC. Enhanced accumulation of theranostic nanoparticles in brain tumor by external magnetic field mediated in situ clustering of magnetic nanoparticles. J Ind Eng Chem. 2017;54:389–97.
16. Shen Z, Liu T, Li Y, Lau J, Yang Z, et al. Fenton-Reaction-Acceleratable magnetic nanoparticles for Ferroptosis therapy of Orthotopic brain tumors. ACS Nano. 2018;12 (11):11355–65.
17. Nasr SH, Kouyoumdjian H, Mallett C, Ramadan S, Zhu DC, et al. Detection of beta-amyloid by Sialic acid coated bovine serum albumin magnetic nanoparticles in a mouse model of Alzheimer's disease. Small. 2018;14(3):1701828.
18. Zarebkohan A, Najafi F, Moghimi HR, Hemmati M, Deevband MR, et al. Synthesis and characterization of a PAMAM dendrimer nanocarrier functionalized by SRL peptide for targeted gene delivery to the brain. Eur J Pharm Sci. 2015;78:19–30.
19. Kanazawa T, Akiyama F, Kakizaki S, Takashima Y, Seta Y. Delivery of siRNA to the brain using a combination of nose-to-brain delivery and cell-penetrating peptide-modified nano-micelles. Biomaterials. 2013;34(36):9220–6.
20. Xie YT, Du YZ, Yuan H, Hu FQ. Brain-targeting study of stearic acid-grafted chitosan micelle drug-delivery system. Int J Nanomedicine. 2012;7:3235–44.
21. Qu M, Lin Q, He S, Wang L, Fu Y, et al. A brain targeting functionalized liposomes of the dopamine derivative N-3,4-bis(pivaloyloxy)-dopamine for treatment of Parkinson's disease. J Control Release. 2018;277:173–82.
22. Boche M, Pokharkar V. Quetiapine Nanoemulsion for intranasal drug delivery: evaluation of brain-targeting efficiency. AAPS Pharm Sci Tech. 2017;18(3):686–96.
23. Hao J, Zhao J, Zhang S, Tong T, Zhuang Q, et al. Fabrication of an ionic-sensitive in situ gel loaded with resveratrol nanosuspensions intended for direct nose-to-brain delivery. Colloids Surf B Biointerfaces. 2016;147:376–86.
24. Liu Y, Ma Y, Xu J, Chen Y, Xie J, et al. Apolipoproteins adsorption and brain-targeting evaluation of baicalin nanocrystals modified by combination of Tween80 and TPGS. Colloids Surf B Biointerfaces. 2017;160:619–27.

25. Kumar MNVR. A review of chitin and chitosan applications. React Funct Polym. 2000;46(1):1–27.
26. Liu Z, Jiang M, Kang T, Miao D, Gu G, et al. Lactoferrin-modified PEG-co-PCL nanoparticles for enhanced brain delivery of NAP peptide following intranasal administration. Biomaterials. 2013;34(15):3870–81.
27. Loureiro JA, Gomes B, Fricker G, Coelho MAN, Rocha S, et al. Cellular uptake of PLGA nanoparticles targeted with anti-amyloid and anti-transferrin receptor antibodies for Alzheimer's disease treatment. Colloids Surf B Biointerfaces. 2016;145:8–13.
28. Hu X, Yang F, Liao Y, Li L, Zhang L. Cholesterol-PEG comodified poly (N-butyl) cyanoacrylate nanoparticles for brain delivery: in vitro and in vivo evaluations. Drug Deliv. 2017;24(1):121–32.
29. Chertok B, David AE, Yang VC. Polyethyleneimine-modified iron oxide nanoparticles for brain tumor drug delivery using magnetic targeting and intra-carotid administration. Biomaterials. 2010;31(24):6317–24.
30. Rassu G, Soddu E, Posadino AM, Pintus G, Sarmento B, et al. Nose-to-brain delivery of BACE1 siRNA loaded in solid lipid nanoparticles for Alzheimer's therapy. Colloids Surf B Biointerfaces. 2017;152:296–301.
31. Gothwal A, Nakhate KT, Alexander A, Ajazuddin GU. Boosted memory and improved brain bioavailability of Rivastigmine: targeting effort to the brain using covalently tethered lower generation PAMAM Dendrimers with Lactoferrin. Mol Pharm. 2018;15(10):4538–49.
32. Shao K, Wu J, Chen Z, Huang S, Li J, et al. A brain-vectored angiopep-2 based polymeric micelles for the treatment of intracranial fungal infection. Biomaterials. 2012;33(28):6898–907.
33. Tanifum EA, Dasgupta I, Srivastava M, Bhavane RC, Sun L, et al. Intravenous delivery of targeted liposomes to amyloid-β pathology in APP/PSEN1 transgenic mice. PLoS One. 2012;7(10):e48515.
34. Dordevic SM, Cekic ND, Savic MM, Isailovic TM, Randelovic DV, et al. Parenteral nanoemulsions as promising carriers for brain delivery of risperidone: design, characterization and in vivo pharmacokinetic evaluation. Int J Pharm. 2015;493(1–2):40–54.
35. Chen T, Li C, Li Y, Yi X, Wang R, et al. Small-sized mPEG-PLGA nanoparticles of Schisantherin A with sustained release for enhanced brain uptake and anti-Parkinsonian activity. ACS Appl Mater Interfaces. 2017;9(11):9516–27.
36. Sanchez-Lopez E, Ettcheto M, Egea MA, Espina M, Calpena AC, et al. New potential strategies for Alzheimer's disease prevention: pegylated biodegradable dexibuprofen nanospheres administration to APPswe/PS1dE9. Nanomed Nanotechnol Biol Med. 2017;13(3):1171–82.
37. Shi B, Du X, Chen J, Fu L, Morsch M, et al. Multifunctional hybrid nanoparticles for traceable drug delivery and intracellular microenvironment-controlled multistage drug-release in neurons. Small. 2017;13(20):1603966.
38. Jasim A, Abdelghany S, Greish K. Current update on the role of enhanced permeability and retention effect in cancer nanomedicine. In: Mishra V, Kesharwani P, Mohd Amin MCI, Iyer A, editors. Nanotechnology-based approaches for targeting and delivery of drugs and genes. Amsterdam: Academic; 2017. p. 62–109.
39. Cheng Y, Morshed RA, Auffinger B, Tobias AL, Lesniak MS. Multifunctional nanoparticles for brain tumor imaging and therapy. Adv Drug Deliv Rev. 2014;66:42–57.
40. Feng Q, Shen Y, Fu Y, Muroski ME, Zhang P, et al. Self-assembly of gold nanoparticles shows microenvironment-mediated dynamic switching and enhanced brain tumor targeting. Theranostics. 2017;7(7):1875–89.
41. Kozler P, Riljak V, Jandova K, Pokorny J. CT imaging and spontaneous behavior analysis after osmotic blood-brain barrier opening in Wistar rat. Physiol Res. 2014;63(Suppl 4):S529–34.

42. Foley CP, Nishimura N, Neeves KB, Schaffer CB, Olbricht WL. Flexible microfluidic devices supported by biodegradable insertion scaffolds for convection-enhanced neural drug delivery. Biomed Microdevices. 2009;11(4):915–24.
43. Chen PY, Ozawa T, Drummond DC, Kalra A, Fitzgerald JB, et al. Comparing routes of delivery for nanoliposomal irinotecan shows superior anti-tumor activity of local administration in treating intracranial glioblastoma xenografts. Neuro-Oncology. 2013;15(2):189–97.
44. Khan AR, Yang X, Fu M, Zhai G. Recent progress of drug nanoformulations targeting to brain. J Control Release. 2018;291:37–64.
45. Gao H. Progress and perspectives on targeting nanoparticles for brain drug delivery. Acta Pharm Sin B. 2016;6(4):268–86.
46. Aulston BD, Schapansky J, Huang Y, Odero GL, Glazner GW. Secreted amyloid precursor protein alpha activates neuronal insulin receptors and prevents diabetes-induced encephalopathy. Exp Neurol. 2018;303:29–37.
47. Tsutsumi K, Inoue Y, Yoshida C. Suppression of hyperlipidemia-associated cataracts in diabetic rats with the lipoprotein lipase activator NO-1886. Biol Pharm Bull. 2008;19(12):1570–3.
48. Qian ZM, Li H, Sun H, Ho K. Targeted drug delivery via the transferrin receptor-mediated endocytosis pathway. Pharmacol Rev. 2002;54(4):561–87.
49. Mahajan SD, Roy I, Xu GX, Yong KT, Ding H, et al. Enhancing the delivery of anti retroviral drug "Saquinavir" across the blood brain barrier using nanoparticles. Curr HIV Res. 2010;8(5):396–404.
50. Bao H, Jin X, Li L, Lv F, Liu T. OX26 modified hyperbranched polyglycerol-conjugated poly (lactic-co-glycolic acid) nanoparticles: synthesis, characterization and evaluation of its brain delivery ability. J Mater Sci Mater Med. 2012;23(8):1891–901.
51. Kim SS, Rait A, Kim E, Pirollo KF, Nishida M, et al. A nanoparticle carrying the p53 gene targets tumors including cancer stem cells, sensitizes glioblastoma to chemotherapy and improves survival. ACS Nano. 2014;8(6):5494–514.
52. Salvati E, Re F, Sesana S, Cambianica I, Sancini G, et al. Liposomes functionalized to overcome the blood–brain barrier and to target amyloid-β peptide: the chemical design affects the permeability across an in vitro model. Int J Nanomedicine. 2013;8(8):1749–58.
53. Hu K, Shi Y, Jiang W, Han J, Huang S, et al. Lactoferrin conjugated PEG-PLGA nanoparticles for brain delivery: preparation, characterization and efficacy in Parkinson's disease. Int J Pharm. 2011;415(1–2):273–83.
54. Molino Y, David M, Varini K, Jabes F, Gaudin N, et al. Use of LDL receptor-targeting peptide vectors for in vitro and in vivo cargo transport across the blood-brain barrier. FASEB J. 2017;31(5):1807–27.
55. Dieu LH, Wu D, Palivan CG, Balasubramanian V, Huwyler J. Polymersomes conjugated to 83-14 monoclonal antibodies: in vitro targeting of brain capillary endothelial cells. Eur J Pharm Biopharm. 2014;88(2):316–24.
56. Agarwal M, Sahoo AK, Bose B. Receptor-mediated enhanced cellular delivery of nanoparticles using recombinant receptor-binding domain of diphtheria toxin. Mol Pharm. 2016;14(1):23–30.
57. Wei X, Zhan C, Shen Q, Fu W, Xie C, et al. A D-peptide ligand of nicotine acetylcholine receptors for brain-targeted drug delivery. Angew Chem Int Ed Engl. 2015;127(10):3066–70.
58. Tortorella S, Karagiannis TC. Transferrin receptor-mediated endocytosis: a useful target for cancer therapy. J Membr Biol. 2014;247(4):291–307.
59. Alahmady ZS. Selective drug delivery approaches to lesioned brain through blood brain barrier disruption. Expert Opin Drug Deliv. 2018;15(4):335–49.
60. Zhu Y, Zhang J, Meng F, Deng C, Cheng R, et al. cRGD-functionalized reduction-sensitive shell-sheddable biodegradable micelles mediate enhanced doxorubicin delivery to human glioma xenografts in vivo. J Control Release. 2016;233:29–38.

61. Fang Y, Jiang Y, Zou Y, Meng F, Zhang J, et al. Targeted glioma chemotherapy by cyclic RGD peptide-functionalized reversibly core-crosslinked multifunctional poly(ethylene glycol)- b -poly(ε-caprolactone) micelles. Acta Biomater. 2017;50:396–406.
62. Wang F, Zhang W, Shen Y, Huang Q, Zhou D, et al. Efficient RNA delivery by integrin-targeted glutathione responsive polyethyleneimine capped gold nanorods. Acta Biomater. 2015;23:136–46.
63. Gao H, Cao S, Yang Z, Zhang S, Zhang Q, et al. Preparation,-characterization and anti-Glioma effects of Docetaxel-incorporated albumin-lipid nanoparticles. J Biomed Nanotechnol. 2015;11(12):2137–47.
64. Araque A, Navarrete M. Glial cells in neuronal network function. Philos Trans R Soc Lond Ser B Biol Sci. 2010;365(1551):2375–81.
65. Anderson MA, Ao Y, Sofroniew MV. Heterogeneity of reactive astrocytes. Neurosci Lett. 2014;565(17):23–9.
66. Overman JJ, Clarkson AN, Wanner IB, Overman WT, Eckstein I, et al. A role for ephrin-A5 in axonal sprouting, recovery, and activity-dependent plasticity after stroke. Proc Natl Acad Sci U S A. 2012;109(33):E2230–9.
67. Haughey NJ, Nath A, Chan SL, Borchard AC, Rao MS, et al. Disruption of neurogenesis by amyloid β-peptide, and perturbed neural progenitor cell homeostasis, in models of Alzheimer's disease. J Neurochem. 2010;83(6):1509–24.
68. Agyare EK, Jaruszewski KM, Curran GL, Rosenberg JT, Grant SC, et al. Engineering theranostic nanovehicles capable of targeting cerebrovascular amyloid deposits. J Control Release. 2014;185(1):121–9.
69. Rip J, Chen L, Hartman R, van den Heuvel A, Reijerkerk A, et al. Glutathione PEGylated liposomes: pharmacokinetics and delivery of cargo across the blood-brain barrier in rats. J Drug Target. 2014;22(5):460–7.
70. Rotman M, Welling MM, Bunschoten A, de Backer ME, Rip J, et al. Enhanced glutathione PEGylated liposomal brain delivery of an anti-amyloid single domain antibody fragment in a mouse model for Alzheimer's disease. J Control Release. 2015;203:40–50.
71. Milletti F. Cell-penetrating peptides: classes, origin, and current landscape. Drug Discov Today. 2012;17(15–16):850–60.
72. Saw PE, Ko YT, Jon S. Efficient liposomal Nanocarrier-mediated Oligodeoxynucleotide delivery involving dual use of a cell-penetrating peptide as a packaging and intracellular delivery agent. Macromol Rapid Commun. 2010;31(13):1155–62.
73. Wang H, Su W, Wang S, Wang X, Liao Z, et al. Smart multifunctional core-shell nanospheres with drug and gene co-loaded for enhancing the therapeutic effect in a rat intracranial tumor model. Nanoscale. 2012;4(20):6501–8.
74. Veiseh O, Kievit FM, Mok H, Ayesh J, Clark C, et al. Cell transcytosing poly-arginine coated magnetic nanovector for safe and effective siRNA delivery. Biomaterials. 2011;32(24):5717–25.
75. Xia H, Gao X, Gu G, Liu Z, Zeng N, et al. Low molecular weight protamine-functionalized nanoparticles for drug delivery to the brain after intranasal administration. Biomaterials. 2011;32(36):9888–98.
76. Qin Y, Zhang QY, Chen HL, Yuan WM, Kuai R, et al. Comparison of four different peptides to enhance accumulation of liposomes into the brain. J Drug Target. 2012;20(3):235–45.
77. Lu W, Wan J, Zhang Q, She Z, Jiang X. Aclarubicin-loaded cationic albumin-conjugated pegylated nanoparticle for glioma chemotherapy in rats. Int J Cancer. 2010;120(2):420–31.
78. Lu W, Sun Q, Wan J, She Z, Jiang XG. Cationic albumin–conjugated Pegylated nanoparticles allow gene delivery into brain tumors via intravenous administration. Cancer Res. 2006;66(24):11878.
79. Wei L. Adsorptive-mediated brain delivery systems. Curr Pharm Biotechnol. 2012;13(12):2340–8.
80. Tabatabaei SN, Duchemin S, Girouard H, Martel S. Towards MR-navigable Nanorobotic carriers for drug delivery into the brain. Paper presented at The IEEE international conference on robotics and automation. 2012.

81. Tabatabaei SN, Girouard H, Carret AS, Martel S. Remote control of the permeability of the blood-brain barrier by magnetic heating of nanoparticles: a proof of concept for brain drug delivery. J Control Release. 2015;206:49–57.
82. Huang HY, Liu HL, Hsu PH, Chiang CS, Tsai CH, et al. A Multitheragnostic Nanobubble system to induce blood–brain barrier disruption with magnetically guided focused ultrasound. Adv Mater. 2015;27(4):655.
83. Vykhodtseva N, McDannold N, Hynynen K. Progress and problems in the application of focused ultrasound for blood-brain barrier disruption. Ultrasonics. 2008;48(4):279–96.
84. Lynn JG, Zwemer RL, Chick AJ, Miller AE. A new method for the generation and use of focused ultrasound in experimental biology. J Gen Physiol. 1942;26(2):179–93.
85. Medel R, Monteith SJ, Elias WJ, Eames M, Snell J, et al. Magnetic resonance-guided focused ultrasound surgery: part 2: a review of current and future applications. Neurosurgery. 2012;71 (4):755–63.
86. Airan RD, Meyer RA, Ellens NP, Rhodes KR, Farahani K, et al. Noninvasive targeted transcranial Neuromodulation via focused ultrasound gated drug release from Nanoemulsions. Nano Lett. 2017;17(2):652–9.
87. Lamsam L, Johnson E, Connolly ID, Wintermark M, Hayden Gephart M. A review of potential applications of MR-guided focused ultrasound for targeting brain tumor therapy. Neurosurg Focus. 2018;44(2):E10.
88. Liu C, Liu XN, Wang GL, Yu H, Meng S, et al. A dual-mediated liposomal drug delivery system targeting the brain: rational construction, integrity evaluation across the blood–brain barrier, and the transporting mechanism to glioma cells. Int J Nanomedicine. 2017;12:2407–25.
89. Byeon HJ, Thao le Q, Lee S, Min SY, Lee ES, et al. Doxorubicin-loaded nanoparticles consisted of cationic- and mannose-modified-albumins for dual-targeting in brain tumors. J Control Release. 2016;225:301–13.
90. Zhang J, Chen N, Wang H, Gu W, Liu K, et al. Dual-targeting superparamagnetic iron oxide nanoprobes with high and low target density for brain glioma imaging. J Colloid Interface Sci. 2016;469:86–92.
91. Ruan S, Xiao W, Hu C, Zhang H, Rao J, et al. Ligand-mediated and enzyme-directed precise targeting and retention for the enhanced treatment of glioblastoma. ACS Appl Mater Interfaces. 2017;9(24):20348–60.
92. Macdonald J, Henri J, Goodman L, Xiang D, Duan W, et al. Development of a bi-functional aptamer targeting the transferrin receptor and EpCAM for the treatment of brain cancer metastases. ACS Chem Neurosci. 2017;8(4):777–84.
93. Gao H. Perspectives on dual targeting delivery systems for brain tumors. J Neuroimmune Pharmacol. 2017;12(1):6–16.
94. Quader S, Liu X, Chen Y, Mi P, Chida T, et al. cRGD peptide-installed epirubicin-loaded polymeric micelles for effective targeted therapy against brain tumors. J Control Release. 2017;258:56–66.
95. Ruan S, Qin L, Xiao W, Hu C, Zhou Y, et al. Acid-responsive transferrin dissociation and GLUT mediated exocytosis for increased blood–brain barrier Transcytosis and programmed Glioma targeting delivery. Adv Funct Mater. 2018;28(30):1802227.
96. Shen Y, Cao B, Snyder NR, Woeppel KM, Eles JR, et al. ROS responsive resveratrol delivery from LDLR peptide conjugated PLA-coated mesoporous silica nanoparticles across the blood–brain barrier. J Nanobiotechnol. 2018;16(1):13.
97. Li Y, Li Y, Ji W, Lu Z, Liu L, et al. Positively charged Polyprodrug Amphiphiles with enhanced drug loading and ROS-responsive release ability for traceable synergistic therapy. J Am Chem Soc. 2018;140(11):jacs.8b01641.
98. Abouhussein DMN, Khattab A, Bayoumi NA, Mahmoud AF, Sakr TM. Brain targeted rivastigmine mucoadhesive thermosensitive in situ gel: optimization, in vitro evaluation, radiolabeling, in vivo pharmacokinetics and biodistribution. J Drug Delivery Sci Technol. 2018;43:129–40.

99. Sridhar V, Wairkar S, Gaud R, Bajaj A, Meshram P. Brain targeted delivery of mucoadhesive thermosensitive nasal gel of selegiline hydrochloride for treatment of Parkinson's disease. J Drug Target. 2018;26(2):150–61.
100. Ma M, Gao N, Sun Y, Ren J, Qu X. A near-infrared responsive drug sequential release system for better eradicating amyloid aggregates. Small. 2017;13(46):1701817.
101. Zhang TT, Li W, Meng G, Wang P, Liao W. Strategies for transporting nanoparticles across the blood-brain barrier. Biomater Sci. 2016;4(2):219–29.
102. Khan AR, Liu M, Khan MW, Zhai G. Progress in brain targeting drug delivery system by nasal route. J Control Release. 2017;268:364–89.
103. Pardeshi CV, Belgamwar VS. Direct nose to brain drug delivery via integrated nerve pathways bypassing the blood-brain barrier: an excellent platform for brain targeting. Expert Opin Drug Deliv. 2013;10(7):957–72.
104. Fonseca FN, Betti AH, Carvalho FC, Gremião MPD, Dimer FA, et al. Mucoadhesive Amphiphilic Methacrylic copolymer-functionalized poly(ε-caprolactone) Nanocapsules for nose-to-brain delivery of olanzapine. J Biomed Nanotechnol. 2015;11(8):1472–81.
105. Zhang C, Chen J, Feng C, Shao X, Liu Q, et al. Intranasal nanoparticles of basic fibroblast growth factor for brain delivery to treat Alzheimer's disease. Int J Pharm. 2014;461 (1–2):192–202.
106. Dhuria SV, Hanson LR, Frey WH 2nd. Intranasal delivery to the central nervous system: mechanisms and experimental considerations. J Pharm Sci. 2010;99(4):1654–73.
107. Kozlovskaya L, Abou-Kaoud M, Stepensky D. Quantitative analysis of drug delivery to the brain via nasal route. J Control Release. 2014;189(10):133–40.
108. Tibbitt MW, Dahlman JE, Langer R. Emerging Frontiers in drug delivery. J Am Chem Soc. 2016;138(3):704.
109. Shi J, Votruba AR, Farokhzad OC, Langer RJNL. Nanotechnology in drug delivery and tissue engineering: from discovery to applications. Nano Lett. 2010;10(9):3223–30.

# Chapter 4
# The Strategies of Nanomaterials for Therapy

**Yang Du, Shuying Wang, Fangyuan Li, and Daishun Ling**

**Abstract** Brain diseases, including brain tumor, Alzheimer's and Parkinson's diseases, and stroke, are becoming increasingly widespread in the aging population. However, due to our limited knowledge of the pathogenesis of these diseases and inefficient drug delivery system, the treatments for these disorders are still major challenges in modern medicine. Nanomaterials have recently emerged as an effective tool in biomedical fields, which have unique properties and structures conferring improved efficacy, safety, sensitivity, and the potential to be personalized. With these advantages, nanomaterial-based therapies are rapidly progressing and show promise to revolutionize the way we treat brain diseases. In this chapter, the state of the art of current nanomaterials in the treatment of brain diseases will be described, with discussions on the prospects and ongoing challenges that must be overcome.

**Keywords** Brain diseases · Neurosciences · Nanomaterials · Nanotechnology · Therapeutics

## 4.1 Introduction

With the aging of the population becoming a global issue, the incidence of central nervous system (CNS) diseases has been growing over the years. Some brain diseases are particularly notorious for the high mortality rate and/or low life quality of the patients. For example, cerebrovascular diseases such as stroke are the world's second leading cause of mortality. Neurodegenerative diseases, like Alzheimer's diseases (AD) and Parkinson's diseases (PD), are closely related to movement and memory disorder, as well as dementia attributable to the gradual loss of neurons, affecting the quality of life in the middle-aged and elderly population. In addition, brain tumor remains a severe clinical problem that accounts for a big part of cancer-

Y. Du · S. Wang · F. Li (✉) · D. Ling (✉)
Institute of Pharmaceutics, College of Pharmaceutical Sciences, Zhejiang University, Hangzhou, Zhejiang, People's Republic of China
e-mail: lfy@zju.edu.cn; lingds@zju.edu.cn



X. Xue (ed.), *Nanomedicine in Brain Diseases*,
https://doi.org/10.1007/978-981-13-8731-9_4

related death worldwide. All these brain diseases severely threaten the public health of modern society and have become an increasingly serious medical burden and social problem [1].

Although brain diseases have been investigated for many years, thus far, effective clinical treatment strategies are still highly limited due to the pathophysiological environment of CNS, where the most known one is the blood-brain barrier (BBB) which restricts access of most therapeutic agents to the brain [2]. Besides, the heterogeneity of cellular and molecular environments and the complexity of the anatomy and information processing of the brain all pose challenges for the successful treatment of brain diseases [3]. To solve these problems and achieve better therapeutic effect, novel strategies are urgently needed.

In recent years, nanomaterials have played an important role in the diagnosis and treatment of various brain disorders. Engineered nanomaterials can not only be used as carriers for the effective delivery of therapeutic agents into the brain but also exhibit therapeutic effect via the inherent antioxidant properties of themselves for neuron protection [3]. In this chapter, we summarize the recent nanomaterial-based strategies for the treatment of brain disease.

## 4.2 Nanomaterials for Brain Tumor Treatment

Brain tumors are highly aggressive with very high mortality rates. However, effective treatments for brain tumors are still very limited as aforementioned. Up to date, surgical resection, radiotherapy, and chemotherapy are three common methods for the treatment of brain tumor in clinic, such as the typical gliomas [4]. Compared with other categories of tumors, the shield of the BBB imposes additional restrictions on the treatment of gliomas. Moreover, because of the strong infiltration of glioma and indistinct tumor boundaries, the cancer cells are difficult to be completely resected by surgery, increasing the risk of tumor recurrence [5]. Nowadays, nanotechnology has been extensively investigated for drug delivery, gene delivery, hyperthermia therapy, and photodynamic therapy in brain tumor therapy, which has provided compelling evidence for that therapeutic strategies taking advantages of unique properties of nanomaterials are promising to achieve more accurate and effective treatments of brain tumors.

### *4.2.1 Nanomaterial-Based Chemotherapy*

It is well known that the severe side effect of chemotherapy seriously affects the life quality of patients [6]. Therefore, it is critical to improve the therapeutic effect and reduce the side effect of chemotherapy after the operation. However, the BBB acts as a fortress to protect the pivotal position such as CNS from potentially detrimental foreign materials [7, 8]. Thus, almost all large-molecule drugs and a majority of small molecules cannot pass it [9]. Moreover, the success penetration is not enough for chemotherapy. For example, the use of temozolomide, the first-line chemotherapy drug, is significantly limited due to the non-specific distribution in the brain despite of a certain degree of brain penetration capability [10]. What's more, the high-dose injection may further promote the drug resistance of tumors. Given the above, current chemotherapeutic strategies are undesirable when applied to gliomas compared with other malignancies.

Fortunately, with the development of the nanotechnology, a variety of nanoparticles have been constructed with rational surface engineering and precise size control, enabling the delivery of therapeutic agents pass through the BBB [11]. Normally, small molecular chemotherapy drugs are integrated with nanoparticles via a variety of physicochemical methods, such as encapsulation, adsorption, or covalent linkage within biocompatible materials. Then, some special proteins or other artificial macromolecules can be conjugated to the surface of nanoparticles to promote the specific targeting. For example, Gao et al. constructed a kind of dual-targeting doxorubicin (DOX) liposome that conjugated with both folate and transferrin [12]. These hybrid liposomes significantly improved the transport of DOX across the BBB as well as to the brain glioma for treatment (Fig. 4.1a). Similarly, Ying et al. also reported a type of dual-targeting DOX-loaded liposome which combined p-aminophenyl-α-D-mannopyranoside (MAN) and transferrin for the drug transportation across the BBB and further brain glioma targeting [13]. The ligand MAN played a key role for liposomes to overcome the BBB, and the transferrin was important for targeting glioma cells after entering the brain. The inhibitory rate to glioma cells after these dual-targeting liposomes crossing the BBB was significantly enhanced up to 64.0%, suggesting that dual-targeting DOX-loaded liposomes were available to improve the therapeutic outcome of brain glioma both in vitro and in vivo.

To sum up, nanoparticle-loading chemotherapy drugs are preferable to cross the BBB and could be efficiently internalized by the tumor tissues. Hence, the nanoparticle-based delivery strategy is promising to considerably improve the intracerebral delivery efficiency of chemotherapy drugs with minimal side effects for the treatment of glioma.

### *4.2.2 Nanomaterial-Based Gene Therapy*

The conventional surgical resection assisted with chemotherapy of glioma can improve the clinical prognosis of patients. However, the complete elimination of the tumor is quite difficult because of the existence of cancer cells tolerance and the insufficient drug concentration in the tumor site. Thus, new strategies to treat glioma are urgently demanded. Gene therapy, as emerging means of cancer treatment, has shown great potential in curing the glioma [14, 15]. In principle, gene therapy is to put the corrective genetic material into specific cells to alleviate the symptoms of disease [16]. For example, cytosine deaminase/5-fluorocytosine (CD/5-FC) can

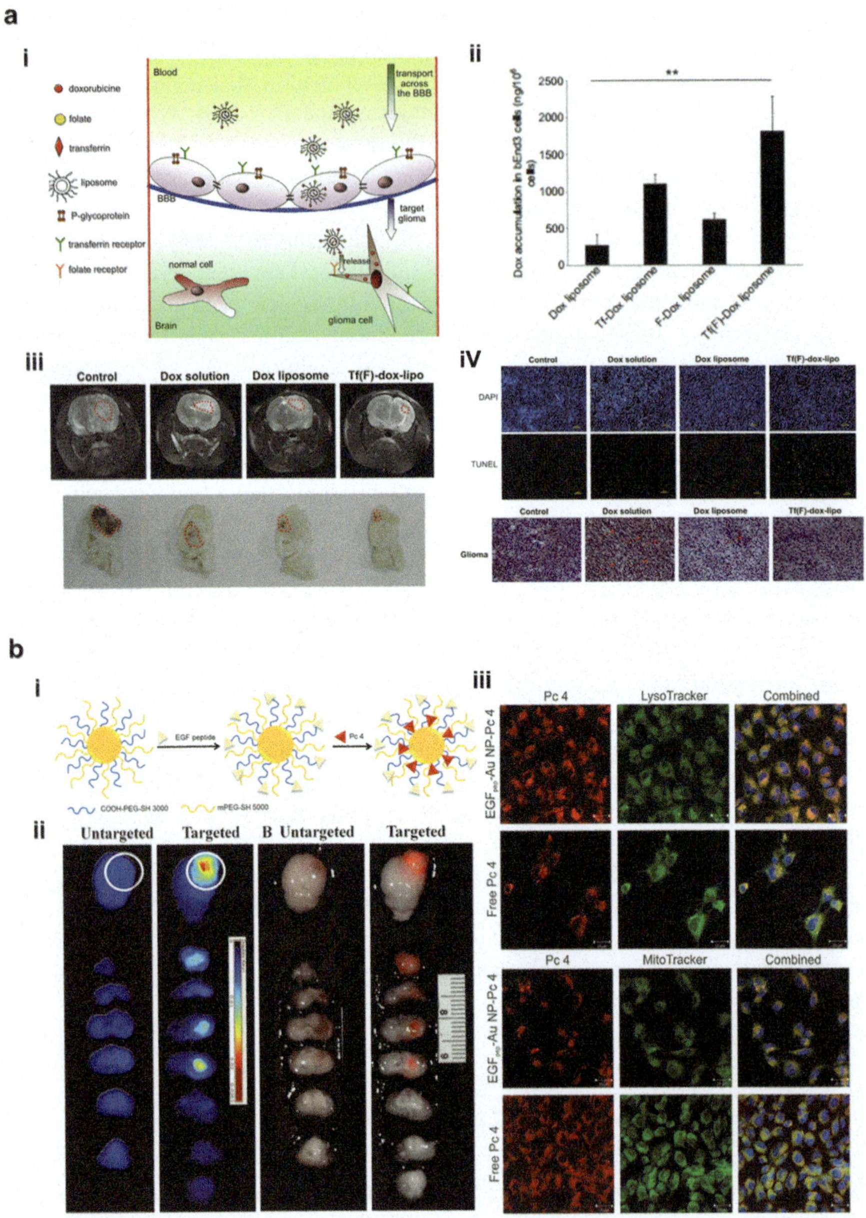


**Fig. 4.1** Nanomaterials for brain tumor treatment. (**a**) Dual-targeting liposome for nanomaterial-based chemotherapy. (*i*) Representative scheme of DOX liposome functionalized with both transferrin and folate for brain tumor targeting; (*ii*) DOX accumulation in cancer cells; (*iii*) Anti-glioma efficacy of dual-targeting liposome observed by MRI and brain column and (*iv*) analyzed by TUNEL and p-gp expression (from Ref. [12]). (**b**) Au NP-based nanoplatform for brain tumor PDT. (*i*) Design of EGFpep-Au NPs; (*ii*) Targeting efficiency of EGFpep-Au NPs-Pc 4 compared with unmodified Au NPs-Pc 4 in model mice; (*iii*) Laser scanning confocal microscopy images of 9L. E29 cells incubated with EGFpep-Au NP-Pc 4 and free Pc 4, respectively, and counterstained by LysoTracker (lysosomes) or MitoTracker (mitochondria). Scale bar = 20 μm. (*i–ii* from Ref. [36], *iii* from Ref. [37])

translate the antifungal drug to 5-fluorouracil (5-FU) which is able to irreversibly inhibit the thymidine synthase and DNA synthesis, and further resist the cancer cells [17]. Polyethyleneimine (PEI) is a widely used gene delivery carrier [18]. However, PEI is incapable of delivery of genes to the brain for the treatment of glioma as it cannot penetrate the BBB. Therefore, novel alternatives urgently await development.

To this regard, nanotechnology provides various opportunities to enhance the transportation of therapeutic genes into the brain. For example, Jung et al. constructed a nano-delivery system of multifunctional siRNA-quantum dots (QDs) for selectively inhibiting the expression of epidermal growth factor receptor variant III (EGFR vIII) in glioma cells and further downregulating signaling pathway with high efficiency [19]. In vitro analysis revealed that the EGFR vIII-overexpressed cancer cells (U87-EGFRvIII) decreased when increasing the siRNA-QDs co-incubation time. In addition, Li et al. established a gene delivery system for brain tumor targeting by the conjugation of myristic acid (MC) with a low-molecular-weight PEI [20]. This delivery system (MC-PEI) exhibited an efficient antitumor ability on the intracranial glioblastoma model. Compared with the control group, MC-PEI significantly prolonged the survival time from 22 to 28 days.

Recently, the gene therapy method has achieved some initial results, but most of these researches are the exploratory tests and unable to be translated in the clinic immediately. Thus, how to manipulate the gene therapy for clinical applications and develop more efficient gene therapy agents should be the focus in the future.

### *4.2.3 Nanomaterial-Based Thermotherapy*

Thermotherapy is a therapeutic method that facilitates the increment of body tissue temperature to cause cell dysfunction, including apoptosis induction [21], cellular structure alteration [22], and DNA conformational change [23]. The mechanism is based on the fact that temperature increment (~ 41–42 °C) could induce severe cell damages as tumor cells are more susceptible to sudden temperature increases than normal counterparts [24]. Thermotherapy has been extensively used for cancer treatment and is particularly attractive for treating brain tumor which is stubborn for current treatments [25]. Meanwhile, more and more nanomaterials were discovered to possess extraordinary optical and magnetic properties, such as gold nanoparticles and iron oxide nanoparticles, which could be investigated for thermotherapy. Here, we briefly introduce the nanomaterial-based thermotherapy induced by laser and magnetic field and their great potential for brain tumor treatment.

Stereotactic injection of magnetically active nanomaterials into the tumor site followed by exposure to an alternating magnetic field (AMF) could produce substantial heat within the tumor, thereby achieving thermotherapy. The feasibility of this strategy has been demonstrated both in preclinical [26] and clinical studies [27]. Klaus et al. [28] investigated the efficacy of iron oxide nanoparticles-mediated thermotherapy combining with radiotherapy on treating recurrent glioblastoma multiforme. The approach showed moderate side effect with the absence of serious complications, demonstrating the conjunction of thermotherapy and radiotherapy was safer and more effective than conventional therapy.

Apart from magnetic hyperthermia, photothermal therapy can also be utilized for brain tumor treatment. Accordingly, Wang et al. [29] designed a multifunctional magnetic graphene-based mesoporous silica with the payload of DOX. Based on the intrinsic thermal property of graphene nanosheets, the magnetic property of superparamagnetic iron oxide nanoparticles, and the surface modification of targeting ligands, this nanoplatform integrated dual targeting (active targeting and magnetic targeting), high biocompatibility, magnetic resonance (MR) imaging, and chemo-photothermal therapy to realize the MRI-monitored synergistic therapy of glioma. Notably, the heat produced by the nanosystem could promote not only photothermal therapy but also enhanced the release of DOX.

### 4.2.4 Nanomaterial-Based Photodynamic Therapy

Photodynamic therapy (PDT) is a noninvasive therapeutic method using photosensitizers activated by light to generate reactive oxygen species (ROS) and eventually inducing tumor cell death [30]. PDT is a promising way to improve cancer treatment and reduce side effects by localized light activation [31, 32]. However, there exist some intrinsic challenges when using PDT for brain tumors. For example, due to their poor solubility, it is hard to deliver PDT agents into the brain and to be used for systemic administration [33]. Besides, traditional PDT drugs remain to be restricted by the multidrug resistance (MDR) of tumor cells [34].

Nanomaterial-based photodynamic therapy possesses both selective delivery of PDT drugs into the tumor tissue and effective improvement of PDT efficacy. Ross et al. described a new kind of photodynamic nanoparticle for brain tumor treatment by coating the polyacrylamide (PAA) core with PEG and molecular targeting groups [35]. The photosensitizers and MRI contrast agents are encapsulated in the core, which enable the simultaneous cancer diagnosis, therapy, and real-time monitoring. This multifunctional platform provides a promising and versatile approach to treat the brain tumor. Basilion et al. [36] have designed a novel drug delivery platform which was composed of epidermal growth factor-targeted Au NPs (EGFpep-Au NPs) for PDT of the brain tumor. They used PEGylated Au NPs to encage photosensitizer (Pc 4) and transported it to the target tissue. The PEG ligands in this platform provide excellent water solubility, biocompatibility, and long blood circulation time. EGF peptides conjugated to the PEG layer could achieve tumor targeting by recognizing EGFRs on the surface of the glioma tumor cell. Besides, the proper size of Au NPs (5 nm in diameter) was able to be excreted by renal at the same time. This strategy was proved to significantly improve the brain tumor targeting efficiency. Further, they found that this platform led to decreased uptake of Au NPs by the reticuloendothelial system (RES) and increased the uptake in cancer cells which result in an enhanced therapeutic outcome in brain cancer (Fig. 4.1b) [37].

### 4.2.5 *Nanomaterial-Based Immunotherapy*

Tumor immunotherapy has been an attractive strategy for cancer treatments [38]. Immunotherapy could induce tumor death with minimal damage to normal tissue and target to both primary and secondary metastases by systemically triggering antitumor immune responses. More importantly, it can produce immunological memory that helps prevent tumor recurrences [39–41]. Nanotechnology may open a new chapter in cancer immunotherapy and vigorously promote the transformation of cancer immunotherapy to clinical practice.

The previous work [42] has shown that multi-walled carbon nanotubes could carry DNA and siRNA and be used as a novel, nontoxic, and biodegradable nanoplatform for immunogenetic therapy of brain tumors. This study suggests nanoparticles are promising candidates in immunotherapy of brain tumors. Furthermore, antigen-modified nanoparticles could be internalized by antigen-presenting cells to initiate a strong immunostimulatory cascade against cancer cells [43].

Zhang et al. designed an iRGD-modified nanoparticle for delivering both chemotherapeutic agents and immune checkpoint inhibitor into glioma [44]. This nanoplatform could overcome the BBB and accumulate in orthotopic brain glioma sites effectively. Moreover, the robust antitumor immune response against glioma was triggered by this platform. Overall, this novel synergistic platform provided a new chemo-immunotherapy strategy for the treatment of brain tumor.

In another example, Sabel et al. [45] used polybutylcyanoacrylate nanoparticles to delivery of transforming growth factor-β (TGF-β) antisense oligonucleotides to block the TGF-β production and virus-infected tumor cells to produce active specific immunization for an upregulation in the immune system activation. This "double-punch" method combined immunization and gene delivery together, showing a promising way for brain tumor immunotherapy.

## 4.3 Nanomaterials for Alzheimer's Disease Treatment

Alzheimer's disease (AD), a kind of progressive neurodegenerative brain disorder, accounts for 60–80% of dementia, and the clinical manifestations of AD are memory loss, aphasia, cognitive decline, etc. Since Alois Alzheimer's description of Auguste D.'s brain in 1907, scientists around the world have paid tremendous attention to search for the underlying mechanism of AD [46]. Some investigators proposed the amyloid hypothesis that the pathological aggregation of amyloid protein β (Aβ) may contribute to the induction of AD [47]. As developed from amyloid precursor protein (APP) via two-step proteolysis reaction, Aβ further self-assembles into neurotoxic fibrils, and the depositing fibrils in the hippocampus and other regions of the brain are defined as an iconic plaque of AD [48]. In addition, neurofibrillary tangles

(NFTs), formulated with hyperphosphorylated tau protein, are widely reported to play unneglected role in the development of AD [49]. Generally, microtube forming by the combination of tau protein and tubulin is responsible for the transport of nutrient in the brain. As an important component, tau protein is unfolded under the normal physiological state while experiences hyperphosphorylation once the PKN, a serine/threonine kinase, is activated. Consequently, the stability of microtube may be affected by generated NTFs, leading to neurodegeneration and neurotoxicity. Besides, there are other factors that may increase the risk of AD, including oxidative stress [50], neuroinflammation [51], genic mutation [52], etc.

Enormous efforts have been devoted to the drug discovery for AD, but the results are far from expected. There only exist four small molecule drugs approved by FDA for the treatment of AD, including donepezil, galantamine, rivastigmine, and memantine [48]. All of them either act on CNS cholinergic pathways or target the N-methyl-D-aspartate (NMDA) receptor and glutaminergic pathways, providing symptomatic relief but failing to reverse the pathological process of AD patients. In addition, lots of therapeutic agents are undergoing the preclinical phases but the proceeding seems to be blocked. For example, the famous pharmaceutical company Eli Lilly has recently regrettably announced that the Phase III clinical trial of the new drug research for AD had failed [53]. As the age structure of society is getting old, the increasing prevalence of AD forces us to find an efficient way to cure AD. So far, some nanomaterial-based therapy strategies, which mainly benefit from their better biocompatibility, penetrability, and versatility, have shown certain peculiarity in the treatment of AD [54, 55]. With the in-depth study of the mechanism of Alzheimer's disease, nanotechnology may bring light to the treatment of AD.

## *4.3.1 Nanomaterials for Drug Delivery*

### 4.3.1.1 Cholinesterase Inhibitors and Acetylcholine Nanocarriers

Small-molecule drugs for the treatment of AD are limited not just in the inability to reverse the pathological process but also in the drug concentration reaching the lesion. It was reported that single-walled carbon nanotubes (SWCNTs) could be excellent carriers for acetylcholine, a kind of cholinesterase inhibitor, to relieve the symptoms of AD [56]. Especially, SWCNTs can deliver the drug precisely and efficiently to the lesions via nerve axons; meanwhile, it can target the lysosomes of the neurons to minimize the toxicity to other normal tissues. In addition, BBB is another obstacle for the treatment of AD, but some investigated that using poly (n-butylcyanoacrylate) nanoparticles as carriers could penetrate the BBB and enhance the concentration of rivastigmine in the brain [57]. Hence, nanomaterials loaded with cholinesterase inhibitors can improve the therapeutic effect of AD.

#### 4.3.1.2 Hormone Nanocarrier

Recently, it is found that estradiol may reduce the risk of AD by degenerating cerebral amyloid deposition. Unfortunately, estradiol has poor bioavailability due to the first-pass effect, resulting in limited dose and high toxicity. G. Mittal et al. designed a nano-delivery system to improve the oral bioavailability of estradiol [58]. Interestingly, the drug release can be controlled subtly through adjusting molecular weight and composition of polymers.

#### 4.3.1.3 Curcuminoids Nanocarrier

Curcumin, a kind of polyphenol, can reduce the oxidative stress induced by amyloid. However, its terrible photostability and low bioavailability impose the tighter restriction on its application for the therapy of diseases relevant to oxidative stress, including AD. Rohit S. Mulik et al. concluded that poly(butyl)cyanoacrylate nanoparticles mediated with apolipoprotein E (ApoE3) can deliver curcumin to the brain without damaging the structure of curcumin [59]. They believed that this drug delivery system mediated by nanoparticles presented a great opportunity for the treatment of neurodegenerative disorders.

#### 4.3.1.4 Polyphenol Nanocarrier

Restraining and reversing the forming process of amyloid may be a promising way to cure or mitigate AD. Epigallocatechin-3-gallate (EGCG) is known as an antioxidant agent and can deviate the APP cleavage process from Aβ production. However, when EGCG was injected alone into the body, it would be cleared quickly by the immune organs such as the liver. Adam Smith et al. [60] found that oral bioavailability could be significantly enhanced by forming nanolipidic EGCG particles, and satisfying results were achieved by this way.

In conclusion, nanomaterials provide an excellent approach to improve the application of small-molecule drug candidates in AD.

### *4.3.2 Nanomaterial-Based Metal Chelation Strategy*

Trace metal ions in the brain (such as $Cu^{2+}$, $Zn^{2+}$, $Mn^{2+}$, $Fe^{2+}$) are balanced in the body under normal circumstances and play a vital role in physiology. When the trace elements in the brain are metabolically imbalanced, the ion concentration will increase or decrease, causing disorder of lipid metabolism in the body, eventually producing ROS. For example, certain metal ions ($Cu^{2+}$ and $Fe^{3+}$) may interact with Aβ to produce hydrogen peroxide, which would cause damage to the brain

[61]. Meanwhile, Aβ, as a metalloprotein, can be affected by some metal ions, ultimately resulting in its self-aggregation. In addition, $Cu^{2+}$ and $Zn^{2+}$ can increase the interaction between Aβ and the cell membrane, resulting in the neurotoxicity of Aβ [62].

Some studies demonstrated that metal chelation can effectively reduce the formation of precipitates and slow the progression of AD. In other words, the metal chelation method to remove excess metal ions in brain tissue can not only delay the accumulation of Aβ but also effectively decrease extracellular oxidative stress, thereby controlling the development of AD.

#### 4.3.2.1 Iron Chelators

Liu et al. designed a chelating agent nanoparticle system that could effectively cross the BBB by combining the FDA-approved metal chelating agent, deferoxamine, with nanoparticles [63]. The nanosystem could adsorb apolipoprotein E and apolipoprotein AI, helping to enter or leave the brain by transporting low-density lipoproteins. This approach was employed to overcome the BBB permeability limitation of traditional chelation therapy and thereby provided highly effective and safe chelation therapy of AD. Subsequently, the group designed a chelator-conjugated nanoparticle that reduced cytotoxicity to human cortical neurons by inhibiting Aβ aggregation [64]. In addition, the chelator-conjugated nanoparticle could also reduce the lipophilicity of the chelate and further reduce toxicity. Therefore, this chelator-conjugated nanoparticle could provide a new strategy for the treatment of AD.

#### 4.3.2.2 Copper Chelators

Cui et al. attached D-penicillamine, a copper ion chelating agent, to nanoparticles through a thioether bond or a disulfide bond. The reducing agent in the body can cleave the disulfide bond to release D-penicillamine. The released D-penicillamine can chelate cuprous ion and effectively dissolve $A\beta_{1-42}$, thereby preventing aggregation of Aβ [62]. Therefore, this method can effectively reverse the protein precipitation which is caused by copper ions. Bu et al. combined an upconversion nanoparticle (UCNP) with a copper ion chelating agent, 8-hydroxyquinoline-2-carboxylic acid (HQC), to prepare an intelligent nanoprobe [65]. The nanoprobes can achieve upconversion luminescence imaging, as well as chelate copper ions to inhibit the aggregation of Aβ.

#### 4.3.2.3 Zinc Chelators

Maluta et al. modified different chelating agents (e.g., EDTA, histidine, ZnAc, etc.) on the surface of nanoliposomes [66]. The modified nanoliposome can induce depolymerization of Aβ aggregates caused by zinc ions. In addition, the nanoliposomes have a certain protective effect on PC12 neuronal cells.

## 4.3.3 Nanomaterials for Antioxidant

Another strategy for the treatment of AD is directed toward delivering antioxidant species to the brain, because of their antioxidant sponge's effect to quench the reactivity of ROS in AD. Recently, many nanomaterials with the excellent ROS scavenging ability are promising to achieve the antioxidant therapy of AD.

### 4.3.3.1 Fullerenes

Fullerene, a common nanomaterial with strong antioxidant and free radical scavenging effects, can, therefore, be used as a neuroprotective agent to treat AD. A variety of functionalized fullerenes have been shown neuroprotective effects in vitro [67, 68]. Its protective mechanism is attributed to the fact that fullerenes can effectively reduce free radical levels and inhibit the neurotoxicity of calcium ions [68, 69]. Xie et al. found that the firm interaction between fullerenes and Aβ hindered the fibrillation of Aβ [70]. Therefore, various multifunctional fullerene-based nanomaterials have great application potential in the treatment of AD.

### 4.3.3.2 Nanoceria

Among various kinds of antioxidants, ceria is the most representative one with the pronounced antioxidant property. With a unique atomic structure and the 4f electronic configuration, nanocrystalline ceria possesses special luminescent, magnetic, and electronic properties [71]. Up to now, ceria nanomaterials have been extensively investigated in many different fields of technology and engineering, such as solid-oxide fuel cells, nano-catalysis, solar cells, oxidative coupling of methane, and the water-gas shift reaction. Furthermore, the high redox ability and low toxicity of ceria nanomaterials enable them to be pharmacological antioxidant agents [71]. By means of the redox potential of cerium(III) and cerium(IV), ceria exhibits a flexible circulation between these two states. Cerium(IV) is metastable in water and easily converts to cerium(III) under surrounding conditions [72]. In recent years, it is demonstrated that ceria nanocrystals can react with hydrogen peroxide and serve as antioxidant agents for eliminating excess ROS in living systems [72]. In addition, ceria nanoparticles could protect neurons from the Aβ-induced toxicity [73]. Barbara et al. found that neuroblastoma cells incubated with ceria nanoparticles were successfully protected from Aβ damage and maintained their original cell viability, thus confirming that ceria nanoparticles can act as a regulator of signaling pathways for neuronal survival [74]. Cimini et al. modified polyethylene glycol on the surface of

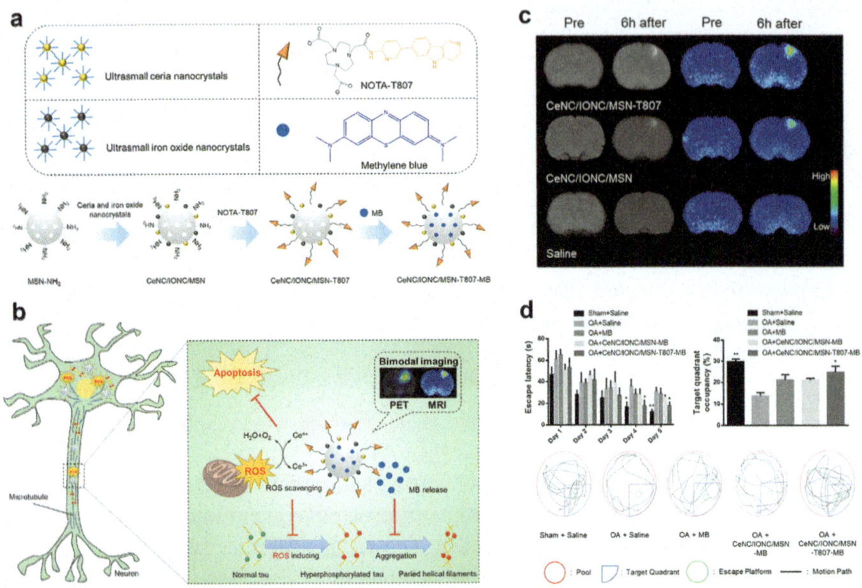


**Fig. 4.2** Nanomaterials for the treatment of AD. (**a**) Schematic diagram of the synthetic process of multifunctional nanocomposite (CeNC/IONC/MSN-T807-MB). (**b**) Schematic illustration of the tau-targeted synergistic treatment of CeNC/IONC/MSN-T807-MB. (**c**) Specific tau targeting ability of CeNC/IONC/MSN-T807-MB confirmed by $T_1$-weighted MR imaging. (**d**) Water maze test of CeNC/IONC/MSN-T807-MB therapy. (From Ref. [77])

ceria nanoparticles and linked anti-Aβ antibodies to specifically target Aβ plaques [75]. The nanocomposite can change the signaling pathway of nerve influence factors in the brain and increase the survival rate of neurons. Kwon et al. designed a triphenylphosphine-modified ceria nanoparticle [76]. It can effectively localize to mitochondria and scavenge ROS, which could relieve mitochondrial damage and neuronal death. Moreover, Chen et al. developed a nanocomposite (CeNC/IONC/MSN-T807) loaded with methylene blue for tau-targeted combinational therapy of AD [77]. This nanocomposite can effectively alleviate mitochondrial oxidative stress and cooperate with methylene blue to inhibit tau hyperphosphorylation (Fig. 4.2). In addition, the nanomaterial was also effective in preventing neuronal death and cell apoptosis. All these results suggest ceria nanoparticles have invaluable potential in the treatment of AD.

### *4.3.4 Nanomaterial-Based Gene Therapy*

Nowadays, gene therapy has been widely applied in genetic disease by altering the behavior of genes. Gene therapy is also a promising strategy for the treatment of AD. Gene delivery is a very important part in gene therapy. A qualified gene carrier should not only deliver the gene to target cells but also protect it from degradation. The commonly used viral vector may lead to safety problems like the immune response, and insufficient specificity hinders its application. Nanomaterials, as a new kind of non-viral vector, have attached much attention due to its high loading capacity and stability for gene transport. Amino-functionalized organically modified silica (ORMOSIL) nanoparticles have been reported, enabling the efficient delivery of gene to CNS. ORMOSIL nanoparticles with positively charged amino groups can bind with the negatively charged gene and protect its activity. The transfection efficiency of these nanoparticles was comparable to a viral vector. The experiment data showed that ORMOSIL nanoparticles were safe and efficient gene carriers for the therapy of brain disease like AD. Besides, they also showed the potential to regulate neural stem cells [78].

### 4.3.5 Nanomaterial-Based Immunotherapy

Recently, most immunotherapies for AD are based on targeting $A\beta_{1-42}$ peptide, by which Aβ antibodies bind and remove amyloid deposits in brain vascular, reducing the toxicity of Aβ. However, challenges come along as this strategy needs the antibodies to cross the BBB to recognize Aβ. To this end, rationally designed nanomaterials with the functionalized surface were employed to overcome biological obstacles to a certain extent and improve the cellular uptake. Kristen et al. reported a kind of immune-nanovehicle for targeted AD therapy. Polylactic-co-glycolic was selected as the core followed by coating with chitosan. These nanovehicle exhibited a cubed shape, and the zeta potential after coating with chitosan and conjugating with anti-amyloid antibody IgG4.1 was $8.0 \pm 1.45$ mV and $16.80 \pm 6.09$ mV, respectively, which had access to electrostatically interact with the negatively charged cell surface and thus promote cellular uptake. Moreover, chitosan could act as a cryoprotectant like mannitol, trehalose, and sucrose to protect the nanocore and antibody conjugated on the surface of nanovehicles as well as increase the water dispersibility. Meanwhile, chitosan coating facilitates the nanovehicles to form cube or rodlike shape, which enhances the cellular penetration. In vitro BBB model constructed by MDCK cell monolayer verified that chitosan-absorbed and IgG4.1-conjugated immune-nanovehicles showed obvious transcytosis across cell monolayer compared to all control groups. After treating the cell with $A\beta_{40}$ proteins, improved cellular uptake of immune-nanovehicles was observed, thereby demonstrating their ability to target cerebrovascular amyloid for AD diagnosis and treatment [79]. Another study designed a smart nanovehicle (SNV) to treat AD and cerebral amyloid angiopathy. The core of SNVs consisted of the biocompatible and biodegradable chitosan, which improves cellular absorption due to its positive charge. Polyamine-modified IgG4.1 fragment was conjugated on the surface of chitosan core to enable the travel of BBB and target amyloid deposits and amyloid plaques, which had higher binding affinity than the whole antibody or unmodified fragment. Therefore, SNVs had a better permeability both in vitro and in vivo than the control nanoparticles and modified antibody fragment alone and accumulated much more in the brain, indicating the great potential for AD treatment [80].

## 4.4 Nanomaterials for Parkinson's Disease Treatment

PD is a progressive degenerative disorder in the CNS, which affects 2–3% of people over 65 years of old [81]. It is characterized by both motor manifestations, including rigidity, bradykinesia, postural instability, and rest tremor, and non-motor symptoms, like depression, cognitive impairment, autonomic features, hallucinations, and sleep problems [82]. Based on current knowledge, the direct cause of PD is the deficiency of dopaminergic cells in the substantia nigra along with the accumulation of intraneuronal α-synuclein aggregates in Lewy neurites and Lewy bodies (namely, Lewy pathology) [82, 83]. However, what triggers the initiation of this pathology and how it spreads across the brain remain unclear. It is originally thought that PD was caused by the exposure to environmental factors. In recent years, a more widely accepted opinion is that PD is caused by the intricate interactions between environmental and genetic factors that impact many fundamental cellular processes [84]. Currently, PD treatment is clinically challenging because of the lack of effective approaches for definitive diagnosis of the disease at the earliest stage and for symptom management at later stages. In fact, there is no disease-modifying but merely symptomatic treatments for PD, which is attributed to the yet elusive etiology of this disease. Available pharmacological therapies for PD can be categorized into five classes, including dopamine agonists, levodopa, catechol-O-methyltransferase inhibitors, monoamine oxidase type B inhibitors, and anticholinergics [85]. These drugs, though to different extents, are effective in controlling the symptoms at initial usage, which largely improve the life quality of early-stage PD patients. However, because of the low selectivity of those drugs and the pathological complexity of PD itself, considerable adverse effects and unsatisfying responses are often seen [82, 84]. In this regard, targeting specific pathways in specific regions and, if necessary, targeting several dysfunctional pathways simultaneously may represent a plausible strategy for better PD treatment. Besides, gene therapies that target genetic factors of PD are on the horizon and hold an optimistic future for PD treatments. To conclude, the final goal of PD research is to develop drugs that slow or reverse the neurodegeneration, but before we excavate deeper into the underlying mechanisms of PD onset, perhaps the best option is to optimize current therapeutics through rational designing. As described below, nanomaterials that can achieve controlled release, targeted therapy, and diverse cargo delivery may hold the key for the treatment of PD.

### 4.4.1 Nanomaterials for Dopamine Replacement

In PD treatment, one major challenge when using dopamine (DA) as the therapeutic agent is the drug penetration through the BBB into the desired tissue. As we know, the BBB is a rate-determining factor for the transport of drugs or agents to the CNS, which results in many treatment failures. Therefore, enhancing the BBB-crossing efficiency of DA is regarded as a promising strategy. The hydrophilic feature and hydrogen bonding potential of DA also limit the delivery of DA into the brain. Levodopa (LD) and bromocriptine 74 (BRC) have been widely used for the treatment of PD. However, the amount of BRC that can cross BBB is limited and largely determined by the efflux system of BBB [86].

One elegant approach to deliver DA and its derivates through BBB is using colloidal carriers, especially nanoparticles (Fig. 4.3a) [87, 88]. Nanoparticles made up of natural lipids, polymers, and some biopolymer materials hold optimistic potential for brain delivery and targeting, since these nanoparticles can be taken up by the brain rapidly with good biodegradability and bioavailability but fewer side effects. There are many nanomaterials that have been reported for dopamine replacement, which have been demonstrated to be useful in PD treatment.

Chitosan (CS) is a cationic polysaccharide applied for DA delivery. Giglio et al. prepared the DA-loaded CS nanoparticles (CSNPsDA) with QCM-D technique; the swelling behavior of nanoparticles and the interaction between DA and nanoparticles were investigated. CSNPsDA was confirmed to be biological available for brain application [89]. Trapani et al. also used chitosan as the vehicle to deliver DA to the brain. The DA-loaded CS nanoparticles exhibited enhanced cellular uptake and reduced ROS production. The nanoparticles can induce the time-dependent increase of DA delivery to the brain and showed low neurotoxicity (Fig. 4.3b, c) [90]. Furthermore, Yang et al. developed levodopa methyl ester (LDME)/benserazide-loaded nanoparticles with poly(lactic-co-glycolic acid) and polylactic acid. Compared with LDME plus benserazide, the nanoparticle significantly reduced the abnormal involuntary movements of PD model mice. The levels of cyclic adenosine monophosphate-regulated phosphoprotein and phosphorylated dopamine, which reflects the PD conditions, were also reduced by using the nanoparticle [91]. In summary, the nanoparticle manifests as a promising platform for engineering DA replacement therapy that performs better than conventional medicine.

### 4.4.2 Nanomaterials for Dopaminergic Agonist Delivery

Though the levodopa has been widely used for PD as the most effective symptomatic treatment for 30 years, long-term usage always results in dyskinesias. Therefore, it is important to minimize the dosage of levodopa and to delay the initiation of dyskinesias. The dopamine agonists have drawn attention for continuous dopaminergic

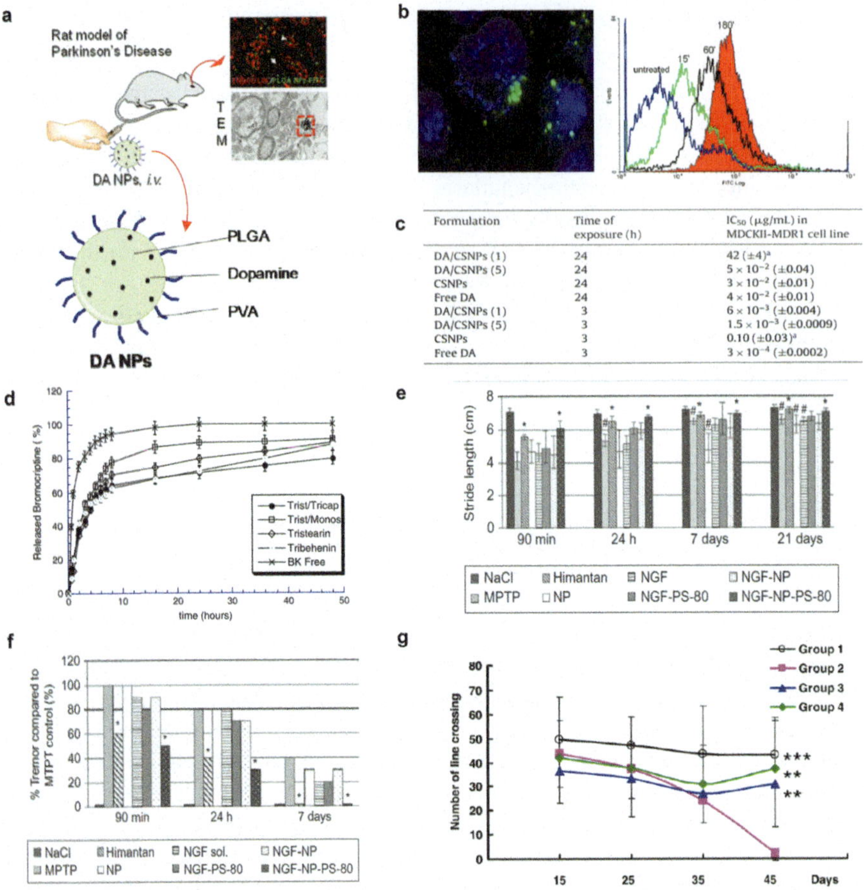


| Formulation | Time of exposure (h) | $IC_{50}$ (μg/mL) in MDCKII-MDR1 cell line |
|---|---|---|
| DA/CSNPs (1) | 24 | 42 (±4)[a] |
| DA/CSNPs (5) | 24 | $5 \times 10^{-2}$ (±0.04) |
| CSNPs | 24 | $3 \times 10^{-2}$ (±0.01) |
| Free DA | 24 | $4 \times 10^{-2}$ (±0.01) |
| DA/CSNPs (1) | 3 | $6 \times 10^{-3}$ (±0.004) |
| DA/CSNPs (5) | 3 | $1.5 \times 10^{-3}$ (±0.0009) |
| CSNPs | 3 | 0.10 (±0.03)[a] |
| Free DA | 3 | $3 \times 10^{-4}$ (±0.0002) |

**Fig. 4.3** Nanomaterials applied in PD treatments. (**a**) Illustrations of trans-BBB delivery of DA-loaded nanoparticles (from Ref. [88]). (**b**) Representative immunofluorescence imaging of cellular uptake of unloaded FITC-CSNP (left) and cell entry dynamics of FITC-DA/CSNPs into MDCKII-MDR1 cells determined by flow cytometry (right) (from Ref. [90]). (**c**) IC50 of free DA, unloaded CSNPs, and DA-loaded NPs in MDCKII-MDR1 cells treated for indicated times. DA/CSNPs showed very low neurotoxicity with a p value $<0.001$ (from Ref. [90]). (**d**) The release curves of bromocriptine form the indicated SLNs (from Ref. [95]). (**e–f**) Effects of indicated drug formulations on horizontal activity (**e**) and vertical activity (**f**) in male C57Bl/6 mice tested for 2 min in the open field, 90 min, 24 h, and 7 and 21 days after administration of MPTP (from Ref. [101]). (**g**) Behavioral changes of rats in line crossing in indicated groups exposed to rotenone for different times. Treatment in different groups are as follows: group 1, oil; group 2, GFP-loaded lactoferrin (Lf)-modified NPs; group 3, hGDNF-loaded Lf-modified NPs; and group 4, hGDNF-loaded Lf-modified NPs (multiple injections) (from Ref. [108])

stimulation. It can stimulate striatal dopamine receptors to induce the production of dopamine for further PD treatment and shows less potential to induce dyskinesia risk [92].

However, most of the dopamine agonists, such as ropinirole and rotigotine, have low bioavailability and short half-life, which are inconvenient for PD patients [93]. Nanomedicine can resolve these problems effectively, as many kinds of nanomaterials have been shown able to significantly prolong the half-life.

Solid lipid nanoparticles (SLNs) are made up of physiological compounds with superior biosafety. More importantly, due to the lipophilic nature, SLNs are considered as useful vehicles for lipophilic drugs [94]. Esposit et al. loaded the dopamine agonist, bromocriptine, into SLN nanoparticles based on tristearin/tricaprin mixture. The SLNs showed up to 6-month long-term stability. Compared with bromocriptine alone, the bromocriptine-loaded SLNs prolonged the drug release duration from 5 h to 48 h and realized better-controlled drug release (Fig. 4.3d) [95]. Tsai et al. utilized SLNs to encapsulate apomorphine and applied them to oral administration. The SLNs showed 12- to 13-fold higher bioavailability, offering an applaudable method for apomorphine delivery by oral routine [96]. Nanomaterials are also investigated for the transdermal delivery. Azeem et al. prepared nanoemulsion gel and observed a 7.5-fold increase in skin permeation rate compared with traditional hydrogel to enhance permeation of ropinirole [97].

### *4.4.3 Nanomaterials for Growth Factor and Peptides Delivery*

As mentioned above, current treatments for PD are mainly dependent on dopamine and dopamine-like drugs, anticholinergic drugs, or their combinations. However, none of them can prevent the loss of neurons. Moreover, these treatments can potentially cause long-term motor complications such as idiosyncratic reactions, on/off effects, and overmedication side effects [98]. Thus, new strategies that can ensure the human body to tolerate a longer treatment cycle or effectively reverse the degeneration are demanded.

#### 4.4.3.1 Growth Factor

It is well known that PD is related to a decreased level of nerve growth factor (NGF) in the substantia nigra and the blood [99]. However, the survival of both and central cholinergic neurons and peripheral ganglion cells in the basal forebrain affects by NGF. Therefore, administration of NGF may reverse the PD progression [100].

The nanoparticle is a promising strategy to deliver NGF across the BBB into the brain. Kurakhmaeva et al. used poly(butylcyanoacrylate) (PBCA) nanoparticles coated with polysorbate 80 as vehicles to transport NGF to the brain. The NGF-loaded PBCA nanoparticles can significantly reduce primary PD symptoms such as rigidity, oligokinesia, and tremor (Fig. 4.3e, f) [101]. In another example, Xie et al. utilized magnetic nanotubes as vehicles for NGF delivery. The magnetic nanotubes with NGF incorporation enabled the differentiation of PC12 cells into neurons exhibiting growth cones and neurite outgrowth [102].

#### 4.4.3.2 Peptides

Peptides have been investigated for the treatment of PD for several years. However, the delivery of these therapeutic peptides is facing many obstacles due to the high enzymatic metabolism, instability, rapid renal elimination, low gastrointestinal absorption, and potential immunogenicity. Recent nanomaterials show the potential possibility to resolve these problems effectively. The nanomaterial-based strategies to delivery of therapeutic peptides are developed.

Urocortin (UCN), a corticotrophin-releasing hormone-related peptide, has been used as a cytoprotectant for GABAergic neurons, cerebellar granule cells, and cultured hippocampal neurons [103, 104]. Hu et al. loaded UCN into lactoferrin (Lf)-conjugated polyethylene glycol polylactide-polyglycolide (PEG-PLGA) nanoparticles (Lf-NPs). The Lf-NPs showed a near threefold higher bioavailability, offering a promising strategy for urocortin delivery for the application in intravenous administration. The administration of urocortin-loaded Lf-NPs effectively relieved the striatum lesion induced by 6-hydroxydopamine [105]. Nanomaterials are also investigated for the intranasal administration. Wen et al. loaded UCN into odorranalectin (OL)-connected PEG-PLGA nanoparticles. The modification of OL improved the delivery of nanoparticles in the brain, thus enhancing the therapeutic effect on PD treatment [106].

### *4.4.4 Nanomaterial-Based Gene Therapy*

Gene therapy has greatly attracted researchers' attention and has been implicated as a disease-modifying strategy directly targeting the etiology of neurodegenerative diseases [107]. Gene therapy transports therapeutic genes to the lesion area, and the expression of genic products maintains the health of neurons or repair-degenerated ones [108]. This strategy might ideally slow or reverse the degenerative process of PD fundamentally.

Currently, most of the therapeutic genes were delivered to the diseased region via the injection of viral vectors [109]. However, this administration approach is not only highly invasive and unsafe but also inconvenient for repetitive or continuous treatments [15].

Nanomaterials, potential non-viral vectors, are efficient gene therapy carriers to cross the BBB and can be used for long-lasting treatments. Huang et al. used Lf-conjected PAMAM/PEG nanoparticles (Lf-NPs) as vehicles to carry the hGDNF gene to the brain via intravenous injection. Multiple injections of Lf-NPs could significantly reduce dopaminergic neuronal loss, enhance monoamine neurotransmitter levels, and improve locomotor activity (Fig. 4.3g) [108, 110]. Huang et al. also used angiopep-conjugated nanoparticles as carriers to transport the hDNF gene to the brain. The improved apparent dopaminergic neuron recovery and locomotor activity could be observed after intravenous injections of nanoparticles for five times [111]. What's more, Zhang et al. loaded glial-derived neurotrophic factor (GDNF) plasmid DNA into Trojan horse liposomes. Nearly complete abrogation of the neurotoxic effect was achieved by multiple intravenous dosing of GDNF plasmid DNA [112].

Overall, nanomaterials provide a powerful toolkit allowing us to maximize the utility of currently available PD therapies. With future advances in better materials, smarter engineering, and, most importantly, the etiology of PD, we may finally beat PD.

## 4.5 Nanomaterials for Stroke Treatment

Stroke, defined as a neurological deficit due to the poor blood flow to the brain results in cell death, is the third leading cause of potential life loss worldwide [113], with a severe morbidity in patients suffering from high blood pressure and diabetes mellitus [114]. It is an acute cerebrovascular disease caused by either the sudden vascular rupture in the brain or the vascular occlusion resulting in poor blood flow to the brain. The insufficient blood supply results in a battery of pathophysiological reactions including the overproduction of free radicals, inflammation, the overexpression of related genes, the degradation of enzymes, and the eventual cell death [115]. Accordingly, it can be divided to two main types of stroke, i.e., ischemic and hemorrhagic stroke, in which ischemia dominated the cause of stroke. [116, 117].

The neuroprotection and thrombolysis (early recanalization) are two major strategies developed for treating acute ischemic stroke. However, brain herniation or intracranial hemorrhage may happen after thrombolysis, and ischemia-reperfusion can be aggravated by recanalization [118, 119]. Many clinical trials of neuroprotectors are carrying on evaluating their therapeutic effects but achieve negligible progresses. [120]. Some reasons responsible for the failure of these drugs are listed as follows. First and foremost, the obstruction of BBB and the complexity of human brain structure are formidable. Second, the effect of drugs would be decreased by the existence of concomitant diseases. Third, the type of drug and dosage depends on the location and the degree of cerebral ischemia-reperfusion process, and the misuse of the drug may cause potential side effects. For example, although tissue plasminogen activator (tPA) has been approved by FDA, few people use it due to the possibility of cerebral hemorrhage [121]. Thus, there is an urgent demand to find treatment strategies to enhance the delivery efficiency and safety.

Nanotechnology has shown great promise in biomedical applications, particularly for enhancing the effectiveness of diagnosis and therapy [122]. In addition to the universal benefits, nanomaterials can achieve dynamic changes in vivo, play a role in a fixed time and place, and thus are extremely suitable for the treatment of such kind of disease that needs to monitor the disease process and respond accordingly [123, 124]. In conclusion, the multifunctional and highly specific nanoparticles provide hope to make significant progress in the stroke treatment.

### 4.5.1 *Nanomaterials for Thrombolysis*

Thrombolysis is the decomposition of blood clots by pharmacological methods, commonly known as broken pieces. Clearing the cross-linked fibrin network promotes the dissolution of clots and subsequent proteolysis by enzymes to restore blood flow in the blocked blood vessels. Nanomaterials have been investigated in the treatment of stroke thrombolytic therapy for several years. Nanoparticles have been successfully applied in the treatment of acute ischemic stroke by transportation of tissue plasminogen activator to thrombi. They exhibit preferential position in thrombosis, effective thrombolysis, and increased safety due to a significant reduction in the dose of fibrinolytic agents and reduced downstream adverse effects [125–127].

#### 4.5.1.1 Tissue Plasminogen Activator (tPA) Delivery

Although there are many thrombolytic agents available, tissue plasminogen activator (tPA) is the only FDA-approved drug for the treatment of fibrinolytic clots in ischemic stroke. tPA is a serine protease which transforms proenzyme plasminogen to plasminogen, initiating the cleavage process of fibrin clots. Since the half-life of tPA in plasma is very short ($\approx$5 min), long-term high dose would be used in the thrombolytic treatment to maintain a sufficient drug level, leading to clotting factor degradation and bleeding. Therefore, it will be highly desirable if the nanoparticles were able to target the thrombus and release thrombolytic agents. Bi et al. [128] developed urokinase (UK)-conjugated magnetic nanoparticles (UK/MNP) for targeted thrombolysis. The treatment showed a reduction in thrombus size compared to control group. The thrombolytic efficacy of magnetic field-mediated UK/MNP was 5.0 times than that of the UK alone and 2.6 times than that of the UK/MNP alone. Moreover, by detecting the residual fibrinogen and counting the number of bleedings, it showed that the magnetic targeting-mediated nanoparticles significantly reduced the systemic activation of plasminogen, thus significantly reducing the number of bleedings [129]. Another effect of nanoparticles is to reduce bleeding, which is the most destructive and severe side effect of thrombolytic drugs. Ma et al. studied the thrombolytic effect of magnetic nanoparticles binding with rtPA and coated by polyacrylic acid [130]. They used magnetic nanoparticles to increase the retention in target site and reduce cerebral hemorrhage. The nanoparticles have the similar fibrinolytic activity and amylolytic effect compared with free rtPA in vitro. The nanoparticles exhibited a fivefold enhancement in thrombolytic activity compared to that without magnetic targeting in the hind limb ischemia model rats. Uesugi Y et al. developed a novel tPA agent delivery nanosystem, which had a suppressed thrombolytic activity of tPA at normal conditions and recovered activity when exposed to ultrasound. The results showed that the responsive nanosystem is a promising tPA delivery system that increases therapeutic activity by localized ultrasound exposure [131].

#### 4.5.1.2 Heparin Delivery

Heparin has been demonstrated to have important and significant effects on the treatment of thrombosis in decades of clinical practice. Heparin prevents new thrombosis and promotes the activation of the patient's own fibrinolytic system to eliminate thrombosis. Tang et al. reported the heparin-modified chitosan (CS)/poly (glutamic acid) (g-PGA) nanoparticles (HP-CS/g-PGA) for the delivery of basic fibroblast growth factor (bFGF) and heparin. The nanoparticles may be a potential therapeutic agent to enhance the regeneration of ischemic tissue and prevent vascular thrombosis [132] (Fig. 4.4). Bai et al. demonstrated that PEG-modified dendritic micelles could extend the half-life of low-molecular-weight heparin (LMWH) and enhance the pulmonary accumulation. It is concluded that PEG-modified PAMAM dendritic micelles are promising to delivery of LMWH into the lung for long-term therapy of deep vein thrombosis [133].

### *4.5.2 Nanomaterials for Neuroprotection*

#### 4.5.2.1 Antioxidant and Anti-inflammatory Agents Delivery

Neuroprotection can lead to the rescue, recovery, or regeneration of the nervous system. Reducing oxidative stress is a promising neuroprotective strategy, because oxidative stress is known as a vital neuropathological process that causes central nervous system ischemia. Therefore, recent researches have centered on the use of nanomaterials with antioxidant ability to scavenge ROS in CNS [134]. For instance, nanomaterials consisting of cerium or yttrium oxide can serve as antioxidants to inhibit the production of ROS. The remarkable neuroprotective activity of these nanomaterials was proven in the HT22 hippocampal neuronal cells [135]. What's more, the fullerenes exhibit high reactivity against ROS and show antioxidative activity and ROS scavenging effect. Particularly, polyhydroxylated $C_{60}$ fullerenol inhibited glutamate receptors, resulting in intracellular calcium elevation and reduced neuronal toxicity [136]. In addition, water-soluble carboxyl fullerenes, $C_{60}$ derivatives of malonate, suppressed the excitotoxicity and apoptosis of cortical neurons, showing the neuroprotective effects in vivo [137]. In another research, the core of fullerene $C_{60}$ was modified by a malonate dendrimer layer to increase the water solubility and the surface area of reaction on the nanoparticles. The modified nanoparticles showed much more antioxidant activity than fullerenes [138].

#### 4.5.2.2 Antioxidant Enzymes Delivery

Antioxidant enzymes are demonstrated to be promising treatments for pathologic conditions accompanied with increased ROS production. However, their efficiency in combating ROS relies on the ability to obtain sufficient levels of active enzymes

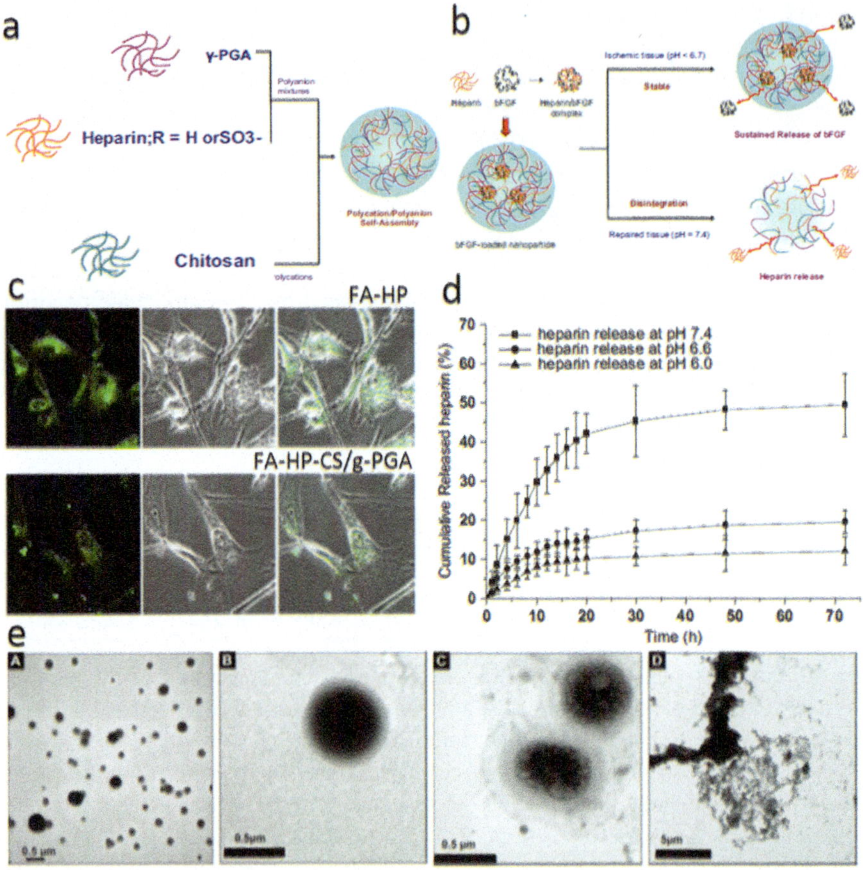


**Fig. 4.4** Nanomaterials for stroke treatment. (**a**) Illustrations of self-assembled HP-CS/g-PGA nanoparticles. (**b**) Illustrations of the mechanism of pH-mediated release of bFGF and heparin from HP-CS/g-PGA nanoparticles. (**c**) Fluorescence microscopy images of FA-HP or FA-HP-CS/g-PGA nanoparticles in HFF cells. (**d**) The cumulative released curve of heparin from HP-CS/g-PGA nanoparticles at different pH values. (**e**) TEM images of HP-CS/g-PGA nanoparticles after storage at different pH values: (A) 2 h at pH 6.0, (B) 10 min at pH 7.4, (C) 30 min at pH 7.4, and (D) 2 h at pH 7.4. (From Ref. [132])

for treatment at injury site. Therefore, the administration of antioxidant enzyme treatment demands a carrier that can not only deliver the drug to the target but also effectively protect the activities of enzymes. Therefore, Chorny et al. developed magnetically sensitive nanoparticles (MNPs) as a novel platform for targeted delivery of antioxidant enzymes. The nanoparticles protect the endothelial cells from ROS-mediated cell death [139]. Klyachko et al. utilize superoxide dismutase 1 and catalase conjugating with cationic block copolymer, PEI-PEG, or poly(l-lysine)-PEG to form antioxidant enzymes nanoplatform. The potential of these nanoplatforms to deliver antioxidant enzymes to CNS to reduce ROS level was demonstrated [140].

#### 4.5.2.3 Nanomaterial-Based Stem Cell Therapy

Recently, stem cells, such as mesenchymal stem cells (MSCs), neural stem cells (NSCs), embryonic stem cells (ESCs), and inducible pluripotent stem cells (iPSCs), exhibit great potential for stroke treatment, due to their extraordinary abilities of self-renewal, proliferation, as well as multipotent differentiation [141–144]. Among them, MSCs are widely investigated in stroke management. On the one hand, there are no serious ethical issues with MSCs. On the other hand, MSCs possess a wide range of sources, which could be obtained from bone marrow, adipose tissues, umbilical cord, and so on [145].

The use of nanomaterials in stem cell therapy offers promising future perspectives for stroke management. The noninvasive tracking of stem cells in vivo is crucial for monitoring the homing ability and therapeutic effect of stem cells after transplantation. Several nanoparticles, such as quantum dots and iron oxide nanoparticles (IONPs), have been reported to label stem cells and track their fate in vivo, but only IONPs detected by magnetic resonance imaging (MRI) are suitable in human medicine [146–148]. Kim et al. [149] transplanted human mesenchymal stem cells (hMSCs) labeled with IONPs into the middle cerebral artery occlusion model. Then, MRI techniques were employed to monitor cells in vivo and found that hMSCs possessed the ability to transplant to the infarcted area extensively in both ipsilateral and contralateral injections. Except for tracking the fate of stem cells in vivo, IONPs also have the potential for improving cell homing efficiency. Huang et al. [150] found that iron-based magnetic nanoparticles actively augmented homing-related chemokine receptor CXCR4 expression of MSCs and consequently improved the homing ability of MSCs to the injured brain.

### *4.5.3 Nanomaterial-Based Gene Therapy*

Gene therapy is a potential therapeutic strategy for stroke in the future. It has been demonstrated that transplantation of stem cells modified with overexpressing neurotrophic factors for stroke treatment, such as VEGF, BDNF, GDNF, PIGF, ANG-1, HGF, NGF, EPO, and noggin, significantly improved the therapeutic effect compared to stem cells only [144]. In addition, there are many kinds of nanomaterials that served as non-viral vectors for delivering genetic materials into stem cells, such as liposomes, dendrimers, and IONPs [151, 152].

## 4.6 Conclusion

Nanomaterial-based strategies have been demonstrated to be the revolutionary therapeutic methods for the treatment of various brain diseases. Nanomaterials make it possible for the delivery of many therapeutic agents across the BBB in brain tumors and neurodegenerative diseases, which is one of the most formidable barriers faced by existing treatments. Moreover, the versatile functionalities of nanomaterials could be achieved by manipulating the physicochemical properties and surface engineering, leading to the personalized treatment of brain diseases. Although promising, the applications of nanomaterials in clinical neuroscience are only in the infancy stage, and there are still many significant issues need us to settle down. Indeed, engineered nanomaterials have shown enhanced penetration into the brain, but how to specifically target lesion locations inside brain? What is the biological fate of nanomaterials remaining in the brain? How about their long-term toxicity problems in CNS? More basic research is urgently needed to fully answer these questions. In addition, the translation from fundamental research to concrete clinical applications also needs much more efforts. Cooperation between pharmaceutical companies, researchers, technologists, and doctors is highly encouraged to facilitate the development of nanomaterial-based strategies for CNS disorders. It is expected that innovative nanomaterials will bring hope for better management of brain diseases and make therapies far more effective.

**Acknowledgments** This work was supported by the National Key Research and Development Program of China (2016YFA0203600), the National Natural Science Foundation of China (31822019, 51703195, 91859116), One Belt and One Road International Cooperation Project from Key Research and Development Program of Zhejiang Province (2019C04024), the Zhejiang Provincial Natural Science Foundation of China (LGF19C100002), the Fundamental Research Funds for the Central Universities (2018QNA7020), and "Thousand Talents Program" for Distinguished Young Scholars.

## References

1. Gilmore JL, Yi X, Quan L, Kabanov AV. Novel nanomaterials for clinical neuroscience. J Neuroimmune Pharmacol. 2008;3(2):83–94.
2. Re F, Gregori M, Masserini M. Nanotechnology for neurodegenerative disorders. Nanomedicine. 2012;8:S51–8.
3. Silva GA. Nanotechnology approaches for the regeneration and neuroprotection of the central nervous system. Surg Neurol. 2005;63(4):301–6.
4. Fernandes C, Costa A, Osório L, Lago RC, Linhares P, Carvalho B, Caeiro C. Current standards of care in Glioblastoma therapy. In: Glioblastoma. Brisbane: Codon Publications; 2017, Chapter 11.
5. Serwer LP, James CD. Challenges in drug delivery to tumors of the central nervous system: an overview of pharmacological and surgical considerations. Adv Drug Deliv Rev. 2012;64 (7):590–7.

6. Liu Y, Lu W. Recent advances in brain tumor-targeted nano-drug delivery systems. Expert Opin Drug Deliv. 2012;9(6):671–86.
7. Cardoso FL, Brites D, Brito MA. Looking at the blood-brain barrier: molecular anatomy and possible investigation approaches. Brain Res Rev. 2010;64(2):328–63.
8. Novakova I, Subileau EA, Toegel S, Gruber D, Lachmann B, Urban E, Chesne C, Noe CR, Neuhaus W. Transport rankings of non-steroidal Antiinflammatory drugs across blood-brain barrier in vitro models. PLoS One. 2014;9(1):e86806.
9. Pardridge WM. The blood-brain barrier: bottleneck in brain drug development. NeuroRx. 2005;2(1):3–14.
10. Agarwal S, Sane R, Oberoi R, Ohlfest JR, Elmquist W. Delivery of molecularly targeted therapy to malignant Glioma, a disease of the whole brain. Expert Rev Mol Med. 2011;13(13): e17.
11. Auffinger B, Thaci B, Nigam P, Rincon E, Yu C, Lesniak MS. New therapeutic approaches for malignant glioma: in search of the Rosetta stone. F1000 Med Rep. 2012;4(18):1–6.
12. Gao JQ, Lv Q, Li LM, Tang XJ, Li FZ, Hu YL, Han M. Glioma targeting and blood-brain barrier penetration by dual-targeting doxorubicin liposomes. Biomaterials. 2013;34 (22):5628–39.
13. Ying X, Wen H, Lu WL, Du J, Guo J, Tian W, Men Y, Zhang Y, Li RJ, Yang TY. Dual-targeting daunorubicin liposomes improve the therapeutic efficacy of brain glioma in animals. J Control Release. 2010;141(2):183–92.
14. Benedetti S, Pirola B, Pollo B, Magrassi L, Bruzzone MG, Rigamonti D, Galli R, Selleri S, Meco FD, Fraja CD. Gene therapy of experimental brain tumors using neural progenitor cells. Nat Med. 2000;6(4):447–50.
15. Schlachetzki F, Zhang Y, Boado RJ, Pardridge WM. Gene therapy of the brain: the trans-vascular approach. Neurology. 2004;62(8):1275–81.
16. Verma IM, Naldini L, Kafri T, Miyoshi H, Takahashi M, Blömer U, Somia N, Wang L, Gage FH. Gene therapy: promises, problems and prospects. Nature. 1997;389(6648):239–42.
17. Kaliberov SA, Markert JM, Gillespie GY, Krendelchtchikova V, Manna DD, Sellers JC, Kaliberova LN, Black ME, Buchsbaum DJ. Mutation of Escherichia coli cytosine deaminase significantly enhances molecular chemotherapy of human glioma. Gene Ther. 2007;14 (14):1111–9.
18. De Smedt SC, Demeester J, Hennink WE. Cationic polymer based gene delivery systems. Pharm Res. 2000;17(2):113–26.
19. Li J, Gu B, Meng Q, Yan Z, Gao H, Chen X, Yang X, Lu W. The use of myristic acid as a ligand of polyethylenimine/DNA nanoparticles for targeted gene therapy of glioblastoma. Nanotechnology. 2011;22(43):435101.
20. Jung J, Solanki A, Memoli K, Kamei K, Kim H, Drahl M, Williams L, Tseng H, Lee K. Selective inhibition of human brain tumor cells through multifunctional quantum-dot-based siRNA delivery. Angew Chem. 2010;49(1):103–7.
21. Sellins KS, Cohen JJ. Hyperthermia induces apoptosis in thymocytes. Radiat Res. 1991;126 (1):88–95.
22. Coss R, Linnemans W. The effects of hyperthermia on the cytoskeleton: a review. Int J Hyperth. 1996;12(2):173–96.
23. Wong RS, Kapp LN, Krishnaswamy G, Dewey WC. Critical steps for induction of chromosomal aberrations in CHO cells heated in S phase. Radiat Res. 1993;133(1):52–9.
24. Cavaliere R, Ciocatto EC, Giovanella BC, Heidelberger C, Johnson RO, Margottini M, Mondovi B, Moricca G, Rossi-Fanelli A. Selective heat sensitivity of cancer cells. Biochemical and clinical studies. Cancer. 1967;20(9):1351–81.
25. Sneed PK, Stauffer PR, Diederich CJ, McDermott MW, Lamborn KR, Weaver KA, Prados MD, Chang S, Malec MK, Spry L. Survival benefit of hyperthermia in a prospective randomized trial of brachytherapy boost ± hyperthermia for glioblastoma multiforme. Int J Radiat Oncol Biol Phys. 1996;36(S1):159.

26. Jordan A, Scholz R, Maier-Hauff K, van Landeghem FK, Waldoefner N, Teichgraeber U, Pinkernelle J, Bruhn H, Neumann F, Thiesen B. The effect of thermotherapy using magnetic nanoparticles on rat malignant glioma. J Neuro-Oncol. 2006;78(1):7–14.
27. Johannsen M, Gneveckow U, Thiesen B, Taymoorian K, Cho CH, Waldöfner N, Scholz R, Jordan A, Loening SA, Wust P. Thermotherapy of prostate cancer using magnetic nanoparticles: feasibility, imaging, and three-dimensional temperature distribution. Eur Urol. 2007;52(6):1653–62.
28. Maier-Hauff K, Ulrich F, Nestler D, Niehoff H, Wust P, Thiesen B, Orawa H, Budach V, Jordan A. Efficacy and safety of intratumoral thermotherapy using magnetic iron-oxide nanoparticles combined with external beam radiotherapy on patients with recurrent glioblastoma multiforme. J Neuro-Oncol. 2011;103(2):317–24.
29. Wang Y, Huang R, Liang G, Zhang Z, Zhang P, Yu S, Kong J. MRI-visualized, dual-targeting, combined tumor therapy using magnetic graphene-based mesoporous silica. Small. 2014;10 (1):109–16.
30. Lucky SS, Soo KC, Zhang Y. Nanoparticles in photodynamic therapy. Chem Rev. 2015;115 (4):1990–2042.
31. Dougherty TJ. An update on photodynamic therapy applications. J Clin Laser Med Surg. 2002;20(1):3–7.
32. Stylli SS, Kaye AH, Macgregor L, Howes M, Rajendra P. Photodynamic therapy of high grade glioma – long term survival. J Clin Neurosci. 2005;12(4):389–98.
33. Koo YE, Reddy GR, Bhojani M, Schneider R, Philbert MA, Rehemtulla A, Ross BD, Kopelman R. Brain cancer diagnosis and therapy with nanoplatforms. Adv Drug Deliv Rev. 2006;58(14):1556–77.
34. Singh G, Wilson BC, Sharkey SM, Browman GP, Deschamps P. Resistance to photodynamic therapy in radiation induced fibrosarcoma-1 and Chinese hamster ovary-multi-drug resistant cells in vitro. Photochem Photobiol. 1991;54(2):307–12.
35. Kopelman R, Koo Y-EL, Philbert M, Moffat BA, Reddy GR, McConville P, Hall DE, Chenevert TL, Bhojani MS, Buck SM. Multifunctional nanoparticle platforms for in vivo MRI enhancement and photodynamic therapy of a rat brain cancer. J Magn Magn Mater. 2005;293(1):404–10.
36. Cheng Y, Meyers JD, Agnes RS, Doane TL, Kenney ME, Broome AM, Burda C, Basilion JP. Addressing brain tumors with targeted gold nanoparticles: a new gold standard for hydrophobic drug delivery? Small. 2011;7(16):2301–6.
37. Meyers JD, Cheng Y, Broome AM, Agnes RS, Schluchter MD, Margevicius S, Wang X, Kenney ME, Burda C, Basilion JP. Peptide-targeted gold nanoparticles for photodynamic therapy of brain cancer. Part Part Syst Charact. 2015;32(4):448–57.
38. Krishnamachari Y, Salem AK. Innovative strategies for co-delivering antigens and CpG oligonucleotides. Adv Drug Deliv Rev. 2009;61(3):205–17.
39. Blattman JN, Greenberg PD. Cancer immunotherapy: a treatment for the masses. Science. 2004;305(5681):200–5.
40. Melief CJ. Cancer: immune pact with the enemy. Nature. 2007;450(7171):803–4.
41. Steinman RM, Banchereau J. Taking dendritic cells into medicine. Nature. 2007;449 (7161):419–29.
42. Kateb B, Van Handel M, Zhang L, Bronikowski MJ, Manohara H, Badie B. Internalization of MWCNTs by microglia: possible application in immunotherapy of brain tumors. NeuroImage. 2007;37:S9–S17.
43. Krishnamachari Y, Geary SM, Lemke CD, Salem AK. Nanoparticle delivery Systems in Cancer Vaccines. Pharm Res. 2011;28(2):215–36.
44. Kuang J, Song W, Yin J, Zeng X, Han S, Zhao YP, Tao J, Liu CJ, He XH, Zhang XZ. iRGD modified chemo-immunotherapeutic nanoparticles for enhanced immunotherapy against glioblastoma. Adv Funct Mater. 2018;28(17):1800025.
45. Schneider T, Becker A, Ringe K, Reinhold A, Firsching R, Sabel BA. Brain tumor therapy by combined vaccination and antisense oligonucleotide delivery with nanoparticles. J Neuroimmunol. 2008;195(1–2):21–7.

46. Graeber MB, Kösel S, Egensperger R, Banati RB, Müller U, Bise K, Hoff P, Möller HJ, Fujisawa K, Mehraein P. Rediscovery of the case described by Alois Alzheimer in 1911: historical, histological and molecular genetic analysis. Neurogenetics. 1997;1(1):73–80.
47. Hardy JA, Higgins GA. Alzheimer's disease: the amyloid cascade hypothesis. Science. 1992;256(5054):184–5.
48. Graham WV, Bonitooliva A, Sakmar TP. Update on Alzheimer's disease therapy and prevention strategies. Annu Rev Med. 2017;68(1):413–30.
49. Goedert M, Eisenberg DS, Crowther RA. Propagation of Tau aggregates and neurodegeneration. Annu Rev Neurosci. 2017;40:189–210.
50. Huang WJ, Zhang X, Chen WW. Role of oxidative stress in Alzheimer's disease. Biomed Rep. 2016;4(5):519–22.
51. Calsolaro V, Edison P. Neuroinflammation in Alzheimer's disease: current evidence and future directions. Alzheimers Dement. 2016;12(6):719–32.
52. Sahni JK, Doggui S, Ali J, Baboota S, Dao L, Ramassamy C. Neurotherapeutic applications of nanoparticles in Alzheimer's disease. J Control Release. 2011;152(2):208–31.
53. Thorgrimsen L. Phase 3 trials of Solanezumab for mild-to-moderate Alzheimer's disease. N Engl J Med. 2014;370(4):311–21.
54. Viola KL, Sbarboro J, Sureka R, De M, Bicca MA, Wang J, Vasavada S, Satpathy S, Wu S, Joshi H. Towards non-invasive diagnostic imaging of early-stage Alzheimer's disease. Nat Nanotechnol. 2015;10(1):91–8.
55. Herran E, Pérezgonzález R, Igartua M, Pedraz JL, Carro E, Hernandez RM. Enhanced hippocampal neurogenesis in APP/Ps1 mouse model of Alzheimer's disease after implantation of VEGF-loaded PLGA nanospheres. Curr Alzheimer Res. 2015;12(10):932–40.
56. Yang Y. Pharmacological and toxicological target organelles and safe use of single-walled carbon nanotubes as drug carriers in treating Alzheimer disease. Nanomed Nanotechnol Biol Med. 2011;6(3):427–41.
57. Wilson B, Samanta MK, Santhi K, Kumar KP, Paramakrishnan N, Suresh B. Poly (n-butylcyanoacrylate) nanoparticles coated with polysorbate 80 for the targeted delivery of rivastigmine into the brain to treat Alzheimer's disease. Brain Res. 2008;1200:159–68.
58. Mittal G, Sahana DK, Bhardwaj V, Ravi Kumar MN. Estradiol loaded PLGA nanoparticles for oral administration: effect of polymer molecular weight and copolymer composition on release behavior in vitro and in vivo. J Control Release. 2007;119(1):77–85.
59. Mulik RS, Mönkkönen J, Juvonen RO, Mahadik KR, Paradkar AR. ApoE3 mediated poly (butyl) cyanoacrylate nanoparticles containing curcumin: study of enhanced activity of curcumin against beta amyloid induced cytotoxicity using in vitro cell culture model. Mol Pharm. 2010;7(3):815–25.
60. Smith A, Giunta B, Bickford PC, Fountain M, Tan J, Shytle RD. Nanolipidic particles improve the bioavailability and alpha-secretase inducing ability of epigallocatechin-3-gallate (EGCG) for the treatment of Alzheimer's disease. Int J Pharm. 2010;389(1):207–12.
61. Bush AI. Drug development based on the metals hypothesis of Alzheimer's disease. J Alzheimers Dis. 2008;15(2):223–40.
62. Cui Z, Lockman PR, Atwood CS, Hsu CH, Gupte A, Allen DD, Mumper RJ. Novel D-penicillamine carrying nanoparticles for metal chelation therapy in Alzheimer's and other CNS diseases. Eur J Pharm Biopharm. 2005;59(2):263–72.
63. Liu G, Men P, Harris PLR, Rolston RK, Perry G, Smith MA. Nanoparticle iron chelators: a new therapeutic approach in Alzheimer disease and other neurologic disorders associated with trace metal imbalance. Neurosci Lett. 2006;406(3):189–93.
64. Liu G, Men P, Kudo W, Perry G, Smith MA. Nanoparticle-chelator conjugates as inhibitors of amyloid-beta aggregation and neurotoxicity: a novel therapeutic approach for Alzheimer disease. Neurosci Lett. 2009;455(3):187–90.
65. Cui Z, Bu W, Fan W, Zhang J, Ni D, Liu Y, Wang J, Liu J, Yao Z, Shi J. Sensitive imaging and effective capture of Cu(2+): towards highly efficient theranostics of Alzheimer's disease. Biomaterials. 2016;104:158–67.

66. Mufamadi MS, Choonara YE, Kumar P, Modi G, Naidoo D, Ndesendo VM, Du LT, Iyuke SE, Pillay V. Surface-engineered nanoliposomes by chelating ligands for modulating the neurotoxicity associated with β-amyloid aggregates of Alzheimer's disease. Pharm Res. 2012;29 (11):3075–89.
67. Dugan LL, Gabrielsen JK, Yu SP, Lin TS, Choi DW. Buckminsterfullerenol free radical scavengers reduce excitotoxic and apoptotic death of cultured cortical neurons. Neurobiol Dis. 1996;3(2):129–35.
68. Huang HM, Ou HC, Hsieh SJ, Chiang LY. Blockage of amyloid beta peptide-induced cytosolic free calcium by fullerenol-1, carboxylate C60 in PC12 cells. Life Sci. 2000;66 (16):1525–33.
69. Kotelnikova RA, Smolina AV, Grigoryev VV, Faingold II, Mischenko DV, Rybkin AY, Poletayeva DA, Vankin GI, Zamoyskiy VL, Voronov II. Influence of water-soluble derivatives of [60]fullerene on therapeutically important targets related to neurodegenerative diseases. MedChemComm. 2014;5(11):1664–8.
70. Xie L, Luo Y, Lin D, Xi W, Yang X, Wei G. The molecular mechanism of fullerene-inhibited aggregation of Alzheimer's β-amyloid peptide fragment. Nanoscale. 2014;6(16):9752–62.
71. Armisén PA, Sancho MB, Almaraz GE, Prieto BG, Polanco AI. Colitis induced by a food allergen. A report of 20 cases. Anales Espanoles De Pediatria. 1996;44(1):21–4.
72. Heckman KL, Decoteau W, Estevez A, Reed KJ, Costanzo W, Sanford D, Leiter JC, Clauss J, Knapp K, Gomez C. Custom cerium oxide nanoparticles protect against a free radical mediated autoimmune degenerative disease in the brain. ACS Nano. 2013;7(12):10582–96.
73. Cimini A, Ceru MP, Amicarelli F, Falone S, Phani RA, Loreto SD, Benedetti E, Santucci S, D'Angelo B. Cerium oxide nanoparticles trigger neuronal survival in a human Alzheimer disease model by modulating BDNF pathway. Curr Nanosci. 2009;5(2):167–76.
74. Xu C, Qu X. Cerium oxide nanoparticle: a remarkably versatile rare earth nanomaterial for biological applications. NPG Asia Mater. 2014;6(3):102–8.
75. Cimini A, D'Angelo B, Das S, Gentile R, Benedetti E, Singh V, Monaco AM, Santucci S, Seal S. Antibody-conjugated PEGylated cerium oxide nanoparticles for specific targeting of Aβ aggregates modulate neuronal survival pathways. Acta Biomater. 2012;8(6):2056–67.
76. Kwon HJ, Cha MY, Kim D, Kim DK, Soh M, Shin K, Hyeon T, Mookjung I. Mitochondria-targeting ceria nanoparticles as antioxidants for Alzheimer's disease. ACS Nano. 2016;10 (2):2860–76.
77. Chen Q, Du Y, Zhang K, Liang Z, Li J, Yu H, Ren R, Feng J, Jin Z, Li F. A Tau-targeted multifunctional nanocomposite for combinational therapy of Alzheimer's disease. ACS Nano. 2018;12(2):1321–38.
78. Bharali DJ, Klejbor I, Stachowiak EK, Dutta P, Roy I, Kaur N, Bergey EJ, Prasad PN, Stachowiak MK. Organically modified silica nanoparticles: a nonviral vector for in vivo gene delivery and expression in the brain. Proc Natl Acad Sci U S A. 2005;102(32):11539–44.
79. Jaruszewski KM, Ramakrishnan S, Poduslo JF, Kandimalla KK. Chitosan enhances the stability and targeting of immuno-nanovehicles to cerebro-vascular deposits of Alzheimer's disease amyloid protein. Nanomedicine. 2012;8(2):250–60.
80. Agyare EK, Curran GL, Ramakrishnan M, Yu CC, Poduslo JF, Kandimalla KK. Development of a smart nano-vehicle to target cerebrovascular amyloid deposits and brain parenchymal plaques observed in Alzheimer's disease and cerebral amyloid angiopathy. Pharm Res. 2008;25(11):2674–84.
81. Zhang ZX, Roman GC, Hong Z, Wu CB, Qu QM, Huang JB, Zhou B, Geng ZP, Wu JX, Wen HB, Zhao H, Zahner GE. Parkinson's disease in China: prevalence in Beijing, Xian, and Shanghai. Lancet. 2005;365(9459):595–7.
82. Kalia LV, Lang AE. Parkinson's disease. Lancet. 2015;386(9996):896–912.
83. Kalia LV, Kalia SK. Alpha-Synuclein and Lewy pathology in Parkinson's disease. Curr Opin Neurol. 2015;28(4):375–81.
84. Kalia LV, Lang AE. Parkinson disease in 2015: evolving basic, pathological and clinical concepts in PD. Nat Rev Neurol. 2016;12(2):65–6.

85. Pires AO, Teixeira FG, Mendes-Pinheiro B, Serra SC, Sousa N, Salgado AJ. Old and new challenges in Parkinson's disease therapeutics. Prog Neurobiol. 2017;156:69–89.
86. Md S, Khan RA, Mustafa G, Chuttani K, Baboota S, Sahni JK, Ali J. Bromocriptine loaded chitosan nanoparticles intended for direct nose to brain delivery: pharmacodynamic, pharmacokinetic and scintigraphy study in mice model. Eur J Pharm Sci. 2013;48(3):393–405.
87. Wu H, Hu K, Jiang X. From nose to brain: understanding transport capacity and transport rate of drugs. Expert Opin Drug Deliv. 2008;5(10):1159–68.
88. Pahuja R, Seth K, Shukla A, Shukla RK, Bhatnagar P, Chauhan LK, Saxena PN, Arun J, Chaudhari BP, Patel DK, Singh SP, Shukla R, Khanna VK, Kumar P, Chaturvedi RK, Gupta KC. Trans-blood brain barrier delivery of dopamine-loaded nanoparticles reverses functional deficits in parkinsonian rats. ACS Nano. 2015;9(5):4850–71.
89. De GE, Trapani A, Cafagna D, Sabbatini L, Cometa S. Dopamine-loaded chitosan nanoparticles: formulation and analytical characterization. Anal Bioanal Chem. 2011;400 (7):1997–2002.
90. Trapani A, De GE, Cafagna D, Denora N, Agrimi G, Cassano T, Gaetani S, Cuomo V, Trapani G. Characterization and evaluation of chitosan nanoparticles for dopamine brain delivery. Int J Pharm. 2011;419(1):296–307.
91. Yang X, Zheng R, Cai Y, Liao M, Yuan W, Liu Z. Controlled-release levodopa methyl ester/ benserazide-loaded nanoparticles ameliorate levodopa-induced dyskinesia in rats. Int J Nanomedicine. 2012;7:2077–86.
92. Kaye CM, Nicholls B. Clinical pharmacokinetics of Ropinirole. Clin Pharmacokinet. 2000;39 (4):243–54.
93. Azeem A, Ahmad FJ, Khar RK, Talegaonkar S. Nanocarrier for the transdermal delivery of an Antiparkinsonian drug. AAPS PharmSciTech. 2009;10(4):1093–103.
94. Dingler A, Gohla S. Production of solid lipid nanoparticles (SLN): scaling up feasibilities. J Microencapsul. 2002;19(1):11–6.
95. Esposito E, Fantin M, Marti M, Drechsler M, Paccamiccio L, Mariani P, Sivieri E, Lain F, Menegatti E, Morari M, Cortesi R. Solid lipid nanoparticles as delivery systems for bromocriptine. Pharm Res. 2008;25(7):1521–30.
96. Tsai MJ, Huang YB, Wu PC, Fu YS, Kao YR, Fang JY, Tsai YH. Oral apomorphine delivery from solid lipid nanoparticles with different monostearate emulsifiers: pharmacokinetic and behavioral evaluations. J Pharm Sci. 2011;100(2):547–57.
97. Wang A, Wang L, Sun K, Liu W, Sha C, Li Y. Preparation of rotigotine-loaded microspheres and their combination use with L-DOPA to modify dyskinesias in 6-OHDA-lesioned rats. Pharm Res. 2012;29(9):2367–76.
98. Wyman T, Rohrer D, Kirigiti P, Nichols H, Pilcher K, Nilaver G, Machida C. Promoter-activated expression of nerve growth factor for treatment of neurodegenerative diseases. Gene Ther. 1999;6(10):1648–60.
99. Rinne JO, Mlic JR, Paljärvi L, Rinne UK. Dementia in Parkinson's disease is related to neuronal loss in the medial substantia nigra. Ann Neurol. 1989;26(1):47–50.
100. Collier DTJ, Sortwell CE. Therapeutic potential of nerve growth factors in Parkinson's disease. Drugs Aging. 1999;14(4):261–87.
101. Kurakhmaeva KB, Djindjikhashvili IA, Petrov VE, Balabanyan VU, Voronina TA, Trofimov SS, Kreuter J, Gelperina S, Begley D, Alyautdin RN. Brain targeting of nerve growth factor using poly(butyl cyanoacrylate) nanoparticles. J Drug Target. 2009;17(8):564–74.
102. Xie J, Chen L, Varadan VK, Yancey J, Srivatsan M. The effects of functional magnetic nanotubes with incorporated nerve growth factor in neuronal differentiation of PC12 cells. Nanotechnology. 2008;19(10):105101.
103. Facci L, Stevens DA, Pangallo M, Franceschini D, Skaper SD, Strijbos PJ. Corticotropin-releasing factor (CRF) and related peptides confer neuroprotection via type 1 CRF receptors. Neuropharmacology. 2003;45(5):623–36.
104. Pedersen WA, Wan R, Zhang P, Mattson MP. Urocortin, but not urocortin II, protects cultured hippocampal neurons from oxidative and excitotoxic cell death via corticotropin-releasing hormone receptor type I. J Neurosci. 2002;22(2):404–12.

105. Hu K, Shi Y, Jiang W, Han J, Huang S, Jiang X. Lactoferrin conjugated PEG-PLGA nanoparticles for brain delivery: preparation, characterization and efficacy in Parkinson's disease. Int J Pharm. 2011;415(1):273–83.
106. Wen Z, Yan Z, Hu K, Pang Z, Cheng X, Guo LR, Zhang Q, Jiang X, Fang L, Lai R. Odorranalectin-conjugated nanoparticles: preparation, brain delivery and pharmacodynamic study on Parkinson's disease following intranasal administration. J Control Release. 2011;151(2):131–8.
107. Björklund A, Björklund T, Kirik D. Gene therapy for dopamine replacement in Parkinson's disease. Sci Transl Med. 2009;1(2):2ps2.
108. Huang R, Ke W, Liu Y, Wu D, Feng L, Chen J, Pei Y. Gene therapy using lactoferrin-modified nanoparticles in a rotenone-induced chronic Parkinson model. J Neurol Sci. 2010;290 (1):123–30.
109. Chtarto A, Yang X, Bockstael O, Melas C, Blum D, Lehtonen E, Abeloos L, Jaspar JM, Levivier M, Brotchi J. Controlled delivery of glial cell line-derived neurotrophic factor by a single tetracycline-inducible AAV vector. Exp Neurol. 2007;204(1):387–99.
110. Huang R, Han L, Li J, Ren F, Ke W, Jiang C, Pei Y. Neuroprotection in a 6-hydroxydopamine-lesioned Parkinson model using lactoferrin-modified nanoparticles. J Gene Med. 2009;11 (9):754–63.
111. Huang R, Ma H, Guo Y, Liu S, Kuang Y, Shao K, Li J, Liu Y, Han L, Huang S. Angiopep-conjugated nanoparticles for targeted long-term gene therapy of Parkinson's disease. Pharm Res. 2013;30(10):2549–59.
112. Zhang Y, Pardridge WM. Near complete rescue of experimental Parkinson's disease with intravenous, non-viral GDNF gene therapy. Pharm Res. 2009;26(5):1059–63.
113. Kim AS, Cahill E, Cheng NT. Global stroke belt: geographic variation in stroke burden worldwide. Stroke. 2015;46(12):3564–70.
114. Shcharbina N, Shcharbin D, Bryszewska M. Nanomaterials in stroke treatment: perspectives. Stroke. 2013;44(8):2351–5.
115. Fernandes LF, Bruch GE, Massensini AR, Frézard F. Recent advances in the therapeutic and diagnostic use of liposomes and carbon nanomaterials in ischemic stroke. Front Neurosci. 2018;12:453–67.
116. Mozaffarian D, Benjamin EJ, Go AS, Arnett DK, Blaha MJ, Cushman M, Ferranti SD, Despres JP, Fullerton HJ, Howard VJ. American Heart Association Statistics Committee and Stroke Statistics Subcommittee. Circulation. 2007;115(5):e69–e245.
117. Donnan GA, Fisher M, Macleod M, Davis SM. Stroke. Lancet. 2008;371(9624):1612–23.
118. Mullen MT, Pisapia JM, Tilwa S, Messé SR, Stein SC. Systematic review of outcome after ischemic stroke due to anterior circulation occlusion treated with intravenous, intra-arterial, or combined intravenous+intra-arterial thrombolysis. Stroke. 2012;43(43):2350–5.
119. Mutoh T, Mutoh T, Taki Y, Ishikawa T. Therapeutic potential of natural product-based oral nanomedicines for stroke prevention. J Med Food. 2016;19(6):521–7.
120. Tomillero A, Moral MA. Gateways to clinical trials. Methods Find Exp Clin Pharmacol. 2008;30(4):313–31.
121. Adeoye O, Hornung R, Khatri P, Kleindorfer D. Recombinant tissue-type plasminogen activator use for ischemic stroke in the United States: a doubling of treatment rates over the course of 5 years. Stroke. 2011;42(7):1952–5.
122. Silvacandal A, Argibay B, Iglesiasrey R, Vargas Z, Vieitesprado A, Lópezarias E, Rodríguezcastro E, Lópezdequidt I, Rodríguezyáñez M, Piñeiro Y. Vectorized nanodelivery systems for ischemic stroke: a concept and a need. J Nanobiotechnol. 2017;15(1):30–44.
123. Sarmah D, Saraf J, Kaur H, Pravalika K, Tekade R, Borah A, Kalia K, Dave K, Bhattacharya P. Stroke management: An emerging role of nanotechnology. Micromachines. 2017;8 (9):262–74.
124. Chen L, Gao X. The application of nanoparticles for neuroprotection in acute ischemic stroke. Ther Deliv. 2017;8(10):915–28.

125. Erdem SS, Sazonova IY, Hara T, Jaffer FA, JR MC. Detection and treatment of intravascular thrombi with magnetofluorescent nanoparticles. Methods Enzymol. 2012;508:191–209.
126. Chen JP, Yang PC, Ma YH, Tu SJ, Lu YJ. Targeted delivery of tissue plasminogen activator by binding to silica-coated magnetic nanoparticle. Int J Nanomedicine. 2012;7:5137–49.
127. Kim BH, Lee N, Kim H, An K, Park YI, Choi Y, Shin K, Lee Y, Kwon SG, Na HB, Park JG, Ahn TY, Kim YW, Moon WK, Choi SH, Hyeon T. Large-scale synthesis of uniform and extremely small-sized iron oxide nanoparticles for high-resolution T1 magnetic resonance imaging contrast agents. J Am Chem Soc. 2011;133(32):12624–31.
128. Bi F, Zhang J, Su Y, Tang YC, Liu JN. Chemical conjugation of urokinase to magnetic nanoparticles for targeted thrombolysis. Biomaterials. 2009;30(28):5125–30.
129. McCarthy JR. Nanomedicine and cardiovascular disease. Curr Cardiovasc Imaging Report. 2010;3(1):42–9.
130. Ma Y-H, Wu S-Y, Wu T, Chang Y-J, Hua M-Y, Chen J-P. Magnetically targeted thrombolysis with recombinant tissue plasminogen activator bound to polyacrylic acid-coated nanoparticles. Biomaterials. 2009;30(19):3343–51.
131. Uesugi Y, Kawata H, Jo J, Saito Y, Tabata Y. An ultrasound-responsive nano delivery system of tissue-type plasminogen activator for thrombolytic therapy. J Control Release. 2010;147 (2):269–77.
132. Tang D-W, Yu S-H, Ho Y-C, Mi F-L, Kuo P-L, Sung H-W. Heparinized chitosan/poly (γ-glutamic acid) nanoparticles for multi-functional delivery of fibroblast growth factor and heparin. Biomaterials. 2010;31(35):9320–32.
133. Bai S, Ahsan F. Synthesis and evaluation of pegylated dendrimeric nanocarrier for pulmonary delivery of low molecular weight heparin. Pharm Res. 2009;26(3):539–48.
134. Gary-Bobo M, Mir Y, Rouxel C, Brevet D, Basile I, Maynadier M, Vaillant O, Mongin O, Blanchard-Desce M, Morere A, Garcia M, Durand JO, Raehm L. Mannose-functionalized mesoporous silica nanoparticles for efficient two-photon photodynamic therapy of solid tumors. Angew Chem. 2011;50(48):11425–9.
135. Schubert D, Dargusch R, Raitano J, Chan SW. Cerium and yttrium oxide nanoparticles are neuroprotective. Biochem Biophys Res Commun. 2006;342(1):86–91.
136. Jin H, Chen WQ, Tang XW, Chiang LY, Yang CY, Schloss JV, Wu JY. Polyhydroxylated C (60), fullerenols, as glutamate receptor antagonists and neuroprotective agents. J Neurosci Res. 2015;62(4):600–7.
137. Dugan LL, Lovett EG, Quick KL, Lotharius J, Lin TT, O'Malley KL. Fullerene-based antioxidants and neurodegenerative disorders. Parkinsonism Relat Disord. 2001;7(3):243–6.
138. Hirsch A. Dendrimeric fullerene derivatives, process for their preparation, and use as neuroprotectants. U.S. Pat. 2003;6(506):928.
139. Chorny M, Hood E, Levy RJ, Muzykantov VR. Endothelial delivery of antioxidant enzymes loaded into non-polymeric magnetic nanoparticles. J Control Release. 2010;146(1):144–51.
140. Klyachko NL, Manickam DS, Brynskikh AM, Uglanova SV, Li S, Higginbotham SM, Bronich TK, Batrakova EV, Kabanov AV. Cross-linked antioxidant nanozymes for improved delivery to CNS. Nanomedicine. 2012;8(1):119–29.
141. Tu J, Yang F, Wan J, Liu Y, Zhang J, Wu B, Liu Y, Zeng S, Wang L. Light-controlled astrocytes promote human mesenchymal stem cells toward neuronal differentiation and improve the neurological deficit in stroke rats. Glia. 2014;62(1):106–21.
142. Pittenger MF, Mackay AM, Beck SC, Jaiswal RK, Douglas R, Mosca JD, Moorman MA, Simonetti DW, Craig S, Marshak DR. Multilineage potential of adult human mesenchymal stem cells. Science. 1999;284(5411):143–7.
143. Jiang Y, Jahagirdar BN, Reinhardt RL, Schwartz RE, Keene CD, Ortiz-Gonzalez XR, Reyes M, Lenvik T, Lund T, Blackstad M. Pluripotency of mesenchymal stem cells derived from adult marrow. Nature. 2002;418(6893):41–9.
144. Hao L, Zou Z, Tian H, Zhang Y, Zhou H, Liu L. Stem cell-based therapies for ischemic stroke. Bio Med Res Int. 2014;2014:468748.

145. Marquez-Curtis LA, Janowska-Wieczorek A, Mcgann LE, Elliott JAW. Mesenchymal stromal cells derived from various tissues: biological, clinical and cryopreservation aspects. Cryobiology. 2015;71(2):181–97.
146. Ellis-Behnke R, Teather L, Schneider G, So K-F. Using nanotechnology to design potential therapies for CNS regeneration. Curr Pharm Des. 2007;13(24):2519–28.
147. Borlongan CV, Masuda T, Walker TA, Maki M, Hara K, Yasuhara T, Matsukawa N, Emerich DF. Nanotechnology as an adjunct tool for transplanting engineered cells and tissues. Curr Mol Med. 2007;7(7):609–18.
148. Kubinová Š, Syková E. Nanotechnology for treatment of stroke and spinal cord injury. Nanomedicine. 2010;5(1):99–108.
149. Roberson ED, Scearcelevie K, Palop JJ, Yan F, Cheng IH, Wu T, Gerstein H, Yu GQ, Mucke L. Reducing endogenous tau ameliorates amyloid beta-induced deficits in an Alzheimer's disease mouse model. Science. 2007;316(5825):750–4.
150. Huang X, Zhang F, Wang Y, Sun X, Choi KY, Liu D, Choi J-S, Shin T-H, Cheon J, Niu G. Design considerations of iron-based nanoclusters for noninvasive tracking of mesenchymal stem cell homing. ACS Nano. 2014;8(5):4403–14.
151. Park W, Yang HN, Ling D, Yim H, Kim KS, Hyeon T, Na K, Park K-H. Multi-modal transfection agent based on monodisperse magnetic nanoparticles for stem cell gene delivery and tracking. Biomaterials. 2014;35(25):7239–47.
152. Nair S, Dileep A, Rajanikant G. Nanotechnology based diagnostic and therapeutic strategies for neuroscience with special emphasis on ischemic stroke. Curr Med Chem. 2012;19 (5):744–56.

# Chapter 5
# Overcoming the Physiopathologic Barriers: Nanoprobes-Mediated Intracranial Glioma Imaging

**Heng Liu, Yu Liu, Fengyuan Man, and Gang Liu**

**Abstract** Malignant glioma is characterized by active angiogenesis, high invasiveness and infiltration, and extremely rapid growth. Accurate visualization of glioma is crucial to the early diagnosis, preoperative localization, intraoperative guidance, and therapeutic evaluation and thus facilitates the clinical decision-making and improves the clinical outcomes of patients. However, conventional contrast agents directed toward intracranial glioma remain challenging, largely attributed to the existence of physiopathologic barriers unique to brain tumors. Remarkable advancements in nanotechnology and nanomedicine open a multidisciplinary field to design various nanoprobes for overcoming the physiopathologic barriers and for improved glioma imaging. This chapter starts with the critical biological challenges facing intracranial glioma. The innovative approaches for enhancing blood-brain barrier permeability and improved glioma targeting ability are presented. It then provides an overview of the unique advantages of nanomaterials for glioma imaging. The advanced applications of nanoprobes in intracranial glioma imaging are reviewed in detail, including magnetic resonance imaging, photoacoustic imaging, fluorescence imaging, multimodality imaging, and intraoperative glioma margin delineation. Finally, the current challenges and perspectives of this field are also discussed.

**Keywords** Malignant glioma · Nanoprobe · Imaging · Blood-brain barrier

---

Heng Liu and Yu Liu contributed equally with all other contributors.

---

H. Liu · F. Man
Department of Radiology, PLA Rocket Force Characteristic Medical Center, Beijing, China

Y. Liu
Department of Ultrasound, Southwest Hospital, Army Medical University, Chongqing, China

G. Liu (✉)
State Key Laboratory of Molecular Vaccinology and Molecular Diagnostics & Center for Molecular Imaging and Translational Medicine, School of Public Health, Xiamen University, Xiamen, China
e-mail: gangliu.cmitm@xmu.edu.cn


X. Xue (ed.), *Nanomedicine in Brain Diseases*,
https://doi.org/10.1007/978-981-13-8731-9_5

## 5.1 Introduction

Malignant glioma, characterized by active angiogenesis, high invasiveness and infiltration, and extremely rapid growth, remains one of the most daunting intracranial malignancies in clinic. Currently, there have been extensive and unprecedented efforts in conventional antitumor therapeutic modalities including advanced neurosurgery techniques, improved adjuvant chemoradiation regimens, and sometimes innovative therapeutic strategies (e.g., gene therapy, immunotherapy, phototherapy, magnetic hyperthermia, and tumor-treating field). Unfortunately, the prognosis remains extremely poor for patients undergoing glioblastoma multiforme, the most frequently occurring primary malignant brain tumor in adults [1]. The median survival and 5-year survival rate after diagnosis are less than 14 months and 10%, respectively [2, 3]. Early and accurate visualization of malignant glioma is crucial to facilitate the clinical decision-making and improve the clinical outcomes of patients.

Currently, gadolinium-based contrast agent (GBCA)-enhanced $T_1$-weighted magnetic resonance examination is the preferred option for early diagnosis, preoperative localization, intraoperative guidance, and therapeutic evaluation of intracranial glioma in clinic. The small molecular GBCAs can diffuse into the tumor extravascular space through the disrupted blood-brain barrier (BBB) and induce positive contrast on $T_1$-weighted MR images. In clinical practice, more than 10% glioblastoma multiforme and 30% anaplastic astrocytoma exhibit no obvious signal enhancement due to the uncompromised and heterogeneous BBB [4]. Moreover, commercially available GBCAs encounter short circulation lifetime, transient imaging time window, non-targeting specificity, and rapid renal elimination. These drawbacks greatly impede the accurate tumor boundary delineation and tumor volume quantification owing to the inherent tumor infiltration, extensive peritumoral edema, and contrast agents' leakage. The lack of definite image margins inevitably results in incomplete surgical resection and consequent tumor recurrence. Concerns with GBCAs also happen in evaluating the therapeutic responses, such as pseudo-progression and pseudo-response. Moreover, despite believed to be exceedingly safe, GBCAs have the possibility to cause rare but serious adverse events such as nephrogenic systemic fibrosis and brain deposition [5, 6]. Therefore, there is an urgent demand to develop novel safe and efficient contrast agents for improved glioma visualization.

Remarkable advancements in nanotechnology and nanomedicine open a multidisciplinary field to design various nanoprobes for neuroimaging, such as magnetic nanoparticles, quantum nanodots, upconversion nanoparticles, and hybrid nanomaterials. Compared with GBCAs, these nanoprobes own unique advantages including tunable physicochemical characteristics, easy functionalization, prolonged circulation time, favorable safety profile, and satisfactory biocompatibility and biodegradability [7]. The properties of nanoprobes can be well tailored by manipulating the fabrication procedures and parameters, such as shape, size, surface charge, surface functional group, coating layer, and surface area-volume ratio. Following appropriate surface engineering, systemically administered nanoprobes with suitable size can exhibit enhanced penetration across the BBB and improved accumulation within the tumor region, mediated by the well-known enhanced permeability and retention (EPR) effect or active targeting. The nanoprobes can provide improved tumor delineation owing to the elevated cellular internalization behavior by tumor cells as well as reactive phagocytic cells and more durable retention at the tumor sites. These distinct features make biotechnology-based nanoprobes as ideal candidates alternative to GBCAs for improved glioma visualization.

This chapter starts with the critical biological challenges facing intracranial glioma. The innovative approaches for enhancing blood-brain barrier permeability and improved glioma targeting ability are presented. It then provides an overview of the unique advantages of nanomaterials for glioma imaging. The advanced applications of nanoprobes in intracranial glioma imaging are reviewed in detail, including magnetic resonance imaging, photoacoustic imaging, fluorescence imaging, multimodality imaging, and intraoperative glioma margin delineation. Finally, the current challenges and perspectives of this field are also discussed.

## 5.2 Critical Biological Challenges Facing Intracranial Glioma

Imaging and therapy strategies directed toward intracranial glioma remain challenging, largely attributed to the existence of two critical physiopathologic barriers that are unique to the brain tumors: blood-brain barrier (BBB) and blood-brain tumor barrier (BBTB). The BBB complex is characterized by continuous tight junction structures coordinated by cerebral endothelium, basement membrane, astrocytic foot processes, and pericytes (Fig. 5.1) [8, 9]. It isolates the systemic blood circulation from the brain parenchyma and creates a physiological barrier for maintaining the brain homeostasis. The pinocytotic vesicles and mitochondria in the endothelial cells and efflux transporters (e.g., P-glycoprotein and various multidrug resistance-associated proteins) together contribute to the controlled BBB permeability, allowing the entry of biologically essential substances into the brain [10]. The BBB strictly excludes hydrophobic compounds and toxic substances from the circulatory system, and it also prevents the effective entry of the vast majority of small molecule drugs against malignant gliomas [11]. Actually, only lipophilic small molecules (<500 Da) and required nutrients can readily pass across the BBB in normal condition, driven by passive diffusion, facilitated diffusion transporters, or active transporters [12, 13].

The rapid growth of malignant glioma makes the BBB integrity compromised. The tumor causes a local and heterogeneous disruption of the selective BBB and creates possible approaches for enhanced delivering various imaging and therapeutic agents. The immature angiogenesis induces abnormal structures of the endothelial lining, contributing to the relative hyperpermeability and leakiness of tumor vasculature [14]. In general, the glioma vasculature permeability dynamically changes at different phases of tumor development [15]. At an early stage, the tumor cells grow

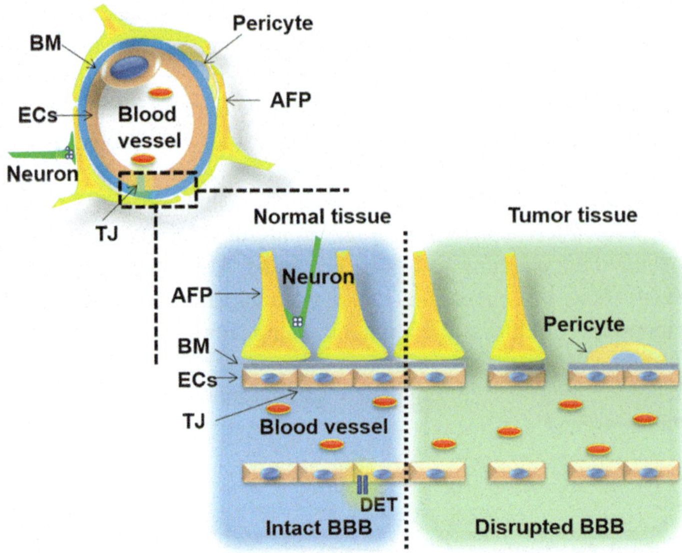


**Fig. 5.1 Illustration of the BBB components.** *BM* basement membrane, *AFP* astrocytic foot processes, *DET* drug efflux transporters, *ECs* endothelial cells, *TJ* tight junctions. (Adapted with permission from Ref. [9])

without inducing pathological angiogenesis, suggesting the intact BBB within the tumor. When the tumor volume increases larger than 0.2 mm [3], the BBB integrity becomes damaged and the BBTB is formed. The fenestrated microvessels composing BBTB prevents effective transvascular delivery of theranostic agents to the brain tumors. After these two barriers are both impaired, the EPR effect comes into action, benefiting from increased vascular permeability along with compromised lymphatic drainage induced by various mediators [16].

## 5.3 Strategies for Bypassing and Crossing BBB

The fundamental pathophysiologic characteristics of the BBB provide possibilities for improved delivery efficacy of imaging and therapeutic components into the brain. To date, various approaches have been exploited, aimed at overcoming this physiopathologic barrier, including circumventing the BBB by local invasive approaches, targeting the BBB by physiological approaches, and manipulating the BBB integrity by modifying the BBB physiology.

To bypass the BBB, local invasive approaches are utilized with the assistance of catheters, pumps, or implantable devices, such as intracerebroventricular injection, intraparenchymal injection, convection-enhanced delivery, intrathecal injection, and intranasal delivery. The invasive approaches have several advantages over intravenous administration, including minimized systemic toxicity, no hepatic first-pass metabolism, and no protein opsonization effect. However, the passive diffusion pattern often induces insufficient penetration distance from the injection site, resulting in inadequate and uneven intratumoral distribution. And the associated risks (e.g., tissue injury, infection, and increased intracranial pressure) need to be considered due to the invasive feature.

The mechanisms of physiological approaches for transporting various imaging and therapeutic components across the BBB mainly include receptor-mediated transcytosis, adsorption-mediated transcytosis, carrier-mediated transcytosis, and transcellular lipophilic diffusion, in which the receptor-mediated transcytosis is the most representative [15]. The dynamically changing vascular features offer possibilities for a myriad of targeting strategies aimed at glioma with different development stages. By incorporating certain tumor-targeting modules (e.g., aptamers, peptides, peptidomimetic agents, recombinant proteins and protein domains of target ligands, and antibodies) aimed at corresponding receptors overexpressed within the tumor, specific tumor targeting and subsequently improved brain penetration can be achieved. Such potential receptors include the transporters localized on the BBB and the biomarkers overexpressed on tumor vasculatures, tumor cells, tumor stroma, and tumor metabolic processes. Cell-penetrating peptides (CPPs), which have been well-recognized to facilitate the penetration into the anionic cell membrane likely mediated by direct electrostatic interactions, is another feasible strategy for crossing the BBB. The CPPs typically contain different sequences with less than 30 amino acids, deriving from natural proteins such as trans-activator of transcription, penetratin, the Syn-B vectors, and engineered peptides.

Besides bypassing and crossing BBB, the BBB permeability can be manipulated through two different approaches: by temporarily opening the tight junctions and by inhibiting the efflux pumps [17]. The former can be achieved by several pharmacological and physical methods such as hyperosmotic agents, zonula occludens toxin, biological and chemical stimuli, high-intensity focused ultrasound, microwave field, and electromagnetic field. The latter can be realized by using various efflux inhibitors (e.g., endothelin-1, verapamil, tariquidar, elacridar, quinine, and quinidine) and siRNA silence technique. Despite promising, manipulation of the BBB may induce the upregulated influx of other undesired components and exacerbate the BBB-associated complications.

## 5.4 Advantages of Nanomaterials for Glioma Imaging

Remarkable advancements in nanomedicine and nanotechnology have created seductive possibilities for engineering various nanoprobes with distinct properties for glioma imaging. The engineered nanoprobes have significant advantages over the available small molecular contrast agents including the following: (i) The size, shape, surface charge, and other physicochemical characteristics of the nanoprobes can be readily modulated by manipulating the fabrication procedures and design parameters. Following appropriate surface engineering, nanoprobes with suitable size can exhibit enhanced penetration across the BBB and improved accumulation within the tumor even without excess modification, mediated by the well-known EPR effect. (ii) Given the distinct pathophysiological biomarkers of brain tumors at different stages, specific tumor targeting can be achieved by conjugating certain tumor-targeting modules onto the nanoprobes or using environment stimulation-responsive guidance (e.g., chemical, thermal, optical, electrical, and magnetic). This will contribute to sufficient accumulation of the nanoprobes at the tumor sites and significantly decrease the background interference. (iii) Multimodal imaging can be realized by integrating different imaging modalities into one nanostructure. It provides complementary information that allows for early and accurate delineation of the tumor molecular and metabolic features. (iv) Compared with GBCAs, nanoprobes own more favorable safety profile and biodegradability by adopting biocompatible and degradable precursor components. This will minimize potential side effects to the living body. (v) Benefiting from abundant active groups on the nanoprobes and physical absorption effect, nanoprobes can represent optimal drug delivery systems for versatile drug loading, achieving diagnosis and therapy simultaneously (theranostics). The loading capacity of therapeutic compounds can be considerably high, preserving their therapeutic potency and improving their bioavailability. The drug distribution and release within the tumor can be modulated in a controlled and effective manner. By co-loading multiple therapeutic agents onto a single nanoplatform, combination therapy can be achieved following administration only once.

## 5.5 Applications of Nanoprobes for Intracranial Glioma Imaging

Noninvasive neuroimaging techniques can provide detailed anatomical, functional, and metabolic information of brain tumors. Different imaging modalities differ in terms of sensitivity and specificity, spatiotemporal resolution, soft tissue contrast, imaging penetration depth, background noise, and data acquisition time. Systemically administered nanomaterials are vulnerable to the opsonization processes and elimination by the reticuloendothelial system, restricting their targeting ability toward intracranial glioma. To realize precise diagnosis, the efficacy of nanoprobes overcoming the BBB must be taken into consideration. Converging progresses in biotechnology-based nanoprobes and improved understanding of the tumor molecular biology urge the emergence of numerous novel nanoprobes for improved diagnostic sensitivity, specificity, and accuracy toward glioma.

### 5.5.1 Magnetic Resonance Imaging

Magnetic resonance imaging (MRI) is the most prominent and powerful neuroimaging modality owning excellent soft tissue contrast, high spatiotemporal resolution, and unlimited penetration depth. MR contrast agents can remarkably promote the imaging sensitivity and diagnostic accuracy by shortening the longitudinal or transverse relaxation time of neighboring water protons. Iron oxide nanoparticles (IONPs) have represented promising MR contrast agents alternative to GBCAs, benefiting from their tunable magnetic properties, easy modification, and excellent biocompatibility. However, IONPs often encounter insufficient accumulation at the glioma site via passive targeting and thus lead to an unconspicuous contrast effect for glioma imaging. To address this, Liu et al. [18] reported recombinant epidermal growth factor-like domain-1 (EGF1)-decorated IONPs for tissue factor-targeted MR visualization of glioma, with EGF1 serving as the targeting module and IONPs as the imaging compound, respectively. The as-designed EGF1-EGFP-IONPs exhibited efficient and persistent MR contrast enhancement with high targeting specificity up to 12 h for tissue factor-positive U87MG glioma in vivo (Fig. 5.2). Similarly, numerous tumor-specific IONPs have been strategically designed for more accurate delineation of glioma through attaching tumor-specific recognition moieties onto the

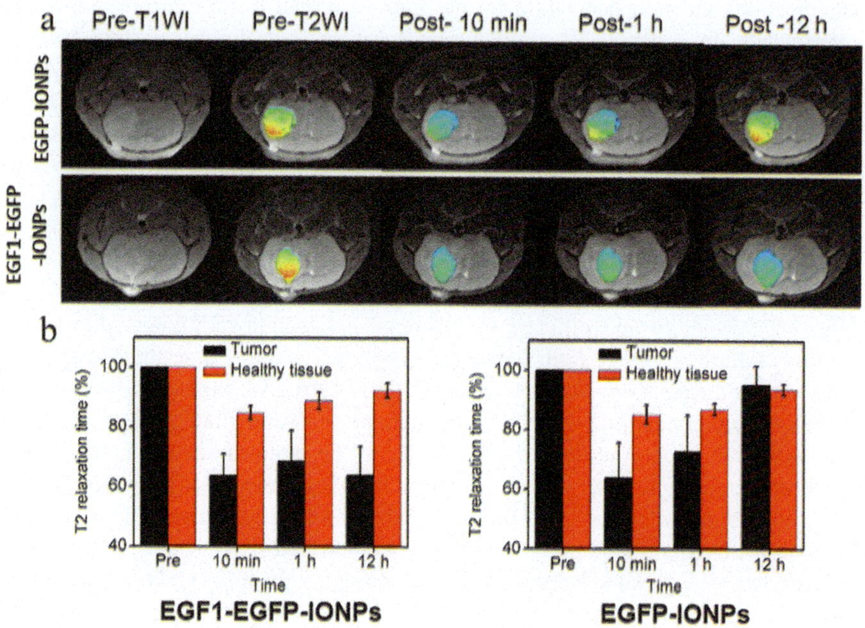


**Fig. 5.2 Recombinant EGF1 domain VII-functionalized IONPs for targeted MR imaging of glioma.** Representative in vivo $T_2$-weighted images (**a**) and normalized $T_2$ relaxation time (**b**) of U87MG glioma-bearing mouse before and at different time points postinjection of magnetic nanoprobes. (Adapted with permission from Ref. [18])

surface of IONPs (Table 5.1). All these nanoprobes demonstrated high imaging specificity and contrast enhancement in intracranial glioma mouse models.

Developing novel nanoprobes derived from natural components in organisms is of great interest for biomedical imaging, as they can undergo disintegration into nontoxic metabolites and thus possess minimal biological side effects [19]. This is crucial to guarantee their biosafety in vivo. Cao et al. reported a ferritin-based nanoprobe for targeted MRI of microscopic glioma in vivo [20]. The recombinant human heavy-chain ferritin protein nanocages were first expressed in Escherichia coli cells and purified, and then the magnetite cores were loaded in the cavity to obtain magnetoferritin (M-HFn). Neither additional surface coating nor targeted ligand modification was performed. The as-prepared M-HFn exhibited an extremely high $r_2$ relaxivity up to 224 $mM^{-1}s^{-1}$. After intravenous injection for 2 h, obvious focal hypointensity was observed on $T_2$-weighted MR images in an orthotopic xenografted model and maintained up to 24 h. The tissue TEM analysis showed that the M-HFn could cross multiple barriers including the endothelium, the epithelium, and the BBB layer. The specific and ultrasensitive detection of M-HFn for microscopic brain glioma was attributed to the transferrin receptor 1-mediated accumulation and internalization into tumor cells.

Although various IONP-based $T_2$ contrast agents have demonstrated satisfactory results for glioma imaging, the high magnetic inhomogeneity and magnetic susceptibility artifacts may lead to poor imaging performance. Moreover, the hemorrhage, calcification, and iron deposition in the brain may lead to false errors and image interference. In contrast, $T_1$ contrast agents can produce bright positive contrast images, providing better visual contrast and imaging performance. Recently, Chen et al. developed folic acid-decorated manganese oxide nanoparticles (MnO-TETT-FA NPs) as glioma-specific $T_1$ contrast agents [21]. The MnO-TETT-FA NPs could efficiently enhance the $T_1$ contrast and showed clearer tumor margin in a tiny glioma model, owing to the folic acid-mediated active targeting effect. Notably, manganese is an essential element for metabolism, without long-term retention and toxicity concerns referred to gadolinium.

Despite being promising, single-targeting IONPs suffer from insufficient targeting specificity and efficacy for accurate delineation of glioma, due to the receptor saturation effect, high tumor heterogeneity, and dynamic molecular characteristics at different developmental stages. Considering multiple types of receptors overexpressed concomitantly in the tumor, multi-targeting strategy is an effective approach to improve the delivery specificity to the targeted sites. Yan et al. first explored the feasibility of two-order targeted imaging strategy for improved glioma visualization, by enhancing the BBB permeability and targeting specificity [22]. In the Den-RGD-Angio nanoprobes, a fifth-generation PAMAM dendrimer served as the nanoscaffold, paramagnetic $Gd^{3+}$-DOTA served as the imaging component, and BBB-permeable angiopep-2 peptide and cyclic [RGDyK] peptide served as the targeting modules, respectively. The nanoprobes first targeted the $\alpha_V\beta_3$ integrin on the tumor neovasculature. After interaction with adjacent LRP receptors, the nanoprobes could cross the BBB via LRPR-mediated endocytosis and then targeted both $\alpha_V\beta_3$ integrin and LRP receptors on tumor cells. In vivo studies revealed that

**Table 5.1** Examples of targeted nanoprobes for crossing the BBB and intracranial glioma MR imaging

| | Targeting module | Nanoprobes | Relaxivity ($mM^{-1}s^{-1}$) | Biological target | Intracranial glioma model | Reference |
|---|---|---|---|---|---|---|
| Receptor-mediated transport | Transferrin (Tf) | Tf-SPIONs | $r_2 = 64.3$ | Tf receptor | Rats, C6 | [24] |
| | Ferritin | M-HFn NPs | $r_2 = 224$ | Tf receptor | Nude mice, U87MG | [20] |
| | Lactoferrin (Lf) | Lf-M-PAEEPPLLA-NPs | $r_2 = 151.3$ | Lf receptor | Rats, C6 | [25] |
| | Lactoferrin (Lf) | Lf-SPIONs | $r_2 = 75.6$ | LRP1 | Rats, C6 | [26] |
| Tumor vasculature-mediated transport | Anti-VEGF | MNP-VEGF | / | VEGF | Rats, C6 | [27] |
| | Anti-VEGFR2 | CLIO-anti-VEGF-R2 | / | VEGF-R2 | Rats, C6 | [28] |
| | GX1 peptide | MNP-HYNIC-PEG4-c(GX1) | / | CD31 | NOD-SCID mice, U87MG | [29] |
| | RGD peptide | $\gamma Fe_2O_3$@PO-PEG-cRGD | $r_2 = 315$ | $\alpha_V\beta_3$ integrin | Nude mice, U87MG | [30] |
| Tumor cell-mediated transport | Anti-IGFBP7 | anti-IGFBP7-IONPs | / | IGFBP7 | Nude mice, U87MG | [31] |
| | EGF | SPION-EGF | $r_2 = 223.7$ | wtEGFR, EGFRvIII | Rats, C6 | [32] |
| | EGFRmAb | EGFRmAb-SPIONs | $r_2 = 92.73$ | EGFR | Rats, C6 | [33] |
| | Hsp70 | Hsp70-SPIONs | $r_2 = 102.7$ | CD40 receptor | Rats, C6 | [34] |
| | EGF1 domain | EGF1-EGFP-IONPs | $r_2 = 177.5$ | TF | Nude mice, U87MG | [18] |
| | IL-1R antagonist | SPION-IL-1Ra | $r_2 = 287.7$ | IL-1R | Rats, C6 | [35] |
| | CREKA peptide | IO nanochains | $r_2 = 103.2$ | Fibronectin | Mice, CNS-1-GFP | [36] |
| | Bio-MAb | Bio-MAb/SAv-Gd-SLs | $r_1 = 8.1$ | Endoglin | Rats, C6 | [37] |
| | Interleukin-13 | IL-13-lip-Gd-DTPA | $r_1 = 4.0$ | IL-13R$\alpha$2 | Nude mice, U251 | [38] |
| | Folic acid | MnO-TETT-FA NPs | $r_1 = 4.84$ | Folate receptor | Nude mice, C6 | [21] |

(continued)

**Table 5.1** (continued)

| | Targeting module | Nanoprobes | Relaxivity ($mM^{-1}s^{-1}$) | Biological target | Intracranial glioma model | Reference |
|---|---|---|---|---|---|---|
| Dual-targeting-mediated transport | [RGDyK], angiopep-2 peptide | Den-RGD-Ang | $r_1 = 7.1$ | $\alpha_V\beta_3$ integrin, LRP | Nude mice, U87MG | [22] |
| | Folic acid, RGD peptide | $Fe_3O_4$-PEG-RGD-$FA_h$ | $r_2 = 200.3$ | Folate receptor, $\alpha_V\beta_3$ integrin | ICR mice, C6 | [23] |

***Abbreviations:*** *VEGF* vascular endothelial growth factor, *GX1* CGNSNPKSC, *RGD* arginine-glycine-aspartic acid, *IGFBP7* insulin-like growth factor-binding protein-7, *EGF* epidermal growth factor, *IL-1R* interleukin-1 receptor, *Bio-Mab* biotinylated monoclonal antibody, *IL-13Rα2* interleukin-13 receptor alpha 2, *angiopep-2* TFFYGGSRGKRNNFKTEEYC, *Hsp70* heat shock protein, *LRP1* low-density lipoprotein receptor-related protein-1 the two-order targeted nanoprobes not only could efficiently pass through the intact BBB in normal condition but also could realize specific intratumoral accumulation and precise boundary delineation with high-contrast capacity in a mouse glioma model. In another study, folic acid and c(RGDyK) peptide co-modified dual-targeting magnetite nanoparticles ($Fe_3O_4$-PEG-RGD-$FA_h$) were reported [23]. Compared with the non-targeted and single-targeted nanoprobes, the synergistic targeting nanoprobes enabled more effective penetration across the BBB, increased internalization by glioma cells, and enhanced MR contrast effect. Furthermore, it was found that the density of dual-target molecules on nanoprobes was critical for controlling the targeting specificity.

### 5.5.2 *Photoacoustic Imaging*

Photoacoustic imaging (PAI) is a burgeoning noninvasive imaging modality based on the light-thermal energy conversion. It integrates the merits of conventional optical imaging and acoustic imaging, possessing high sensitivity and specificity, high spatial resolution to micrometers, and increased tissue penetration to centimeters. As the generated thermal energy is related to the light energy absorbed by specific molecules, the PA signal strength is determined by the optical absorbance at specific wavelengths. For intracranial tumors, the dense skull can severely slash the light energy and decreased the imaging depth. And endogenous molecules such as hemoglobin may make imaging interference in the NIR-I region. To achieve longer imaging depth and minimize the background interference by biological tissues, PA contrast agents with higher absorption coefficient and longer absorption wavelength are urgently demanded. Lu et al. first explored the feasibility of c(KRGDf) peptide-functionalized hollow gold nanospheres (c(KRGDf)-PEG-HAuNS) for integrin-targeted photoacoustic tomography (PAT) imaging of orthotopic glioma [39]. The absorption peak of HAuNS was about at 800 nm. After intravenous injection of c (KRGDf)-PEG-HAuNS into U87MG glioma for 24 h, PAT images showed that the PA signal intensity ratio of tumor-to-contralateral healthy brain tissue was twice higher than that prior to injection. And no change of the PA signal ratio was observed in the non-targeted nanoprobe group. Micro-positron emission tomography results verified that the enhanced PAT contrast of U87MG glioma was attributed to selective accumulation of the nanoprobes mediated by the c(KRGDf) peptide.

Chen et al. fabricated single, few, and multilayer molybdenum disulfide ($MoS_2$) nanosheets via albumin-assisted layer-by-layer exfoliation method [40]. The as-prepared $MoS_2$ nanosheets demonstrated layer number-dependent PA effect. As the layer number decreased, significantly amplified imaging sensitivity was observed, attributed to the enhanced NIR absorption and the improved elastic property. The single-layered $MoS_2$ nanosheets exhibited excellent biocompatibility and satisfactory tumor-targeting ability. In situ glioma sitting at 1.8 mm below the scalp could be clearly visualized by using monolayer $MoS_2$ nanoprobe-enhanced PA imaging. To realize greater imaging depth, this group subsequently covalently

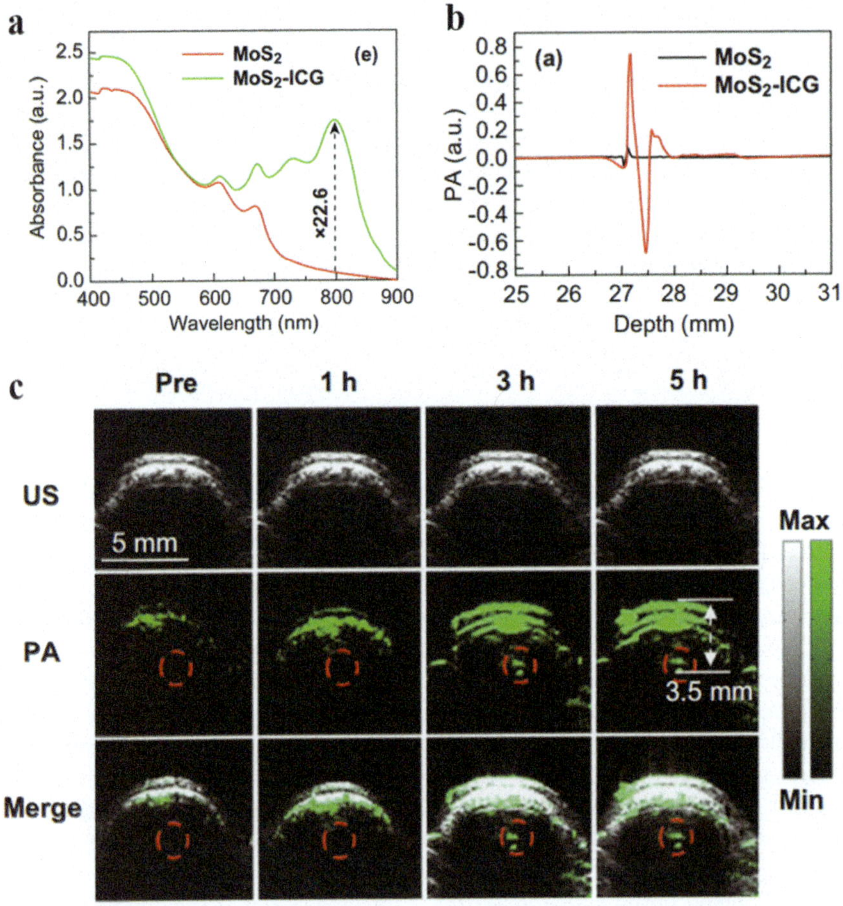


**Fig. 5.3** **$MoS_2$-ICG hybrid for highly sensitive PA imaging of orthotopic glioma at deep site.** (**a**) Absorbance spectra and (**b**) PA signal comparison of $MoS_2$-ICG and $MoS_2$ solutions. (**c**) B-scan ultrasound (US), PA, and the merged images of U87MG glioma-bearing mice before and at different time points postinjection of $MoS_2$-ICG. (Adapted with permission from Ref. [41])

conjugated ICG-Sulfo-NHS onto the monolayer $MoS_2$ nanosheets by facile mixing, obtaining $MoS_2$-ICG hybrid. The conjugation strategy produced strong and broad NIR absorption spectrum and red shifting of the $MoS_2$ absorption peak (800 nm vs. 675 nm), enabling greater PA imaging performance. The $MoS_2$-ICG hybrid demonstrated high PA imaging sensitivity of U87MG orthotopic glioma at 3.5 mm below the mouse scalp, nearly twofold deeper than $MoS_2$ nanosheets only (Fig. 5.3) [41]. The remarkably improved imaging depth warrants this covalent conjugation strategy promising for PA imaging of deep-sitting glioma.

Besides aforementioned inorganic materials, conjugated polymer (CP) nanoparticles have also been developed as PA contrast agents, benefiting from their large extinction coefficient and good photostability. Fan et al. fabricated

DSPE-mPEG5000 micelle-enveloped perylene-3,4,9,10-tetracarboxylic diimide (PDI) organic nanoparticles with water solubility and NIR absorptivity [42]. The tertiary amine group and diimide group in the perylene bisimide molecule could serve as the electron donor and electron acceptor, respectively, forming an effective donor-π-acceptor structure. The PDI NPs exhibited an absorption peak at 675 nm and high PAI contrast capacity. After intravenous injection into C6 glioma mice for 2 days, obvious PA signal was detected within the tumor region with a 4.0 mm depth while negligible signal in the control mice group. It indicated the excellent EPR effect of PDI NPs for penetrating the disrupted BBB and the satisfactory imaging performance.

Although promising, most reported PA nanoprobes for glioma are based on NIR-I (700-900 nm) PAI with low signal-background ratio. Developing NIR-II (1000-1700 nm) absorptive nanoprobes is of great potential for glioma diagnosis with higher sensitivity and deeper penetration. More recently, Guo et al. [43] reported cyclic c-RGD peptide-decorated conjugated polymer nanoparticles (P1RGD NPs) with strong NIR-II absorption for integrin-targeted glioma PA imaging. The planar backbone structure and high donor-acceptor strength of this conjugated polymer resulted in a broad and strong absorption peak centered at 1064 nm. The baseline images prior to injection showed no obvious PA background signal, because of the negligible absorption of biological tissues at 1064 nm. After intravenous injection of P1RGD NPs, the PA signal gradually increased within the tumor region. At 24 h postinjection, a high PA signal-background ratio up to 90 was detected, which was about 1.5-fold higher than that in the non-targeted NPs group, attributed to the active targeting effect for the former. The prominent PA imaging sensitivity of P1RGD NPs through the scalp and skull makes them promising for deep tumor imaging and therapy guidance.

### 5.5.3 Fluorescence Imaging

Fluorescence imaging modality owns significant advantages including high sensitivity and high contrast, real-time monitoring, rapid acquisition, and fast feedback. Unfortunately, currently available small molecular fluorescent organic dyes encounter extremely poor water solubility, susceptible to photobleaching, and short fluorescence lifetime. To address these limitations, various fluorescent nanoprobes have been extensively investigated for sensitive and durable bioimaging. Compared with fluorescent organic dyes, they possess many prominent advantages, including excellent water solubility, superior photo- and chemical stability, high quantum fluorescence yield, broad optical absorption with narrow emission, and long photoluminescence lifetime. Quantum dots (QDs) are emerging semiconductor nanocrystals with sizes ranging from 2 to 10 nm and possess unique optical properties [44]. Tang et al. constructed biotin-aptamer-conjugated streptavidin-PEG-CdSe/ZnS QDs (QD-Apt) for targeted fluorescence imaging of U87MG glioma overexpressing epidermal growth factor receptor variant III [45]. In vivo fluorescence imaging results demonstrated that QD-Apt could effectively penetrate the BBB and accumulate within the tumor, due to the aptamer-mediated active targeting effect, realizing the clear glioma margin visualization. Similarly, this group engineered NGR peptide-decorated PEGylated CdSe/ZnS QDs (NGR-PEG-QDs) and verified their effectiveness for penetrating across the BBB and targeted fluorescence imaging toward both glioma tissue and tumor vasculature [46].

Although promising, the intrinsic safety concerns of QDs limit their practical applications, due to the involvement of toxic heavy metal precursor molecules. Recently, photoluminescent carbon nanodots (CNDs) have been recognized as optimal substitutes to QDs, owing to their favorable biocompatibility. Using 1-methyl-2-pyrrolidinone serving as the solvent, carbon source and nitrogen source simultaneously, Wang et al. synthesized polymer-coated and nitrogen-doped carbon nanodots (pN-CNDs) with good water dispersibility by a simple "direct solvent-derived" method [47]. The resultant pN-CNDs exhibited stable blue fluorescence and a quantum yield up to 8.4%. The in vivo study showed glioma fluorescence imaging with high contrast, owing to the small hydrodynamic size of pN-CNDs and the EPR effect. Similarly, Zheng et al. reported novel photoluminescent carbon nanodots (CD-Asp) via a simple and facile pyrolysis approach by using L-aspartic acid and D-glucose as the precursor molecules (Fig. 5.4) [48]. The as-obtained CD-Asp exhibited excellent biocompatibility, continuous full-color emission, and self-targeting capacity toward C6 glioma cells. After intravenous injection into a C6 glioma model, intensive fluorescent signal was observed within the tumor region rather than in normal brain tissues, indicating their capacity of penetrating across the

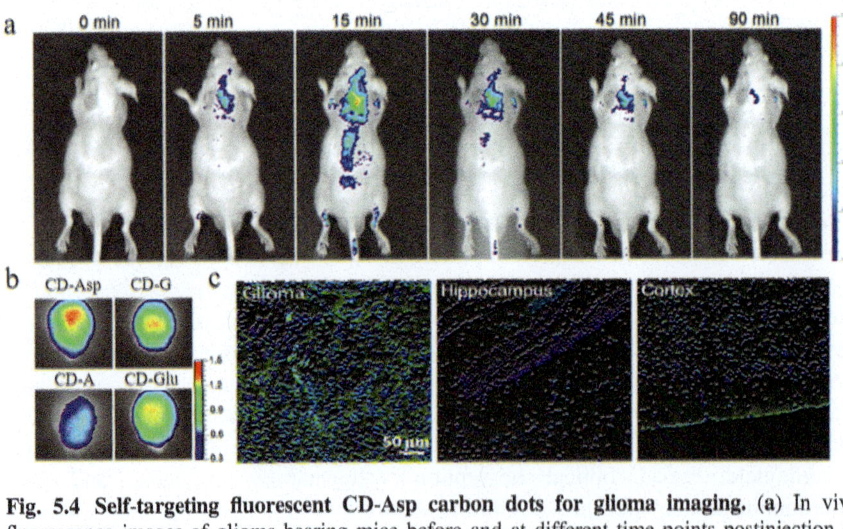


**Fig. 5.4 Self-targeting fluorescent CD-Asp carbon dots for glioma imaging.** (**a**) In vivo fluorescence images of glioma-bearing mice before and at different time points postinjection of CD-Asp. (**b**) Ex vivo fluorescence images of brain at 24 h postinjection of CD-Asp and its counterparts CD-A, CD-G, and CD-Glu. (**c**) Fluorescent distribution of CD-Asp in the brain at 24 h postinjection of CD-Asp (green). Blue represents the cell nucleus. (Adapted with permission from Ref. [48])

BBB and glioma targeting. In comparison, any other counterpart of the CD-Asp (CD-A, CD-G, and CD-Glu) showed scarcely any selectivity toward glioma. It indicated that CD-Asp held great potential as self-targeting contrast agents for precise fluorescence imaging of glioma.

Rare-earth upconversion nanoparticles (UCNPs), absorbing NIR light energy and emit fluorescent in the visible region, have recently been highly attractive for fluorescent bioimaging. Upconversion luminescence (UCL) possesses advantages including superior photostability without blinking, deep tissue penetration, long fluorescence lifetime, and minimal tissue autofluorescence [49]. By covalently conjugating angiopep-2 onto PEGylated UCNPs, UCNP-based nanoprobes (ANG/PEG-UCNPs) were constructed [50]. Both cellular and animal experiments revealed that the as-prepared ANG/PEG-UCNPs could efficiently cross the BBB and subsequently accumulated within the tumor. The UCL imaging in vivo results demonstrated better contrast performance compared with the frequently used fluorescent dye five-aminolevulinic acid in clinic. Considering the high atomic number of elements in UCNPs, they hold great potential as radiosensitizers for effective tumor theranostics.

Similar with PA contrast agents, NIR-II fluorescence nanoprobes can provide deeper penetration and higher spatial resolution than that in the NIR-I region, owing to the significantly decreased photon attenuation and negligible tissue autofluorescence for the former. Very recently, Sheng et al. designed a new donor-acceptor-tailored NIR-II fluorescent molecule (TB1) with aggregation-induced emission features for targeted imaging toward orthotopic glioma [51]. The TB1 dots exhibited a high-fluorescence quantum yield up to 6.2%, much higher than most reported NIR-II contrast agents with quantum yields in the range of 0.3–2%. The dots exhibited a strong absorption peak around 740 nm and a fluorescence emission peak centered at 975 nm, with high NIR-II fluorescence sensitivity and high photostability in various media. After functionalization with c-RGD peptide, the as-obtained TB1-RGD dots demonstrated specific tumor accumulation, with a high signal-background ratio up to 4.4 and high spatial resolution up to 38 μm. It suggested that the TB1-RGD dots could penetrate the BBB and selectively accumulate within the tumor region. The NIR-II dots with excellent imaging performance represented the ideal candidates for precise glioma diagnosis.

### 5.5.4 Multimodal Imaging

Although remarkable advances have been made in developing numerous nanoprobes for glioma diagnosis, conventional single-imaging modality cannot offer adequate information, owing to their inherent limitations. With the rapid development of advanced biomedical imaging techniques, multimodal imaging nanoprobes have attracted intensive interest. They can provide comprehensive and complementary diagnostic information through integrating the advantages of different imaging modalities into one multifunctional nanoplatform [52]. This allows for early, precise,

**Table 5.2** Examples of multifunctional nanoprobes for multimodal imaging of intracranial glioma

| Nanoprobes | Targeting moiety | Intracranial glioma model | Imaging modality | Ref. |
|---|---|---|---|---|
| PEG-GdIO NPs | | Rats, C6 | $T_1$–$T_2$ MRI | [60] |
| $Fe_{0.6}Mn_{0.4}O$ NFs | | SCID mice, U87MG | $T_1$–$T_2$ MRI | [62] |
| CS-DX-SPIONs | | Rats, C6 | $T_1$–$T_2$ MRI | [61] |
| $NaGdF_4$@PLL NDs | | Rats | $T_1$ MRI, CEST MRI | [63] |
| Cy5.5-Lf-MPNA nanogels | Lactoferrin (Lf) | Rats, C6 | MRI, FLI | [53] |
| ANG/PEG-UCNPs | Angiopep-2 | Nude mice, U87MG | MRI, UCL | [50] |
| UCNPs@$SiO_2$-CX-Lf | Chlorotoxin, Lf | Rats, C6 | MRI, UCL | [64] |
| MnO-PEG-Cy5.5 NPs | | Nude mice, C6-GFP | MRI, FLI | [56] |
| NCD-DTPA-Gd | | Nude mice, U87MG | MRI, FLI | [54] |
| Gd-G5-DL680 | | Nude mice, U87MG | MRI, FLI | [55] |
| Gold/SPION micelles (GSMs) | | Nude mice, U251 | MRI, CT | [57] |
| Au-AZ/Au-AK mixture | Angiopep-2 | Nude mice, U87MG | MRI, SERRS | [65] |
| [64]Cu-c(KRGDf)-PEG-HAuNS | c(KRGDf) peptide | Nude mice, U87MG-TGL | PAT, PET | [39] |
| TB1-RGD dots | c-RGD peptide | Nude mice, U87MG | FLI, PAI | [51] |
| SapC-DOPS nanovesicles | | Nude mice, U87ΔEGFR-Luc | PET, FLI | [58] |
| SERRS-MSOT-nanostars | | Transgenic mice | SERRS, MSOT | [66] |
| MPRs | | Nude mice, U87MG-eGFP | MRI, PAI, Raman | [67] |
| Au@MIL-88(Fe) | | Nude mice, U87MG | MRI, CT, PAI | [59] |

***Abbreviations:*** *FLI* fluorescence imaging, *UCL* upconversion luminescence, *SERRS* surface-enhanced resonance Raman scattering, *PAT* photoacoustic tomography, *PET* positron emission tomography, *MSOT* multispectral optoacoustic tomography

and consolidated characterization of the tumor molecular and metabolic features. Examples of nanoprobes for multimodal visualization of glioma are shown in Table 5.2.

Among the currently available dual-modal contrast agents, multifunctional nanoprobes combining MRI with fluorescence imaging capacities have received the most extensive attention for glioma imaging, due to the integrated advantages of high spatial resolution and high imaging sensitivity. These nanoprobes mainly include Cy5.5-labled and lactoferrin-conjugated magnetic nanogels (Cy5.5-Lf-MPNA nanogels) [53], angiopep-2-conjugated PEGylated upconversion nanoparticles (ANG/PEG-UCNPs) [50], $Gd^{3+}$-doped polymer-coated carbon nanodots (NCD-DTPA-Gd) [54], Gd-DOTA and Dylight680-incorporated G5 dendrimers (Gd-G5-DL680) [55], and PEG-Cy5.5-modified MnO nanoparticles (MnO-PEG-Cy5.5 NPs) [56]. Other dual-modal imaging nanoprobes toward glioma include gold and SPION-loaded micelles (GSMs) for MRI/CT [57], SapC-DOPS nanovesicles for PET/FLI [58], cRGD peptide-decorated aggregation-induced emission dots (TB1-RGD dots) for FLI/PAI [51], and 64Cu-labeled and c(KRGDf)-functionalized hollow gold nanospheres for PAT/PET [39]. All these nanoprobes were able to penetrate the BBB, resulting in clear tumor visualization in orthotopic glioma models. And good co-registration images were acquired by using the combination of different imaging modalities.

To obtain more abundant and synergetic information, several triple-modal imaging nanoprobes have also been developed for glioma imaging. For example, Shang et al. synthesized PEGylated core-shell-structured Au@MIL-88(Fe) nanocomposites with starlike morphology via a simple and robust microemulsion method, through controllable growth of metal-organic framework (MOF) shell layer onto gold nanorod core in N,N-dimethylformamide phase [59]. The as-designed Au@MIL-88(Fe) possess PA and CT contrast properties from the Au core and MR-negative contrast property from the MOF shell simultaneously. After intravenous injection for 12 h, the nanoprobes showed high-contrast performance for CT/PAI/MRI triple-modality imaging in orthotopic U87 MG glioma-bearing mice. The mild synthetic condition and compositional tunability of the nanocomposites make them great potential in clinical settings.

Inevitably, each imaging technique aforementioned must be realized on separated equipment, owing to the default of integrated multimodal imaging machines. The impossible image fusion and complicated practical operation greatly hamper their further application. One feasible solution is to develop MRI-based dual-modal contrast agents and perform sequential scanning on a single MR scanner, which is highly attractive and remains challenging. To overcome the IONP-induced negative contrast limitations and obtain synergetic information, PEGylated gadolinium-doped iron oxide nanoparticles (PEG-GdIO NPs) [60], chitosan-dextran superparamagnetic nanoparticles (CS-DXSPIONs) [61], and ferromagnetic $Fe_{0.6}Mn_{0.4}O$ nanoflowers (FIMO-NFs) [62] have been designed. These hybrid nanoprobes possessed suitable longitudinal relaxivity and transverse relaxivity, indicating their applicability as $T_1$–$T_2$ dual-modal MR contrast agents. Compared with the baseline images, in vivo studies showed positive contrast on $T_1$-weighted MR images and simultaneous negative contrast on $T_2$-weighted MR images within the tumor region. Recently, MR contrast agents integrating anatomic and functional imaging capability have gained extensive interest as they can acquire abundant complementary information and facilitate the understanding of pathological processes. For instance, ultrasmall core-shell-structured $NaGdF_4$@PLL nanodots were designed as novel $T_1$/chemical exchange saturation transfer (CEST) dual-modal MR contrast agents (Fig. 5.5) [63]. The Gd core and the poly-L-lysine shell contributed to high $r_1$ relaxivity (6.42 $mM^{-1}s^{-1}$) and sensitive CEST effect at +3.7 ppm, respectively. In vivo MRI studies verified the effectivity and specificity of $NaGdF_4$@PLL nanodots for glioma imaging and pH mapping.

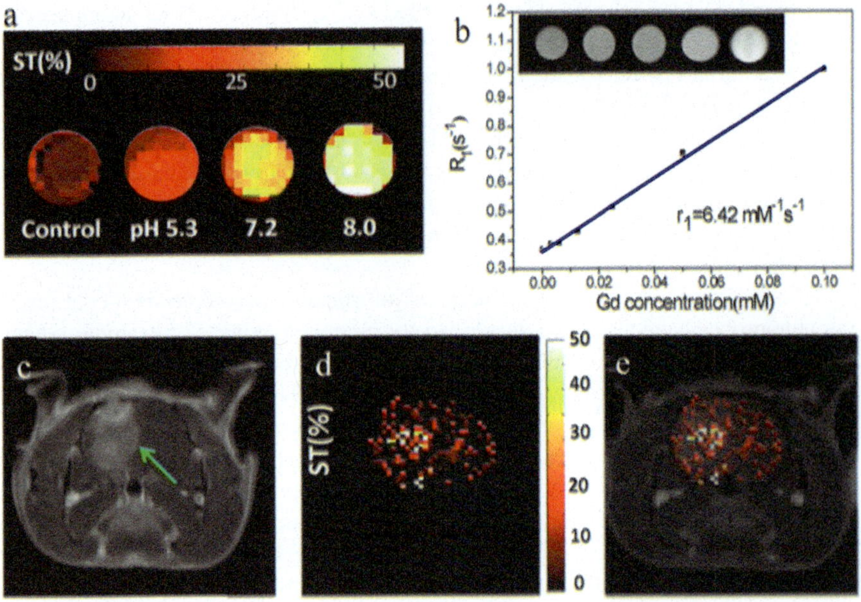


**Fig. 5.5 Ultrasmall core-shell-structured NaGdF4@PLL nanodots for $T_1$/CEST dual-modal imaging of glioma.** (**a**) CEST phantom images of NaGdF4@PLL nanodots solution with different pH. (**b**) $r_1$ relaxivity of NaGdF4@PLL nanodots based on gadolinium concentrations. (**c**) In vivo $T_1$-weighted MR images of glioma after intravenous injection of NaGdF4@PLL nanodots. (**d**) CEST ST difference map between pre- and postinjection at 3.0 μT. (**e**) The merged image of Figure (**c**) and Figure (**d**). (Adapted with permission from Ref. [63])

## 5.5.5 *Intraoperative Glioma Margin Delineation*

Difficulty in intraoperative visualization of infiltrating glioma margins remains a crucial challenge in achieving precise complete surgical resection and thus improved clinical outcomes. Raman imaging holds great potential in guiding tumor resection owing to its ultrahigh-sensitivity, high-resolution, and fingerprint-like spectra. To date, several Raman imaging-incorporated hybrid nanoprobes have been developed for glioma margin delineation and guiding tumor resection. Kircher and coworkers first designed gold-silica-based SERS nanoparticles modified with DOTA-$Gd^{3+}$ (MPRs) as an ultrasensitive PA/MR/Raman triple-modal imaging nanoprobe [67]. After intravenous injection into glioma-bearing mice, the as-developed MPRs led to efficient accumulation and retention within the tumor, enabling noninvasive and accurate delineation of brain glioma margins by these three modalities with picomolar sensitivity. Notably, the histological analysis confirmed that the clear tumor margins delineated by Raman imaging realized precise intraoperative tumor resection (Fig. 5.6). Subsequently, this group developed novel gold-silica nanostars as surface-enhanced resonance Raman scattering (SERRS) and multispectral optoacoustic tomography (MSOT) dual-modal contrast agents for intraoperative

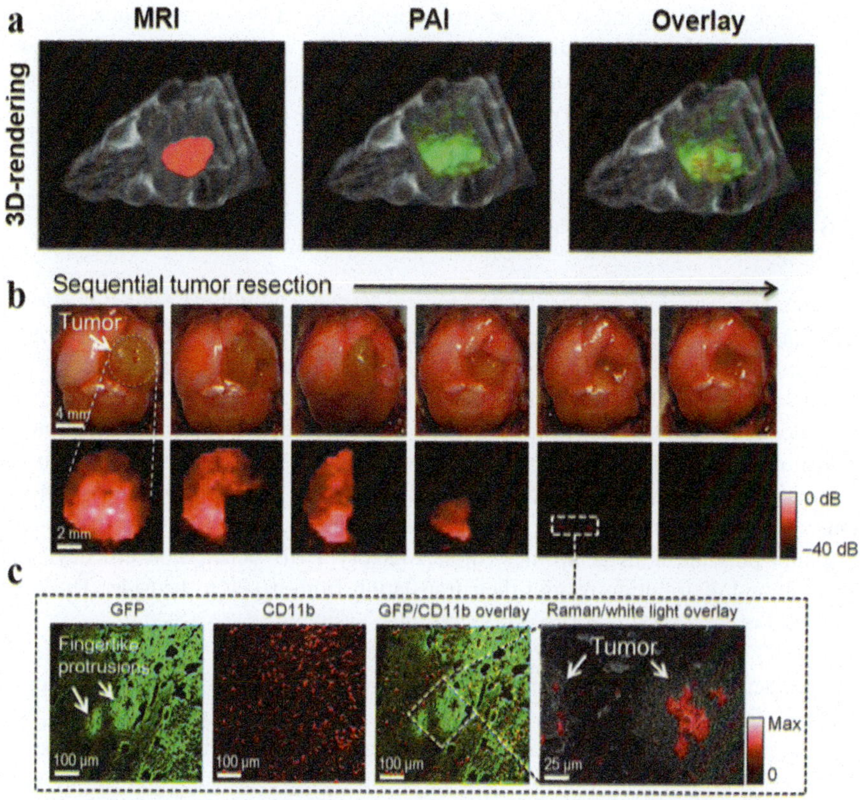


**Fig. 5.6 Triple-modality MRI-Photoacoustic-Raman nanoparticles (MPRs) for delineating brain tumor margins.** (**a**) 3D-rendering of MR, PA, and overlay images of glioma. (**b**) Quarters of the tumor tissue were removed sequentially, and intraoperative Raman imaging was performed after each resection step, until the entire tumor was removed via visual inspection. Several tiny foci of Raman signals were detected in the resection bed (white dashed box). (**c**) Histological analysis of these foci showed an infiltrative pattern, appearing as fingerlike protrusions into surrounding tissues. The Raman signal within these protrusions suggests the specific and selective localization of MPRs. (Adapted with permission from Ref. [67])

tumor delineation in a transgenic mouse glioma model [66]. After intravenous injection of nanostars, MSOT could accurately delineated the extent of infiltrating tumor with high specificity, which correlated well with the SERRS signals. Similarly, Gao et al. designed a pair of gold nanoprobes modified with BBB-permeable angiopep-2 peptide, Au-AZ, and Au-AK [65]. After the nanoprobes enter the tumor interstitium via LRP1-involved receptor-mediated transcytosis, the acidic tumor environment etches the outer shielding layer to expose the azide and alkyne groups, triggers the self-assembly into spherical nanoclusters, and induces remarkable activation of both SERRS and MR signals. They could not only define glioma via MRI with high sensitivity preoperatively but also could guide tumor excision intraoperatively using a handheld Raman scanner, confirmed by histological analysis. Considering the satisfactory excretion behavior of the nanoprobes and the universal extracellular acidification of the tumor environment, these acidic environment-responsive nanoprobes hold a great promise in guiding glioma surgical resection with high specificity and high safety.

## 5.6 Challenges and Perspectives

Conventional contrast agents directed toward intracranial glioma remain challenging, largely attributed to the existence of two critical physiopathologic barriers unique to the brain tumors. In addition, systemic-administrated nanomaterials are always susceptible to the non-specific opsonization processes and elimination by the reticuloendothelial system. In this chapter, the strategies for overcoming the physiopathologic barriers and the applications of versatile nanoprobes for improved glioma visualization are systematically summarized. Despite tremendous efforts and considerable progresses in nanoprobes-mediated glioma imaging, the uncontrollable pharmacokinetic profiles, dissatisfactory BBB penetration efficacy, and intratumoral distribution restrict their translation from bench to bedside. The concerns about long-term biodistribution, possible systemic adverse events and immunological response, exact degradation and metabolic routes, and the elimination mechanisms remain unclear. Much detailed investigations are required in order to push the field forward.

Currently, one major technological challenge on multifunctional nanoprobes is large-scale production with fine stability and good reproducibility in a well-controlled manner. To this end, novel green synthesis approaches with biocompatible precursors, mild synthetic conditions, and less procedures are urgently needed. Multifunctional nanoprobes need to be carefully optimized because multifarious modification may change their physicochemical properties and pharmacokinetic profiles. In-depth understanding of the interactions between different components on one nanostructure is gaining increasing interests in the design of multifunctional nanoprobes. For nanoprobes with multimodal imaging abilities, the impact of one component on the contrast performance of another should be carefully investigated in the integrated design. Moreover, the imaging sensitivity and detection thresholds for each imaging modality should be taken into consideration to ensure their imaging performance.

Ideal nanoprobes should be effective enough to cross or bypass the physiopathologic barriers and target glioma. Both glioma and the physiological barriers are highly heterogeneous in nature, preventing the homogeneous and widespread intratumoral distribution of nanoprobes. Exclusively targeting tumor-specific biomarkers remains challenging since these tumor cell-associated markers may be present on normal cells to some extent as well. Going forward, more potent targeting modules should be routinely incorporated based on the continuously changing molecular characteristics of glioma. A new generation of high-throughput microarray-based tumor gene and protein expression profiles will facilitate the identification of novel whole-process overexpressed biomarkers toward glioma. Stimulus-responsive nanoprobes with multi-targeting and multimodal imaging capacities will realize more personalized glioma delineation and more accurate therapeutic assessment. Further evaluation and standardization through interdisciplinary collaborations are required before being translated into clinical practice. Despite in its infancy, nanoprobes-mediated neuroimaging hold immense potential in making exciting breakthroughs for early and precise visualization of intracranial glioma.

**Acknowledgments** This work was supported by the Major State Basic Research Development Program of China (2017YFA0205201), the National Natural Science Foundation of China (81901872, 81422023, U1705281, and U1505221), the Fundamental Research Funds for the Central Universities (20720160065 and 20720150141), and the Program for New Century Excellent Talents in University, China (NCET-13-0502).

## References

1. Nano R, Lascialfari A, Corti M, Paolini A, Pasi F, Corbella F, et al. New frontiers for astrocytic tumours. Anticancer Res. 2012;32(7):2755–8.
2. Wen PY, Kesari S. Malignant gliomas in adults. N Engl J Med. 2008;359(5):492–507.
3. Linz U. Commentary on effects of radiotherapy with concomitant and adjuvant temozolomide versus radiotherapy alone on survival in glioblastoma in a randomised phase III study: 5-year analysis of the EORTC-NCIC trial (Lancet Oncol. 2009;10:459–466). Cancer. 2010;116 (8):1844–6.
4. Scott JN, Brasher PM, Sevick RJ, Rewcastle NB, Forsyth PA. How often are nonenhancing supratentorial gliomas malignant? A population study. Neurology. 2002;59(6):947–9.
5. Prince MR, Zhang H, Morris M, MacGregor JL, Grossman ME, Silberzweig J, et al. Incidence of nephrogenic systemic fibrosis at two large medical centers. Radiology. 2008;248(3):807–16.
6. Quattrocchi CC, van der Molen AJ. Gadolinium retention in the body and brain: is it time for an international joint research effort? Radiology. 2017;282(1):12–6.
7. Sun C, Lee JS, Zhang M. Magnetic nanoparticles in MR imaging and drug delivery. Adv Drug Deliv Rev. 2008;60(11):1252–65.
8. Cardoso FL, Brites D, Brito MA. Looking at the blood-brain barrier: molecular anatomy and possible investigation approaches. Brain Res Rev. 2010;64(2):328–63.
9. Liu H, Zhang J, Chen X, Du XS, Zhang JL, Liu G, et al. Application of iron oxide nanoparticles in glioma imaging and therapy: from bench to bedside. Nanoscale. 2016;8(15):7808–26.
10. Lesniak MS, Brem H. Targeted therapy for brain tumours. Nat Rev Drug Discov. 2004;3 (6):499–508.
11. Pardridge WM. Crossing the blood-brain barrier: are we getting it right? Drug Discov Today. 2001;6(1):1–2.
12. Pardridge WM. Drug and gene delivery to the brain: the vascular route. Neuron. 2002;36 (4):555–8.
13. Koo YE, Reddy GR, Bhojani M, Schneider R, Philbert MA, Rehemtulla A, et al. Brain cancer diagnosis and therapy with nanoplatforms. Adv Drug Deliv Rev. 2006;58(14):1556–77.
14. Vajkoczy P, Menger MD. Vascular microenvironment in gliomas. J Neuro-Oncol. 2000;50 (1–2):99–108.
15. Liu Y, Lu W. Recent advances in brain tumor-targeted nano-drug delivery systems. Expert Opin Drug Deliv. 2012;9(6):671–86.

16. Sarin H, Kanevsky AS, Wu H, Brimacombe KR, Fung SH, Sousa AA, et al. Effective transvascular delivery of nanoparticles across the blood-brain tumor barrier into malignant glioma cells. J Transl Med. 2008;6:80.
17. Furtado D, Bjornmalm M, Ayton S, Bush AI, Kempe K, Caruso F. Overcoming the blood-brain barrier: the role of nanomaterials in treating neurological diseases. Adv Mater. 2018:e1801362.
18. Liu H, Chen X, Xue W, Chu C, Liu Y, Tong H, et al. Recombinant epidermal growth factor-like domain-1 from coagulation factor VII functionalized iron oxide nanoparticles for targeted glioma magnetic resonance imaging. Int J Nanomedicine. 2016;11:5099–108.
19. Liu H, Chu C, Liu Y, Pang X, Wu Y, Zhou Z, et al. Novel Intrapolymerization doped manganese-eumelanin coordination nanocomposites with ultrahigh relaxivity and their application in tumor theranostics. Adv Sci. 2018; https://doi.org/10.1002/advs.201800032.
20. Cao C, Wang X, Cai Y, Sun L, Tian L, Wu H, et al. Targeted in vivo imaging of microscopic tumors with ferritin-based nanoprobes across biological barriers. Adv Mater. 2014;26 (16):2566–71.
21. Chen N, Shao C, Qu Y, Li S, Gu W, Zheng T, et al. Folic acid-conjugated MnO nanoparticles as a T1 contrast agent for magnetic resonance imaging of tiny brain gliomas. ACS Appl Mater Interfaces. 2014;6(22):19850–7.
22. Yan H, Wang L, Wang J, Weng X, Lei H, Wang X, et al. Two-order targeted brain tumor imaging by using an optical/paramagnetic nanoprobe across the blood brain barrier. ACS Nano. 2012;6(1):410–20.
23. Zhang J, Chen N, Wang H, Gu W, Liu K, Ai P, et al. Dual-targeting superparamagnetic iron oxide nanoprobes with high and low target density for brain glioma imaging. J Colloid Interface Sci. 2016;469:86–92.
24. Jiang W, Xie H, Ghoorah D, Shang Y, Shi H, Liu F, et al. Conjugation of functionalized SPIONs with transferrin for targeting and imaging brain glial tumors in rat model. PLoS One. 2012;7(5):e37376.
25. Luo B, Wang S, Rao R, Liu X, Xu H, Wu Y, et al. Conjugation magnetic PAEEP-PLLA nanoparticles with Lactoferrin as a specific targeting MRI contrast agent for detection of brain Glioma in rats. Nanoscale Res Lett. 2016;11(1):227.
26. Xie H, Zhu Y, Jiang W, Zhou Q, Yang H, Gu N, et al. Lactoferrin-conjugated superparamagnetic iron oxide nanoparticles as a specific MRI contrast agent for detection of brain glioma in vivo. Biomaterials. 2011;32(2):495–502.
27. Abakumov MA, Shein SA, Vishvasrao H, Nukolova NV, Sokol'ski-Papkov M, Sandalova TO, et al. Visualization of experimental glioma C6 by MRI with magnetic nanoparticles conjugated with monoclonal antibodies to vascular endothelial growth factor. Bull Exp Biol Med. 2012;154 (2):274–7.
28. Towner RA, Smith N, Asano Y, He T, Doblas S, Saunders D, et al. Molecular magnetic resonance imaging approaches used to aid in the understanding of angiogenesis in vivo: implications for tissue engineering. Tissue Eng Part A. 2010;16(2):357–64.
29. de Oliveira EA, Lazovic J, Guo L, Soto H, Faintuch BL, Akhtari M, et al. Evaluation of Magnetonanoparticles conjugated with new angiogenesis peptides in intracranial Glioma tumors by MRI. Appl Biochem Biotechnol. 2017;183(1):265–79.
30. Richard S, Boucher M, Lalatonne Y, Meriaux S, Motte L. Iron oxide nanoparticle surface decorated with cRGD peptides for magnetic resonance imaging of brain tumors. Biochim Biophys Acta Gen Subj. 2017;1861(6):1515–20.
31. Tomanek B, Iqbal U, Blasiak B, Abulrob A, Albaghdadi H, Matyas JR, et al. Evaluation of brain tumor vessels specific contrast agents for glioblastoma imaging. Neuro-Oncology. 2012;14(1):53–63.
32. Shevtsov MA, Nikolaev BP, Yakovleva LY, Marchenko YY, Dobrodumov AV, Mikhrina AL, et al. Superparamagnetic iron oxide nanoparticles conjugated with epidermal growth factor (SPION-EGF) for targeting brain tumors. Int J Nanomedicine. 2014;9:273–87.

33. Mu K, Zhang S, Ai T, Jiang J, Yao Y, Jiang L, et al. Monoclonal antibody-conjugated superparamagnetic iron oxide nanoparticles for imaging of epidermal growth factor receptor-targeted cells and gliomas. Mol Imaging. 2015;14
34. Shevtsov MA, Yakovleva LY, Nikolaev BP, Marchenko YY, Dobrodumov AV, Onokhin KV, et al. Tumor targeting using magnetic nanoparticle Hsp70 conjugate in a model of C6 glioma. Neuro-Oncology. 2014;16(1):38–49.
35. Shevtsov MA, Nikolaev BP, Yakovleva LY, Dobrodumov AV, Zhakhov AV, Mikhrina AL, et al. Recombinant interleukin-1 receptor antagonist conjugated to superparamagnetic iron oxide nanoparticles for theranostic targeting of experimental glioblastoma. Neoplasia. 2015;17(1):32–42.
36. Perera VS, Covarrubias G, Lorkowski M, Atukorale P, Rao A, Raghunathan S, et al. One-pot synthesis of nanochain particles for targeting brain tumors. Nanoscale. 2017;9(27):9659–67.
37. Qiu LH, Zhang JW, Li SP, Xie C, Yao ZW, Feng XY. Molecular imaging of angiogenesis to delineate the tumor margins in glioma rat model with endoglin-targeted paramagnetic liposomes using 3T MRI. J Magn Reson Imaging: JMRI. 2015;41(4):1056–64.
38. Liu X, Madhankumar AB, Miller PA, Duck KA, Hafenstein S, Rizk E, et al. MRI contrast agent for targeting glioma: interleukin-13 labeled liposome encapsulating gadolinium-DTPA. Neuro-Oncology. 2016;18(5):691–9.
39. Lu W, Melancon MP, Xiong C, Huang Q, Elliott A, Song S, et al. Effects of photoacoustic imaging and photothermal ablation therapy mediated by targeted hollow gold nanospheres in an orthotopic mouse xenograft model of glioma. Cancer Res. 2011;71(19):6116–21.
40. Chen J, Liu C, Hu D, Wang F, Wu H, Gong X, et al. Single-layer MoS2 nanosheets with amplified photoacoustic effect for highly sensitive photoacoustic imaging of orthotopic brain tumors. Adv Funct Mater. 2016; https://doi.org/10.1002/adfm.201603758.
41. Liu C, Chen J, Zhu Y, Gong X, Zheng R, Chen N, et al. Highly sensitive MoS2-Indocyanine green hybrid for Photoacoustic imaging of Orthotopic brain Glioma at deep site. Nano-Micro Lett. 2018;10(3):48.
42. Fan Q, Cheng K, Yang Z, Zhang R, Yang M, Hu X, et al. Perylene-diimide-based nanoparticles as highly efficient photoacoustic agents for deep brain tumor imaging in living mice. Adv Mater. 2015;27(5):843–7.
43. Guo B, Sheng Z, Hu D, Liu C, Zheng H, Liu B. Through scalp and skull NIR-II photothermal therapy of deep Orthotopic brain tumors with precise photoacoustic imaging guidance. Adv Mater. 2018;30(35):e1802591.
44. Nurmikko A. What future for quantum dot-based light emitters? Nat Nanotechnol. 2015;10 (12):1001–4.
45. Tang J, Huang N, Zhang X, Zhou T, Tan Y, Pi J, et al. Aptamer-conjugated PEGylated quantum dots targeting epidermal growth factor receptor variant III for fluorescence imaging of glioma. Int J Nanomedicine. 2017;12:3899–911.
46. Huang N, Cheng S, Zhang X, Tian Q, Pi J, Tang J, et al. Efficacy of NGR peptide-modified PEGylated quantum dots for crossing the blood-brain barrier and targeted fluorescence imaging of glioma and tumor vasculature. Nanomedicine. 2017;13(1):83–93.
47. Wang Y, Meng Y, Wang S, Li C, Shi W, Chen J, et al. Direct solvent-derived polymer-coated nitrogen-doped carbon Nanodots with high water solubility for targeted fluorescence imaging of Glioma. Small. 2015;11(29):3575–81.
48. Zheng M, Ruan S, Liu S, Sun T, Qu D, Zhao H, et al. Self-targeting fluorescent carbon dots for diagnosis of brain Cancer cells. ACS Nano. 2015;9(11):11455–61.
49. Prodi L, Rampazzo E, Rastrelli F, Speghini A, Zaccheroni N. Imaging agents based on lanthanide doped nanoparticles. Chem Soc Rev. 2015;44(14):4922–52.
50. Ni D, Zhang J, Bu W, Xing H, Han F, Xiao Q, et al. Dual-targeting upconversion nanoprobes across the blood-brain barrier for magnetic resonance/fluorescence imaging of intracranial glioblastoma. ACS Nano. 2014;8(2):1231–42.

51. Sheng Z, Guo B, Hu D, Xu S, Wu W, Liew WH, et al. Bright aggregation-induced-emission dots for targeted synergetic NIR-II fluorescence and NIR-I Photoacoustic imaging of Orthotopic brain tumors. Adv Mater. 2018;30:e1800766.
52. Louie A. Multimodality imaging probes: design and challenges. Chem Rev. 2010;110 (5):3146–95.
53. Jiang L, Zhou Q, Mu K, Xie H, Zhu Y, Zhu W, et al. pH/temperature sensitive magnetic nanogels conjugated with Cy5.5-labled lactoferrin for MR and fluorescence imaging of glioma in rats. Biomaterials. 2013;34(30):7418–28.
54. Du Y, Qian M, Li C, Jiang H, Yang Y, Huang R. Facile marriage of Gd(3+) to polymer-coated carbon nanodots with enhanced biocompatibility for targeted MR/fluorescence imaging of glioma. Int J Pharm. 2018;552(1–2):84–90.
55. Gonawala S, Ali MM. Application of Dendrimer-based nanoparticles in Glioma imaging. J Nanomed Nanotechnol. 2017;8(3):444.
56. Chen N, Shao C, Li S, Wang Z, Qu Y, Gu W, et al. Cy5.5 conjugated MnO nanoparticles for magnetic resonance/near-infrared fluorescence dual-modal imaging of brain gliomas. J Colloid Interface Sci. 2015;457:27–34.
57. Sun L, Joh DY, Al-Zaki A, Stangl M, Murty S, Davis JJ, et al. Theranostic application of mixed gold and Superparamagnetic Iron Oxide nanoparticle micelles in glioblastoma Multiforme. J Biomed Nanotechnol. 2016;12(2):347–56.
58. Blanco VM, Chu Z, LaSance K, Gray BD, Pak KY, Rider T, et al. Optical and nuclear imaging of glioblastoma with phosphatidylserine-targeted nanovesicles. Oncotarget. 2016;7 (22):32866–75.
59. Shang W, Zeng C, Du Y, Hui H, Liang X, Chi C, et al. Core-Shell gold Nanorod@Metal-Organic framework nanoprobes for multimodality diagnosis of Glioma. Adv Mater. 2017;29(3). https://doi.org/10.1002/adma.201603917
60. Xiao N, Gu W, Wang H, Deng Y, Shi X, Ye L. T1-T2 dual-modal MRI of brain gliomas using PEGylated Gd-doped iron oxide nanoparticles. J Colloid Interface Sci. 2014;417:159–65.
61. Shevtsov M, Nikolaev B, Marchenko Y, Yakovleva L, Skvortsov N, Mazur A, et al. Targeting experimental orthotopic glioblastoma with chitosan-based superparamagnetic iron oxide nanoparticles (CS-DX-SPIONs). Int J Nanomedicine. 2018;13:1471–82.
62. Liu XL, Ng CT, Chandrasekharan P, Yang HT, Zhao LY, Peng E, et al. Synthesis of ferromagnetic Fe0.6 Mn0.4 O Nanoflowers as a new class of magnetic Theranostic platform for in vivo T1 -T2 dual-mode magnetic resonance imaging and magnetic hyperthermia therapy. Adv Healthc Mater. 2016;5(16):2092–104.
63. Ni D, Shen Z, Zhang J, Zhang C, Wu R, Liu J, et al. Integrating anatomic and functional dual-mode magnetic resonance imaging: design and applicability of a bifunctional contrast agent. ACS Nano. 2016;10(3):3783–90.
64. Yang L, Shao B, Zhang X, Cheng Q, Lin T, Liu E. Multifunctional upconversion nanoparticles for targeted dual-modal imaging in rat glioma xenograft. J Biomater Appl. 2016;31(3):400–10.
65. Gao X, Yue Q, Liu Z, Ke M, Zhou X, Li S, et al. Guiding brain-tumor surgery via blood-brain-barrier-permeable gold Nanoprobes with acid-triggered MRI/SERRS signals. Adv Mater. 2017 Jun;29(21). https://doi.org/10.1002/adma.201603917. Epub 2017 Mar 15.
66. Neuschmelting V, Harmsen S, Beziere N, Lockau H, Hsu HT, Huang R, et al. Dual-modality surface-enhanced resonance Raman scattering and multispectral Optoacoustic tomography nanoparticle approach for brain tumor delineation. Small. 2018;14(23):e1800740.
67. Kircher MF, de la Zerda A, Jokerst JV, Zavaleta CL, Kempen PJ, Mittra E, et al. A brain tumor molecular imaging strategy using a new triple-modality MRI-photoacoustic-Raman nanoparticle. Nat Med. 2012;18(5):829–34.

# Chapter 6
# The Advances of Nanozyme in Brain Disease

**Ruofei Zhang, Xiyun Yan, and Kelong Fan**

**Abstract** Reactive oxygen species (ROS), a class of metabolites produced in biological aerobic metabolism, play a key role in conducting cellular signals and maintaining normal nerve functions in central nerve system. However, under some pathological conditions, oxidative stress caused by excessive ROS may become an important factor in the occurrence and deterioration of neurological disease. Therefore, it is crucial to control the level of ROS in the central nervous system in time. Traditional ROS regulators, such as some natural enzymes and assemblies based on them, have not been well applied in brain diseases due to their instability and limited ability to cross the blood–brain barrier (BBB). Nanozymes, stable inorganic nanomaterials that possess intrinsic enzyme-mimic activities, have attracted wide attention in the scientific community in recent years attributed to their efficient ability to alleviate oxidative stress in the central nervous system. This chapter reviews the advances in the application of nanozymes in the treatment of neurological diseases and discusses the challenges and perspectives of nanozymes for their clinical translations. We hope that this chapter will arouse readers' interest in nanozymes and promote the research and application of nanozymes in brain diseases.

**Keywords** Nanozyme · Central nervous system disease · Neurological disorders · Antioxidative therapy

## 6.1 Introduction

Inflammation induced by oxidative stress and oxidative damage has been considered as the key factor in the occurrence and development of brain injury, aging, or chronic neurodegeneration including Alzheimer's disease (AD), Parkinson's disease (PD),

R. Zhang · X. Yan (✉) · K. Fan (✉)
CAS Engineering Laboratory for Nanozyme, Key Laboratory of Protein and Peptide Pharmaceutical, Institute of Biophysics, Chinese Academy of Sciences, Beijing, China
e-mail: yanxy@ibp.ac.cn; fankelong@ibp.ac.cn



X. Xue (ed.), *Nanomedicine in Brain Diseases*,
https://doi.org/10.1007/978-981-13-8731-9_6

Huntington's disease (HD), multiple sclerosis (MS), amyotrophic lateral sclerosis (ALS), and cerebral ischemia (CI) [1, 2]. Oxidative stress has been defined as "an imbalance in pro-oxidants and antioxidants with associated disruption of redox circuitry and macromolecular damage" [3], in which reactive oxygen species (ROS) play central roles. ROS in cells, mainly including superoxide anion ($O_2^{\bullet -}$), hydrogen peroxide ($H_2O_2$), hydroxyl radical (•OH), and singlet oxygen ($^1O_2$), are usually produced by both exogenous and endogenous sources [4]. The sources of exogenous ROS include ultraviolet (UV), ionizing radiation, pollution, smoke, and other factors [5], while endogenous ROS are mainly produced by mitochondrial electron transport chain (ETC) and NADPH oxidase (NOX) system [2]. Although oxygen is lethal to many life forms and almost all cell macromolecules (e.g., protein, lipid, and nucleic acid) may undergo oxidative damage, aerobic organisms have evolved the ability to use ROS derived from oxygen for efficient energy transfer. In aerobic organisms, appropriate ROS such as $H_2O_2$ can be used as cell signal transduction messenger to regulate the body's function based on their fast-created and short-lived characteristics, while redundant ROS can be removed by antioxidant systems in a timely and effective manner [6]. However, under some pathological conditions, oxidative stress produced by high-level ROS exceeds the threshold that the antioxidant system could withstand; then these aerobic cells will be damaged. In addition, because of its high frequency of aerobic metabolism, the brain is most sensitive to oxidative stress-induced damage in all organs [2]. Therefore, the use of appropriate antioxidants to alleviate oxidative stress in time is essential for the treatment of brain diseases.

Currently, many antioxidants, such as some natural enzymes, small biomolecules, and the synthetic composites based on them, have been used in brain disease-related research. Most of these antioxidants are difficult to be used in clinical research mainly because of their low ROS-scavenging efficiency, instability, toxicity, or inadequate penetration of blood–brain barrier (BBB). Besides, although a few antioxidative drugs (e.g., edaravone, idebenone, dimethyl fumarate) have been approved for treating human neurological disease, most if not all of them have not achieved ideal therapeutic effect [2]. Consequently, new kinds of antioxidants are urgently needed for the treatment of neurological diseases.

Unlike traditional antioxidants, which are mainly composed of organic components, nanozymes are a class of inorganic nanomaterials with intrinsic enzyme-like activities. $Fe_3O_4$ nanoparticles are the first nanozymes found to possess enzyme-like activity, which have the ability to catalyze the oxidation of the substrates of horseradish peroxidase (HRP) in the presence of $H_2O_2$, and their catalytic mechanism and efficiency are all similar to those of HRP [7]. Since then, a series of inorganic nanomaterials have been found to possess intrinsic enzyme-like activities. These nanomaterials perform efficient valence conversion and electron transfer on their surfaces, which determines that the reactions catalyzed by them are most redox-related. That is to say, these nanozymes mainly mimic some redox-related enzymes, such as superoxide dismutase (SOD), catalase, peroxidase, and oxidase [8]. Because SOD and catalase play important/vital roles in scavenging ROS in the brain [2, 9], the nanozymes which possess SOD-like or catalase-like activities (e.g., $CeO_2$, fullerene, $Mn_3O_4$, $Fe_3O_4$, etc.) have stimulated researchers' interest in using them to treat ROS-related brain disorders. Compared with traditional antioxidants, these nanozymes own the advantages of stable structure, adjustable activity, and diverse functions. At present, many nanozymes have been found to effectively protect nerve cells from oxidative damage and alleviate neurological disorders [8, 10, 11], indicating that nanozymes have great potential in the treatment of ROS-related brain diseases.

In order to give readers a deeper understanding of the role of nanozymes in the treatment of neurological diseases, this chapter first introduces the relationship between ROS and neurological disorders and the mechanism of nanozymes in clearing ROS. Then, the research and application of nanozymes in the treatment of neurological diseases are systematically reviewed. Finally, the challenges and perspectives are discussed.

## 6.2 ROS-Mediated Oxidative Stress and Its Role in Neurological Diseases

Regardless of external sources, ROS in brain is mostly produced by ETC and NOX family. ETC consists of four complexes, including NADH-CoQ reductase (complex I), succinate dehydrogenase (complex II), $CoQH_2$-cytochrome c reductase (complex III), and cytochrome c oxidase (complex IV). The generation of various ROS mostly begins with the production of $O_2^{\bullet-}$, which are mainly produced by complex I and complex III [6]. Chance et al. demonstrated for the first time that ETC can produce ROS by detecting $H_2O_2$ in isolated mitochondria in 1966 [12]. Later, it was found that $H_2O_2$ in mitochondria was mainly produced from the dismutation of $O_2^{\bullet-}$ catalyzed by SOD [13, 14]. In addition, studies have shown that complex I plays a more important role in the production of $O_2^{\bullet-}$ in the brain than complex III [2, 15]. Under the normal function of mitochondria, only a small amount of $O_2^{\bullet-}$ is produced. In this case, NADH passes through complex I, CoQ, complex III, cytochrome c, and complex IV to transfer electrons to oxygen, and the protons are discharged into the mitochondrial membrane space to form ATP through ATP synthase (Fig. 6.1 MODE3) [6]. However, when there is a high $NADH/NAD^+$ ratio (Fig. 6.1 MODE1) or a high-proton motive force ($\Delta p$) and a reduced CoQ pool (Fig. 6.1 MODE2) in the mitochondria, significant $O_2^{\bullet-}$ production will be observed. In many acute or chronic neuronal disorders, notable amounts of $O_2^{\bullet-}$ are also produced through the NOX family [2]. The NOX family is a transmembrane complex composed of six subunits, including seven homologs (NOX1, NOX2, NOX3, NOX4, NOX5, DUOX1, and DUOX2), of which the NOX2 complex has been reported to be expressed in the corpus callosum, hippocampus, spinal cord, and medulla of human [16]. Activated NOX complex transfer electrons from the NADPH in the cytoplasm to extracellular space to form $O_2^{\bullet-}$ [17]. The NOX family maintains a certain activity at ordinary times, and when the extracellular stimulus

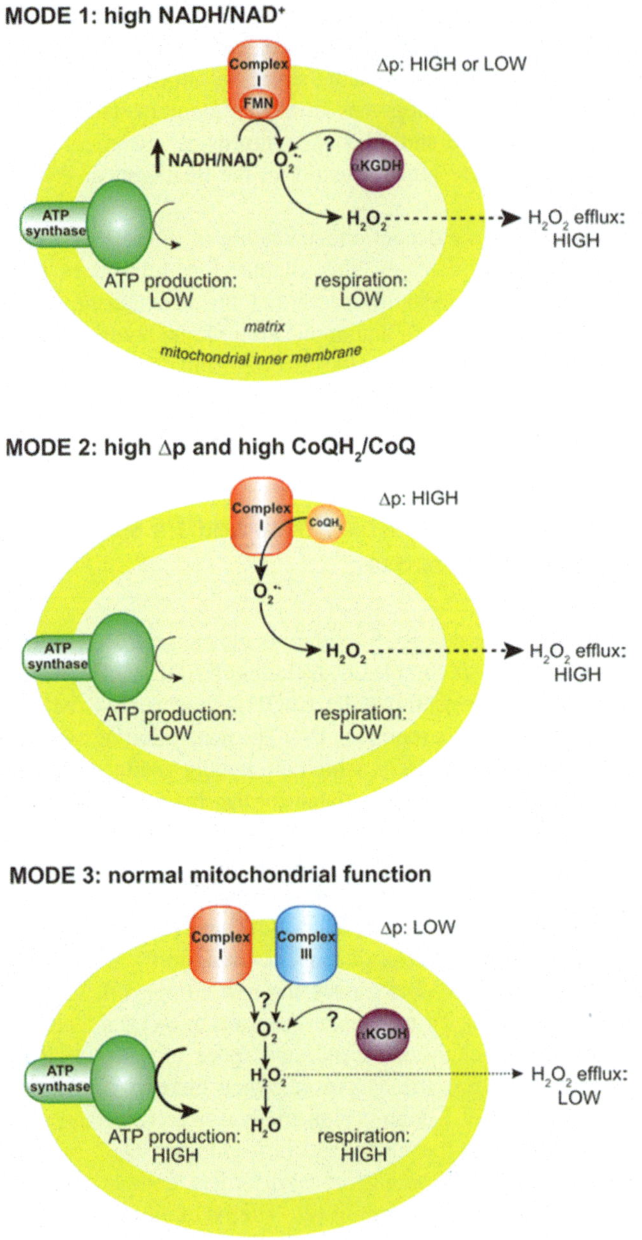


**Fig. 6.1** Modes of mitochondrial operation that lead to $O_2^{\bullet-}$ production [6]. There are three modes of mitochondrial operation that are associated with $O_2^{\bullet-}$ production. In mode 1, the NADH pool is reduced, for example, by damage to the respiratory chain, loss of cytochrome c during apoptosis, or low ATP demand. This leads to a rate of $O_2^{\bullet-}$ formation at the flavin mononucleotide (FMN) of complex I that is determined by the extent of FMN reduction which is in turn set by the NADH/ $NAD^+$ ratio. Other sites such as αKGDH may also contribute. In mode 2, there is no ATP

signal is sensed, the $O_2^{\bullet -}$ level can rapidly increase, resulting in high concentration of ROS in cells. In addition to the two main ROS sources mentioned above, ROS can also be generated by other enzymes in the central nervous system, such as xanthine oxidase/xanthine dehydrogenase (XD/XDH), myeloperoxidase (MPO), nitric oxide synthases (NOSs), cyclooxygenase (COX), xanthine oxidase (Xox), nitric oxide synthase (NOS), phospholipase $A_2$, and cytochrome P450s [1, 2, 18].

In addition to these ROS producers, there is also an antioxidant system in the body that can scavenge ROS. The intracellular antioxidant system consists of antioxidants enzymes (e.g., SOD, catalase, glutathione peroxidase (GPx), hemeoxygenase-1 (HO-1), and thioredoxin) and some nonenzymatic antioxidants (e.g., vitamin C (VC), vitamin E (VE), lipoic acid, and uric acid). Under physiological conditions, the production and removal of ROS are maintained at the level of steady-state equilibrium. When ROS is generated faster than the body's clearance ability, the balance of redox will be disturbed, which will lead to oxidative stress, resulting in the damage of many macromolecules (e.g., DNA, protein, and lipid), causing inflammation, and then lead to the occurrence and development of many neurological diseases (Fig. 6.2) [1].

Although the brain weighs only 2% of the body's weight, it consumes 20% of the body's oxygen [2]. Oxygen metabolism in the brain is very vigorous, so even low levels of oxidative stress in the brain can lead to serious consequences [1]. Oxidative stress and related inflammation have been proved to be closely related to a variety of neurodegenerative diseases (e.g., AD, PD) and cerebrovascular disorders (e.g., stroke, CI). In these diseases, excessive ROS can directly cause oxidative damage to intracellular components or mediate inflammation by activating some signaling pathways. Postmortem tissues of patients with AD and PD have been reported to show oxidative and related inflammatory reactions [19]. The damaged brain areas showed activation of astrocytes and microglia, and high expression of pro-inflammatory cytokines such as TNF-α or IL-1B, IL-6, and IL-8, suggesting the importance of oxidation and related inflammatory reactions in these diseases. In the process of inflammation, the main signal pathways mediated by ROS are receptor tyrosine kinases (RTKs), phosphoinositide-3'-kinase (PI3K)/Akt cascade, transcription factors (e.g., activator protein-1 (AP-1), nuclear factor-κB (NF-κB)), and mitogen-activated protein kinases (MAPKs) [1]. Because the content of ROS-related signaling pathway is sophisticated and does not belong to the main content of this chapter, we will not introduce further information here. There are some related reviews for interested readers to refer to [1, 4, 16, 20]. The relationship between oxidative stress and some major related neurological diseases will be introduced separately in Sect. 6.4.

**Fig. 6.1** (continued) production and there is a high Δp and a reduced CoQ pool which leads to RET through complex I, producing large amounts of $O_2^{\bullet -}$. In mode 3, mitochondria are actively making ATP and consequently have a lower Δp than in mode 2 and a more oxidized NADH pool than in mode 1. Under these conditions, the flux of $O_2^{\bullet -}$ within mitochondria is far lower than in modes 1 and 2, and the $O_2^{\bullet -}$ sources are unclear. (Copyright 2009 The Author(s))

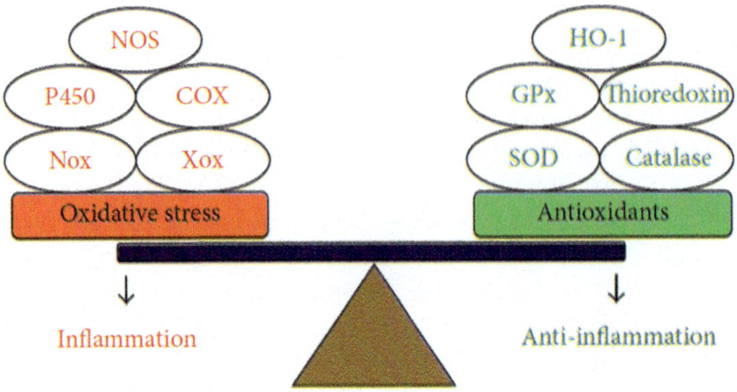


**Fig. 6.2** Oxidative stress and antioxidants imbalance in inflammation [1]. In inflammation, the balance appears to be tipped in favor of increased oxidative stress by various specialized enzymes, including NOX, Xox, P450, COX, or NOS, because of either excessive ROS release or inflammatory mediators leading to the amplification of the pro-inflammatory effects. In contrast, induction of several antioxidants, such as SOD, catalase, GPx, thioredoxin, or HO-1, may reduce ROS generation and attenuate the inflammatory response (anti-inflammation). *NOX* NADPH oxidase, *Xox* xanthine oxidase, *P450* P450 enzyme, *COX* cyclooxygenase, *NOS* nitric oxide synthase, *SOD* superoxide dismutase, *GPx* glutathione peroxidase, *HO-1* heme oxygenase-1. (Copyright 2013 The Author(s))

## 6.3 Nanozymes with Antioxidant Activity

As we mentioned earlier, many natural enzymes and natural biomolecules play important roles in the antioxidant system. Some of these natural organic molecules and some artificial synthetic complex based on these molecules have been used in the treatment of brain diseases. For example, in some studies, catalase and Cu/Zn-SOD have been encapsulated or coupled with some chemical polymers for the treatment of neurological diseases such as PD and CI [21–27]. Although these synthesized enzyme-mimetics have been reported to be effective in alleviating oxidative stress in the brain, they still have many shortcomings, such as poor operational stability, high environmental requirements for catalytic activity, complex preparation process, and high preparation cost [28]. In recent years, some inorganic nanomaterials have been found to have intrinsic enzyme-like catalytic activity without coupling natural enzymes. These nanomaterials have been widely called as nanozymes, which have high catalytic activity and can make up for the defects of traditional mimetic enzymes [8, 28, 29]. In this section, we will introduce the catalytic mechanisms of some typical nanozymes, focusing on their antioxidant ability to scavenge ROS. Although those natural enzyme-based nanoformulations are also called as nanozymes in some studies, they are not within the scope of our discussion in this chapter.

## 6.3.1 Nanozymes with SOD-Like Activity

Superoxide dismutase (SOD) is a kind of metal-containing enzyme widely existing in animals, plants, and microorganisms, which consecutively catalyzes the dismutation of the superoxide free radicals ($O_2^{\bullet-}$) to produce ordinary molecular oxygen ($O_2$) or hydrogen peroxide ($H_2O_2$). Superoxide free radicals are by-products of cellular aerobic metabolism, causing serious oxidative damage if not removed in time. Therefore, SOD plays a significant antioxidant role in organisms. There are three kinds of SOD enzymes in the human body: SOD1, SOD2, and SOD3. SOD1 and SOD3 contain copper and zinc, while SOD2 contains manganese [30]. The reaction catalyzed by copper-containing SOD is as follows [31]:

$$O_2^{\bullet-} + 2H^+ + (Cu^+) - SOD \rightarrow H_2O_2 + (Cu^{2+}) - SOD \tag{6.1}$$

$$O_2^{\bullet-} + (Cu^{2+}) - SOD \rightarrow O_2 + (Cu^+) - SOD \tag{6.2}$$

$$O_2^{\bullet-} + O_2^{\bullet-} + 2H^+ \rightarrow H_2O_2 + O_2 \text{ (overall reaction)} \tag{6.3}$$

Inspired by the catalytic mechanism of natural SOD enzymes, a series of SOD-like nanozymes with similar catalytic abilities have been found, including cerium dioxide ($CeO_2$ or nanoceria) nanoparticles, platinum (Pt) nanoparticles, carbon clusters, fullerenes ($C_{60}$), and their derivatives.

### 6.3.1.1 Nanoceria with SOD-Like Activity

Nanoceria have fluorite structure and variable $Ce^{4+}/Ce^{3+}$ valence pairs on its surface, which can easily and reversibly form a series of nonstoichiometric oxides ($CeO_{2-x}$, $0 < x < 0.5$) in reducing atmosphere [32]. These unique physical properties make nanoceria an efficient oxygen storage agent and catalyst. In industry, nanoceria has been used as one of the main components of three-way catalysts in automobile exhaust treatment [33]. Self et al. reported for the first time the SOD-like activity of nanoceria through a comparative experiment between nanoceria and cytochrome c [34]. Before that, the catalytic activity of nanoceria has been reported to be closely related to the oxygen vacancies and the $Ce^{3+}/Ce^{4+}$ ratio on its surface [35, 36]. Based on previous studies, Self et al. synthesized two kinds of nanoceria with different $Ce^{3+}/Ce^{4+}$ ratios and pointed out that the nanoceria with high $Ce^{3+}/Ce^{4+}$ has higher SOD-like activity [34]. In this study, the ability of nanoceria to scavenge $O_2^{\bullet-}$ was confirmed by electron paramagnetic resonance (EPR) measurements. The production of $H_2O_2$ was also observed, which is a product of SOD enzymatic reaction. In addition, the reaction kinetics experiments showed that the catalytic rate of nanoceria

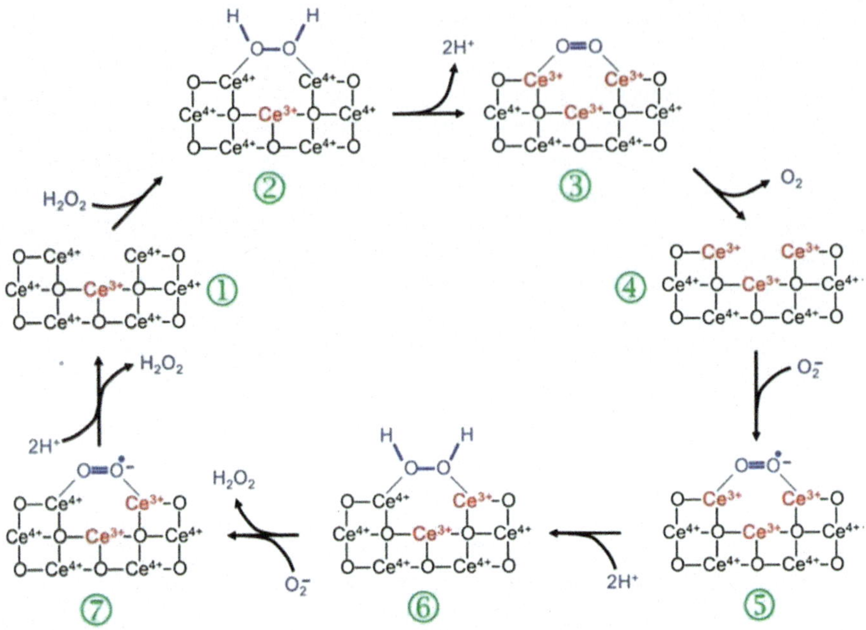


**Fig. 6.3** A model of the reaction mechanism for the oxidation of $H_2O_2$ by nanoceria and the regeneration via reduction by superoxide [37]. An oxygen vacancy site on the nanoceria surface (*1*) presents a 2 $Ce^{4+}$ binding site for $H_2O_2$. (*2*) After the release of protons and two-electron transfer to the two cerium ions, (*3*) oxygen is released from the now fully reduced oxygen vacancy site (*4*). Subsequently, superoxide can bind to this site (*5*), and after the transfer of a single electron from one $Ce^{3+}$, and uptake of two protons from the solution, $H_2O_2$ is formed (*6*) and can be released. After repeating this reaction with a second superoxide molecule (*7*), the oxygen vacancy site returns to the initial 2 $Ce^{4+}$ state (*1*). It is also possible that the third $Ce^{4+}$ indicated, which gives rise to the oxygen vacancy, could participate directly in the reaction mechanism. The square Ce–O matrix is shown here only to illustrate the model and does not correspond to the actual spatial arrangement of the atoms in the crystal structure. (Copyright 2011 RSC Pub)

with particle size of 3–5 nm was higher than that of natural Cu-Zn SOD ($3.6 \times 10^9$ $M^{-1}s^{-1}$ vs. $(1.3\text{–}2.8) \times 10^9$ $M^{-1}s^{-1}$). Referring to the catalytic mechanism of natural SOD, a SOD-like reaction mechanism of nanoceria was proposed [34]:

$$O_2^{\bullet -} + Ce^{3+} + 2H^+ \rightarrow H_2O_2 + Ce^{4+} \tag{6.4}$$

$$O_2^{\bullet -} + Ce^{4+} \rightarrow O_2 + Ce^{3+} \tag{6.5}$$

In addition, Celardo et al. proposed a more specific catalytic reaction model for the SOD-like activity of nanoceria (Fig. 6.3) [37]. However, these reaction mechanisms are all hypotheses based on the existing experimental phenomena, and the detailed reaction process of the SOD-like activity of nanoceria still needs further study.

#### 6.3.1.2 Fullerene and Its Derivatives with SOD-Like Activity

Fullerene is a hollow carbon isotope composed entirely of carbon, which is spherical, ellipsoidal, cylindrical, or tubular in shape. Fullerenes have unique ability to absorb free radicals attributed to their delocalized π double bond system and have been called as free radical "sponge" [38]. Nevertheless, unmodified fullerenes are difficult to be directly used as antioxidants because of their poor water solubility, which makes it difficult for them to interact directly with biological molecules. A series of water-soluble fullerene derivatives can be obtained by chemical modification, and much evidences showed that water-soluble fullerene derivatives could play a neuroprotective role in some cell or animal models, including degeneration of dopaminergic neurons in Parkinson's disease [39–41] and central nervous system ischemia [42, 43].

Duan et al. have done many works in studying the role of buckminsterfullerenes ($C_{60}$) and their derivatives in neuroprotection. $C_{60}$ is a kind of spherical fullerenes which composed of 60 carbon atoms, shaped like a football [44]. In their initial studies, two water-soluble $C_{60}$ derivatives ($C_{60}(OH)_{12}$ and $C_{60}(OH)_{18-20}O_{3-7}$) were found to reduce both excitotoxic and apoptotic neuronal death induced by oxidative stress [45]. Through EPR experiments, these water-soluble $C_{60}$ derivatives were found to be able to effectively scavenge •OH produced by the decomposition of $H_2O_2$, which explained their neuroprotective ability to some extent [45]. Later, similar to the two $C_{60}$ derivatives mentioned above, two isomers of water-soluble malonic acid $C_{60}$ derivatives ($C_{60}[C(COOH)_2]_3$) with $C_3$ or $D_3$ symmetry have also been proved to be effective in scavenging free radicals and protecting cultured cortical neurons from oxidative damage [46]. Moreover, interestingly, the $C_3$ derivative reduced death and functional degradation of transgenic mice carrying human mutation (G93A) SOD gene, which is associated with familial amyotrophic lateral sclerosis (FALS), suggesting that it may play a similar role as SOD [46].

Although in some studies [41] the $C_3$ derivative has been found to eliminate both $O_2^{\bullet -}$ and •OH, it was not proposed that they possess SOD-like activity at that time. Some hypotheses suggested that the antioxidant ability of the water-soluble $C_{60}$ derivatives may be due to the direct transfer of electrons from $O_2^{\bullet -}$ to these fullerene derivatives, but no $C_{60}$ radical intermediates have been found in subsequent studies [47, 48]. Until 2004, Duan et al. found that the $C_3$ derivative is capable of removing the $O_2^{\bullet -}$ through catalytic dismutation of $O_2^{\bullet -}$ like a natural SOD enzyme rather than via stoichiometric scavenging [49]. Without the change of the $C_3$'s structure, the regeneration of oxygen, production of $H_2O_2$, and absence of radical products were all observed, which indicates that $C_3$ plays a SOD-like catalytic role in the $O_2^{\bullet -}$-removing reaction. Besides, by a competitive experiment with cytochrome c for $O_2^{\bullet -}$ dismutation, the rate constant of $C_3$ ($K_{C3}$) to scavenge $O_2^{\bullet -}$ was tested to be 2.2 ($\pm$ 0.1) $\times$ $10^6$ $M^{-1}$ $s^{-1}$, which is approximately 1/100 of that of the SOD. The effect of pH on the reaction between $C_3$ and $O_2^{\bullet -}$ was also measured. The results showed that the $K_{C3}$ decreases with the increase of pH. This phenomenon may be attributed to the fact that the negative charge of carboxyl groups on $C_3$ increases with the increase of pH, which prevents the contact of $O_2^{\bullet-}$ with the main part of the derivatives, thereby decreasing the $K_{C3}$ [49].

In order to elucidate the specific mechanism of the SOD-like activity of $C_3$, these researchers used semiempirical quantum mechanical calculations to determine the electron distribution on $C_3$ surface. A mechanism model of $O_2^{\bullet-}$ dismutation by $C_3$ was proposed as a result (Fig. 6.4). In this model, the susceptibility of $C_3$ surface sites to be attacked by $O_2^{\bullet-}$ is shown in the form of color maps. As can be seen from this map, the malonic acid group is located in the region with the most serious electron defects, which makes it easier to attract $O_2^{\bullet-}$ (Fig. 6.4b). When $O_2^{\bullet-}$ exists in the environment, it binds to the protons of malonic acid group on $C_3$ by hydrogen bond driven by potential difference. Later, with the second $O_2^{\bullet-}$ approaches, these two radicals are dismutated into $O_2$, $H_2O_2$, or •OH with the help of protons in the malonic acid group or in the local water molecules (Fig. 6.4a) [49]. To verify the effect of carboxyl groups on the SOD-like activity and neuroprotective function of $C_{60}$ derivatives, Duan et al. selected six carboxyfullerenes with different quantity and distribution of carboxyl groups. The results showed that the more the amount of carboxyl groups on the surface of the fullerene derivatives and the closer the distribution of carboxyl groups, the stronger the SOD-like activity and the stronger the neuroprotective ability of these derivatives [50]. In addition, Gozin et al. found that when $C_3$ binds to human serum albumin, the original SOD-like activity remains unchanged, indicating that this kind of nanozyme can simultaneously carry SOD mimic activity and biocompatibility, which provides a possibility for its medical application [51].

#### 6.3.1.3 Carbon Clusters with SOD-Like Activity

In addition to fullerenes, another carbon-based nanomaterial, a kind of hydrophilic polyethylene glycol (PEG)-coated carbon clusters (PEG-HCCs), has also been reported to have SOD-like activity [52]. The carbon cores of PEG-HCCs is about 3 nm wide and 30–40 nm long, containing about 2000–5000 carbon atoms in each of them. The results showed that PEG-HCCs specifically eliminate $O_2^{\bullet-}$, while it is inert to peroxynitrite radical ($ONOO^-$) and NO•. X-ray photoelectron spectroscopy (XPS) showed that covalent binding with ROS was not the main mechanism of PEG-HCCs scavenging $O_2^{\bullet-}$. EPR experiments showed that PEG-HCCs act as a catalyst which promoted the transition of $O_2^{\bullet-}$ to $O_2$, rather than as an ROS adsorbent. Besides, the production of $O_2$ and $H_2O_2$ was also detected during the reaction. These experimental evidences indicate that PEG-HCCs are specific SOD-like nanozymes catalyzing the decomposition of $O_2^{\bullet-}$. The state kinetic test showed that the turnover number of $O_2^{\bullet-}$ to $O_2$ catalyzed by PEG-HCCs was more

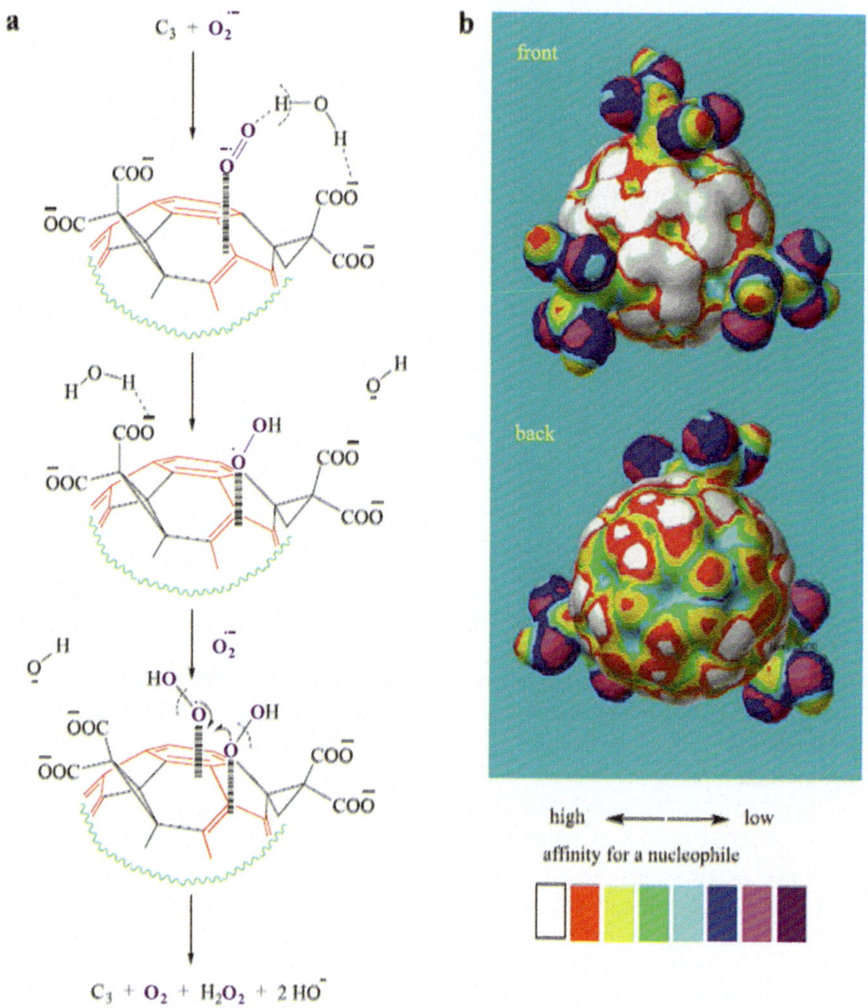


**Fig. 6.4** Proposed SOD-like mechanism of $O_2^{\bullet -}$ dismutation by $C_3$ [49]. (**a**) Schematic representation of the catalytic interaction of $C_3$ and $O_2^{\bullet -}$. Chemical bonds colored in red are associated with electron-deficient areas and predicted through semiempirical quantum calculations (compare with **b**). Incoming superoxide ions and oxygen atoms derived from them are colored in purple to facilitate visualization of the suggested mechanism. Broken lines represent hydrogen bonding between oxygen and hydrogen atoms, and hyphenated lines are used to represent electrostatic attraction between negatively charged oxygen atoms and electron-deficient areas on the $C_{60}$ moiety. In the proposed mechanism, $C_3$ is suggested to electrostatically drive superoxide anions toward electron-deficient areas on its surface until a second $O_2^{\bullet -}$ arrives to undergo dismutation with the help of protons from carboxyl groups and/or surrounding water molecules. (**b**) Map of the electron density reflecting nucleophilic superdelocalizability. On the resulting isosurface, locations that are susceptible to attack by superoxide are designated by a color scale showing decreasing susceptibility, in the order white through cyan (legend). (Copyright 2004 Elsevier)

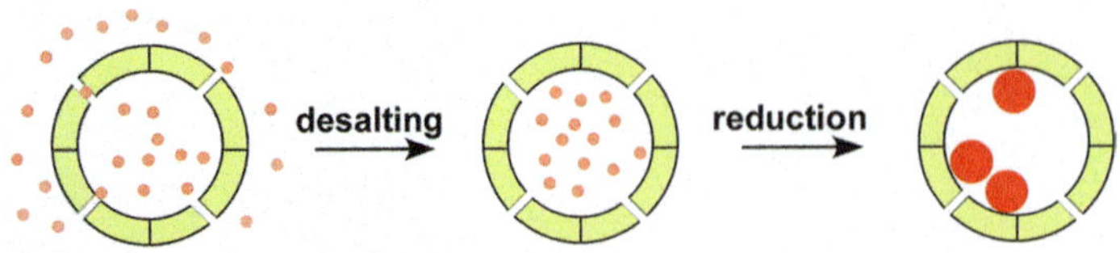


**Fig. 6.5** Preparation of apoferritin-encapsulated platinum nanoparticles (Pt-apo) [55]. (Copyright 2010 American Chemical Society)

than 20,000 $s^{-1}$, which was equivalent to the catalytic activity of Cu/Zn SOD enzyme. Together, the dismutation process that PEG-HCCs catalyze the conversion of $O_2^{\bullet-}$ was given as follows [52]:

$$\mathrm{PEG-HCC^{\bullet} + O_2^{\bullet-} \rightarrow PEG-HCC^{-} + O_2} \tag{6.6}$$

$$\mathrm{PEG-HCC^{-} + O_2^{\bullet-} + 2H_2O \rightarrow PEG-HCC^{\bullet} + H_2O_2 + 2HO^{-}} \tag{6.7}$$

#### 6.3.1.4 Platinum Nanoparticles with SOD-Like Activity

Some studies found that platinum nanoparticles could also effectively clear $O_2^{\bullet-}$ and $H_2O_2$ [53, 54]. Later on, Knez et al. encapsulated platinum nanoparticles within apoferritin (Fig. 6.5), a cage-like protein, and found that the protein–platinum nanoparticles (Pt-apo) showed good SOD-like activity and long-term stability in vitro [55]. In addition, platinum nanoparticles can effectively eliminate the oxidative stress caused by external $H_2O_2$ and improve the survival rate of human intestinal Caco-2 cells after being internalized into cells mediated by apoferritin via its receptor. In addition, a study mentioned that the SOD-like activity of polyacrylic acid-modified platinum nanoparticles (Pt-PAA) was slightly lower than that of ceria nanoparticles in the same mass [56]. No detailed mechanism has been proposed to explain the SOD-like behavior of platinum nanoparticles.

### 6.3.2 *Nanozymes with Catalase-Like Activity*

As mentioned above, $O_2^{\bullet-}$ can be dismutated by SOD into $H_2O_2$ or $O_2$ (Sect. 6.3.1). Compared with other active ROS, $H_2O_2$ is a relatively stable molecule and plays physiological roles in signal transduction in cells. However, although $H_2O_2$ is relatively stable, it can generate damaging •OH through Fenton reaction or other similar processes. Therefore, removal of excessive cellular $H_2O_2$ in a timely manner is crucial to protect cells from oxidative damage. In the body, the most effective way to remove $H_2O_2$ is to decompose it into water and oxygen by catalase. Catalase is an enzyme that exists in almost all aerobic organisms. Although the detailed mechanism has not been thoroughly explained, the reaction process of catalase catalyzing the decomposition of $H_2O_2$ is believed to be as follows [57, 58]:

$$H_2O_2 + Fe\ (III) - OH \rightarrow H_2O + Fe\ (III) - OOH\ (.+) \tag{6.8}$$

$$H_2O_2 + Fe\ (III) - OOH\ (.+) \rightarrow H_2O + Fe\ (III) - OH + O_2 \tag{6.9}$$

$$2H_2O_2 \rightarrow 2H_2O + O_2\ \text{(overall reaction)} \tag{6.10}$$

The Fe(III)-OH in these formulations represents the iron in the heme of the catalase active center, while the (.+) of the Fe(III)-OOH (.+) represents that iron is not completely oxidized to +4 but only accepts some electron density from the heme ligand. When $H_2O_2$ enters the active site, it interacts with the amino acid of catalase, leading to proton transfer between oxygen atoms. Free oxygen atoms coordinate with iron in the active center (Fe(III)-OH) to release newly formed water molecules and Fe(III)-OOH (.+). The active site coordinated with oxygen reacts with the second $H_2O_2$ molecule, reforming Fe(III)-OH and generating water and oxygen. Finally, the two molecules of $H_2O_2$ are decomposed into two molecules of water and one molecule of oxygen.

Like SOD, catalase is essential in protecting nerve cells from oxidative damage. Similar as SOD-like nanozymes, many inorganic nanomaterials (e.g., $CeO_2$, $Fe_3O_4$, $Co_3O_4$, Pt, etc.) have been found to be catalase-like catalysts for $H_2O_2$ decomposition. In this section, some major catalase-like nanozymes and their catalytic mechanisms will be introduced.

#### 6.3.2.1 Nanoceria with Catalase-Like Activity

Section 6.3.1.1 mentions that Self et al. first discovered that nanoceria exhibits SOD-like activity, and the catalase-like activity of nanoceria was also reported by them for the first time [59]. The ability of nanoceria to decompose $H_2O_2$ was confirmed by measuring changes in dissolved oxygen production in this study. Besides, the catalase-like activity of nanoceria was also found to be closely related to its surface redox state. However, unlike nanoceria with SOD-like activity, which has a relatively high $Ce^{3+}/Ce^{4+}$ ratio on their surface, results from the Amplex Red assay showed that nanoceria with a lower proportion of $Ce^{3+}/Ce^{4+}$ exhibited a significant catalase-like activity [59]. Although the specific mechanism has not been clarified, the reaction equation for the decomposition of $H_2O_2$ catalyzed by nanoceria is generally assumed to be as follows [60, 61]:

$$H_2O_2 + 2Ce^{4+} + 2OH^- \rightarrow 2H_2O + O_2 + 2Ce^{3+} \tag{6.11}$$

The regeneration of $Ce^{3+}$ to $Ce^{4+}$ can also be accomplished with the participation of $H_2O_2$, which involves the formation of cerium monoperoxy trihydroxide (Ce $(OH)_2OOH$) intermediates [60]:

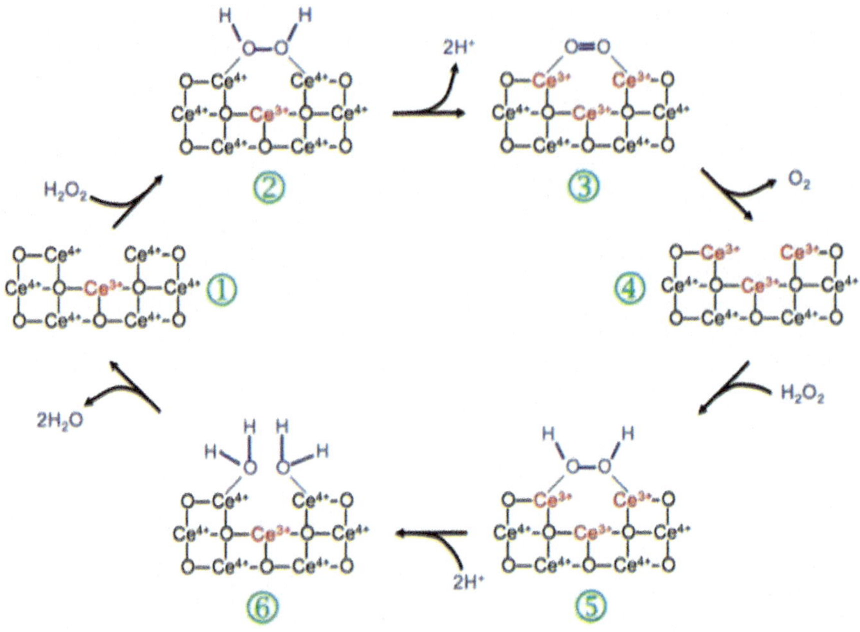


**Fig. 6.6** A model of the reaction mechanism for the complete dismutation of $H_2O_2$ [37]. The oxidative half-reaction is identical to the sequence shown in Fig. 6.3 (1)–(4). The reductive half involves binding of $H_2O_2$ to the 2 $Ce^{3+}$ site (5), uptake of two protons and homolysis of the O–O bond with transfer of electrons to the two $Ce^{3+}$ (6), and release of the water molecules to regenerate the initial $Ce^{4+}$ site (1). This reaction sequence would be analogous to the one found in catalases. (Copyright 2011 RSC Pub)

$$Ce(OH)_3 + H_2O_2 \rightarrow Ce(OH)_2OOH + H_2O \tag{6.12}$$

$$3\,Ce(OH)_2OOH + H_2O \rightarrow 2\,Ce(OH)_3OOH + Ce(OH)_3 \tag{6.13}$$

$$2\,Ce(OH)_3OOH \rightarrow 2\,Ce(OH)_4 + O_2 \tag{6.14}$$

$$2\,Ce(OH)_3 + 3H_2O_2 \rightarrow 2\,Ce(OH)_4 + O_2 + 2H_2O \text{ (overall regeneration reaction)} \tag{6.15}$$

Although $Ce^{3+}$ can be regenerated into $Ce^{4+}$ while $H_2O_2$ is decomposed, this process is limited by many rate-limiting factors (e.g., pH and concentration of $H_2O_2$), so the concentration of $Ce^{3+}$ is not decisive for the catalase-like activity of nanoceria [60]. To better reflect the effect of the redox state and the oxygen vacancy on the catalase-like activity of nanoceria, Celardo et al. have proposed an intuitive reaction mechanism model (Fig. 6.6) [37]. Unlike natural catalase, which has no complete valence transition of iron in its active center, the complete valence transition between $Ce^{3+}$ and $Ce^{4+}$ occurs in all these abovementioned mechanisms. This may explain the difference in catalytic efficiency between nanoceria and natural enzymes, but the detailed mechanism still needs further study.

In addition, although nanoceria can exhibit SOD-like or catalase-like activity under different conditions, it is still unknown whether nanoceria can possess both of these enzyme-like activities under same conditions [34, 59].

#### 6.3.2.2 Iron Oxide Nanoparticles with Catalase-Like Activity

It has been reported that iron oxide nanoparticles (INOPs) can directly catalyze $H_2O_2$ to produce water and oxygen in neutral pH environment, which indicates that INOPs possess catalase activity [62]. Interestingly, in this study, Gu et al. found that INOPs exhibited a pH-dependent double enzyme-like activity. In neutral conditions (pH = 7.4), iron oxide nanoparticles exhibited catalase-like activity, while in acidic conditions (pH = 4.8), they exhibited peroxidase-like activity, which catalyzes oxidation of peroxidase substrates in the presence of $H_2O_2$. They compared the dual enzyme-like activity of two kind of INOPs ($Fe_3O_4$ and $\gamma$-$Fe_2O_3$) with similar size. By measuring the content of soluble oxygen in the reaction environment, they found that the catalase-like activity of $Fe_3O_4$ in neutral conditions was higher than that of $\gamma$-$Fe_2O_3$. Moreover, the peroxidase-like activity of $Fe_3O_4$ in acidic conditions was also higher than that of $\gamma$-$Fe_2O_3$ by observing the color change of substrates. Based on the results of Electron Spin Resonance (ESR) spectroscopy measurements, they proposed a mechanism model to explain the activity of the pH-dependent dual enzyme-like behaviors of INOPs. Under acidic conditions, INOPs can catalyze a Fenton-like reaction of $H_2O_2$. The process is likely to be as follows [62]:

$$Fe^{3+} + H_2O_2 \rightarrow FeOOH^{2+} + H^+ \tag{6.16}$$

$$FeOOH^{2+} \rightarrow Fe^{2+} + HO_2\cdot \tag{6.17}$$

$$Fe^{2+} + H_2O_2 \rightarrow Fe^{3+} + OH^- + OH\cdot \tag{6.18}$$

However, under neutral conditions, because of the low concentration of $H^+$, the reaction rate in formula (6.16) is faster and excessive $FeOOH^{2+}$ and $HO_2\cdot$ is produced. Then, $HO_2\cdot$ can be protonated to form $O_2^{\bullet-}$ (Formula 6.19), and $HO_2\cdot$ or $O_2^{\bullet-}$ can further react with $OH\cdot$ to produce oxygen and water (Formula 6.20). Thus, after a series of reactions, $H_2O_2$ is finally decomposed into oxygen, which means that INOP exhibits catalase activity.

$$HO_2\cdot + H^+ \rightarrow O_2^{\bullet-} \tag{6.19}$$

$$OH\cdot + HO_2\cdot /O_2^{\bullet-} \rightarrow H_2O + O_2 \tag{6.20}$$

Because catalase can protect cells by scavenging $H_2O_2$, peroxidase can catalyze the production of •OH and damage cells. Therefore, these INOPs may play a protective role in neutral cell environment (e.g., in cytoplasm) and a lethal role in acidic cell environment (e.g., in lysosome) (Fig. 6.7) [62]. This characteristic may be

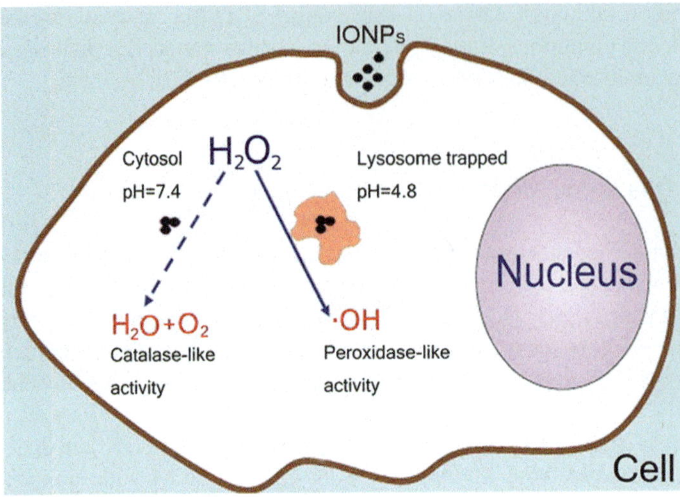


**Fig. 6.7** Schematic illustration of peroxidase-like activity-induced cytotoxicity by IONPs [62]. IONPs are trapped in acidic lysosomes when internalized into cells, so they catalyze $H_2O_2$ to produce hydroxyl radicals through peroxidase-like activity; however, in neutral cytosol, IONPs would decompose $H_2O_2$ through catalase-like activity. (Copyright 2012 American Chemical Society)

used to regulate the ROS level in cells, hence protecting normal cells or killing diseased cells.

#### 6.3.2.3 $Co_3O_4$ Nanozymes with Catalase-Like Activity

Similar to INOPs, $Co_3O_4$ nanoparticles have also been reported to have both peroxidase-like and catalase-like activities [63]. $Co_3O_4$ nanoparticles can exert peroxidase mimic activity by promoting electron transfer between $H_2O_2$ and reductive substrates rather than producing •OH like INOPs. This study suggested that $Co_3O_4$ nanoparticles do not produce •OH, so far as to reduce it (Fig. 6.8a), probably because it catalyzes the decomposition of $H_2O_2$ in a concentration-dependent manner like catalase. After that, the concentration of soluble oxygen in the reaction system was determined to prove that oxygen was produced in the reaction (Fig. 6.8b), which indicated that $Co_3O_4$ nanoparticles did have catalase-like activity, but the reaction mechanism was not studied in depth [63].

#### 6.3.2.4 Platinum Nanoparticles with Catalase-Like Activity

As mentioned in Sect. 6.3.1.4, platinum nanoparticles have been reported to effectively scavenge $O_2^{\bullet-}$ and $H_2O_2$ [53, 54], and studies have confirmed that apoferritin-coated platinum nanoparticles (Pt-apo) possess SOD-like activity [55]. Using a

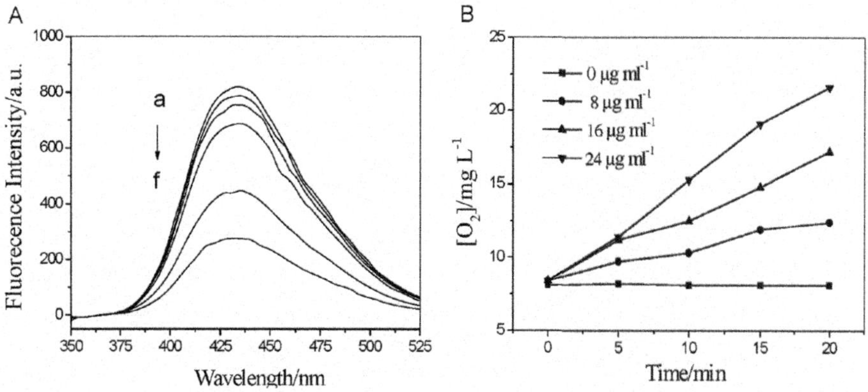


**Fig. 6.8** (**A**) The effect of the $Co_3O_4$ NPs on the formation of hydroxyl radical with terephthalic acid as a fluorescence probe [63]. a–f; 0, 10, 15, 20, 25, 30 μg/mL$^{-1}$. 10 mM $H_2O_2$, 0.5 mM terephthalic acid, and different concentrations of the $Co_3O_4$ NPs were first incubated in 100 mM acetate buffer (pH 5.0) exposed to UV light at 365 nm for 20 min. (**B**) Effect of the $Co_3O_4$ NPs concentration on the generation of $O_2$ by decomposition of $H_2O_2$. Reaction conditions: 50 mM $H_2O_2$ and different concentrations of $Co_3O_4$ NPs in 100 mM NaAc buffer (pH 5.0). (Copyright 2012 Royal Society of Chemistry)

similar method, Nie et al. synthesized platinum nanoparticles with a diameter of 1–2 nm in an apoferritin nanoreactor (Pt-Ft) and found that the Pt-Ft nanoparticles exhibited both catalase-like and peroxidase-like activities [64]. In this study, the formation of oxygen bubbles confirmed that Pt-Ft nanoparticles could catalyze the decomposition of $H_2O_2$ like catalase. In addition, the change of distinctive product color confirmed that Pt-Ft nanoparticles had peroxidase-like activity at the same time. In addition, like natural enzymes, the catalytic activity of Pt-Ft nanozyme is also affected by pH and temperature. The optimum reaction conditions of the peroxidase-like activity of Pt-Ft nanozyme are similar to those of HRP (pH = 4, temperature = 37 °C), but the catalase-like activity of Pt-Ft nanozyme increased gradually with the increase of pH and temperature, which is different from that of catalase (Fig. 6.9) [64]. Although the apoferritin-coated platinum nanoparticles (Pt-apo or Pt-Ft) have been separately reported to have SOD-like or catalase-like activities, whether these nanoparticles can simultaneously display these two enzyme-like activities has not been reported, and further research is needed [55, 64].

### 6.3.3 *Nanozymes with Multiple Antioxidant Enzyme-Like Activities*

In the preceding 6.3.1 and 6.3.2 sections, we have introduced the nanozymes with SOD-like or catalase-like activities, respectively. Although some other enzymes (e.g., HO-1, GPx, glutathione reductase, etc.) also play important roles in the

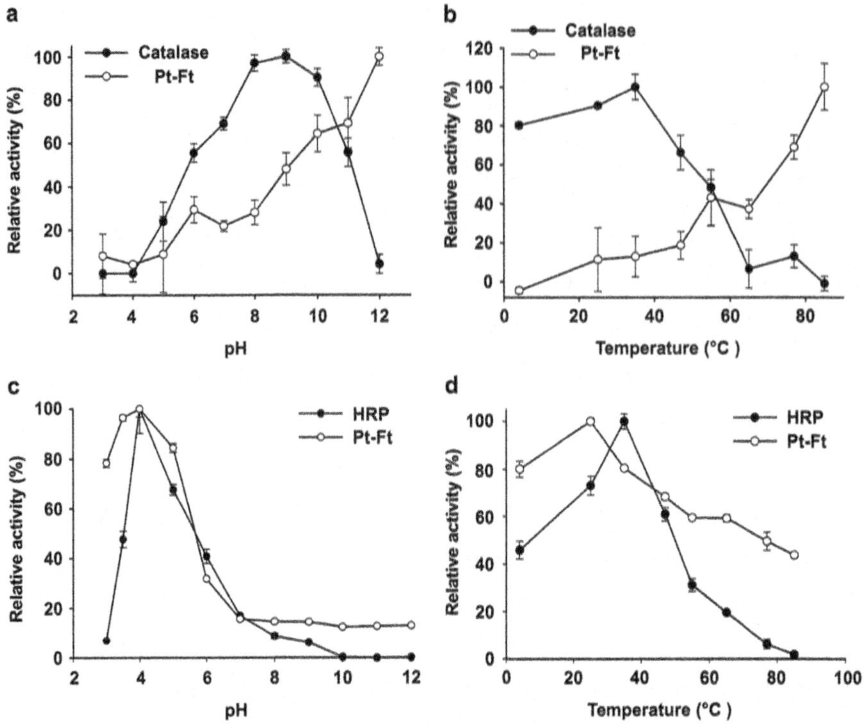


**Fig. 6.9** The catalase-like activity of Pt-Ft is pH and temperature dependent (**a** and **b**) [64]. (**a**) Catalase shows a pH optimum of 8–9, and the activity of Pt-Ft gradually increases as the pH increases. (**b**) Catalase shows a temperature optimum of around 25–35 °C, and the activity of Pt-Ft gradually increases as the temperature increases. HRP-like activity of Pt-Ft is pH and temperature dependent (**c** and **d**). (**c**) The optimal pH values of HRP and Pt-Ft were both at pH 4. (**d**) The optimal temperatures of HRP and Pt-Ft were 37 and 25 °C, respectively. The maximum activity of enzymes in each curve was set as 100%. (Copyright 2011 Elsevier)

body's antioxidant system, there are few studies on their related nanozymes [1, 8]. In addition, some nanozymes (e.g., nanoceria, Pt-apo, or Pt-Ft) have been reported to exhibit SOD-like or catalase-like activities under different conditions or in different studies, but few researches have been done on whether these nanozymes can simultaneously possess both enzyme-like activities under the same condition [34, 55, 59, 64]. However, in addition to the nanozymes mentioned above, in fact, some nanozymes have been reported to possess multiple antioxidative enzyme-like activities under the same condition and show effective neuroprotective effects.

#### 6.3.3.1 $Mn_3O_4$ Nanozyme

Mugesh et al. have synthesized a flowerlike $Mn_3O_4$ nanoparticles (Mnf), which simultaneously showed multiple enzyme-like activities of catalase, GPx, and SOD

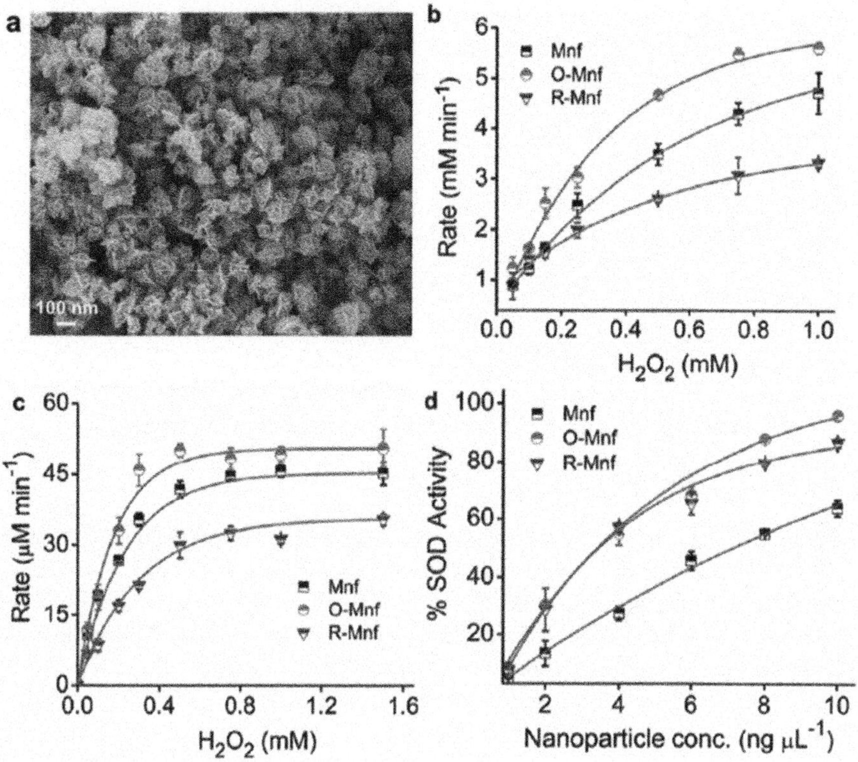


**Fig. 6.10** (**a**) SEM and TEM image of Mnf. (**b–d**) The catalase, GPx, and SOD-like activity of Mnf, O-Mnf, and R-Mnf, respectively. The assay conditions remained same as mentioned previously [65]. (Copyright 2017 John Wiley and Sons)

(Fig. 6.10a) [65]. By monitoring the decrease in $H_2O_2$ absorbance at 240 nm and the formation of oxygen by a gas chromatography experiments, the catalase-like activity of Mnf was determined and found to be accorded with Michaelis–Menten kinetics. GPx-like activity of Mnf was studied by classical GR coupling method, in which the concentration of NADPH was monitored by spectrophotometry at 340 nm. In addition, Mnf can also reduce the formation of formzan, a reduction product of WST-1 (a cell proliferation assay kit) in the presence of $O_2^{\bullet-}$, indicating that Mnf also has SOD-like activity. These researchers believed that multienzyme-like activity of Mnf is mainly related to its rapid valence transition between $Mn^{2+}$ and $Mn^{3+}$. In order to study the effect of two oxidized states on the activity, they synthesized two kinds of Mnf with different ratios of $Mn^{3+}/Mn^{2+}$, namely, oxidized Mnf (O-Mnf) and reduced Mnf (R-Mnf). X-ray photoelectron spectroscopy (XPS) analysis showed that Mnf with higher $Mn^{3+}/Mn^{2+}$ ratio (O-Mnf) exhibited enhanced catalase-like and GPx-like activities as compared to Mnf, while the SOD-like activities of O-Mnf and R-Mnf were both slightly higher than that of Mnf (Fig. 6.10b–d). Besides, the effect of morphology on the activity of $Mn_3O_4$

nanoparticles was also studied. In addition to Mnf, the flakes, cubes, polyhedron, and hexagonal plates also exhibited significant SOD-like activity, but none of them showed the activity of catalase-like or GPx-like comparable to that of Mnf [65].

In addition to the antioxidative nanozymes mentioned above, some other nanozymes have also been reported to have the potential to scavenge ROS and have been studied in the treatment of diabetes, cardiovascular diseases, cancer, and some non-nervous system inflammation. However, due to the lack of research related to the central nervous system diseases, they are not introduced here. If interested, readers can get more comprehensive knowledge from some other excellent reviews [8, 11, 66, 67].

## 6.4 Nanozymes Used for the Treatment of Neurological Diseases

Although nanozymes as a new type of antioxidants have not yet been approved for clinical treatment, many nanozymes have been studied in some oxidative stress-related disease models (e.g., AD, CI, PD, and MS). In this section, based on the existing research, according to the different types of diseases, we will introduce the progress of nanozymes in the treatment of neurological diseases.

### *6.4.1 Application of Nanozymes in Neuroprotection*

The damage and loss of neurons are the key links in the occurrence and deterioration of various types of the central nervous system diseases, including Alzheimer's disease (AD), Parkinson's disease (PD), multiple sclerosis (MS), cerebral ischemia (CI), and so on. Oxidative damage, mitochondrial dysfunction, chronic inflammation and degradation, and misfolding of damaged proteins mediate neuronal damage and loss in neurological diseases [68]. Among them, it is believed that oxidative stress and its mediated inflammatory response in the central nervous system are the main causes of neuron loss [1]. Therefore, timely removal of harmful ROS is essential for neuron protection. At present, some antioxidative nanozymes (e.g., $C_{60}$ derivatives, graphene oxide, etc.) have been found to protect nerve cells from oxidative damage. This section mainly introduces the research progress of nanozymes in nerve cell protection.

Antioxidative nanozymes based on buckminsterfullerene ($C_{60}$) have been shown to be effective in protecting neurons from toxic-induced damage. Duan et al. found that two polyhydroxylated $C_{60}$ derivatives, Fullerenol-1 ($C_{60}(OH)_{18\text{-}20}O_{3\text{-}7}$) and Fullerenol-2 ($C_{60}(OH)_{12}$), can alleviate the excitotoxic neuronal death of mouse neocortical neurons induced by exposing to NMDA (by 80%), AMPA (by 65%), or kainate (by 50%), which can kill cells through their receptors on neurons

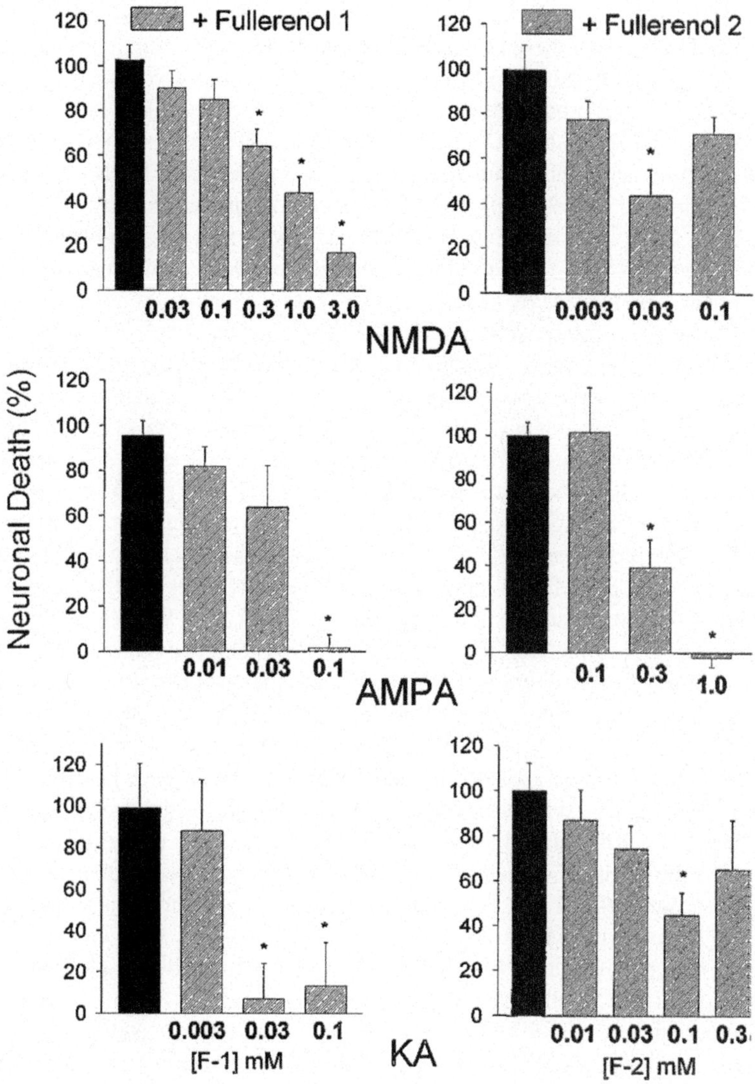


**Fig. 6.11** Neuroprotection by fullerenols in excitotoxic injury [45]. Neurotoxicity produced by exposure to NMDA, AMPA, or kainate (KA) is reduced by coapplication of the fullerenols. (Copyright 1996 Elsevier)

[45]. Among these two fullerenols, Fullerenol-1 with better water solubility exhibited more steady and renewable neuroprotective effects (Fig. 6.11). In addition, electrophysiological and $^{45}Ca^{2+}$-flux experiments showed that the nerve currents and $Ca^{2+}$ uptake induced by NMDA or AMPA were not affected, suggesting that these fullerenols did not act by antagonizing glutamate receptors. These fullerenols can also alleviate the apoptosis of cortical neurons cultured in the absence of serum and

glial cells, which normally occurs within 24–48 h. Based on the results of mass spectrometry and EPR, these researchers speculated that the antioxidant mechanism of these two $C_{60}$ derivatives may be attributed to their ability to absorb a variety of free radicals without producing paramagnetic (radical) adducts [45].

However, contrary to the above studies, Wu et al. showed that polyhydroxylated $C_{60}$ derivatives could antagonize glutamate receptors and protect cells but had no effect on $H_2O_2/Fe^{2+}$-induced neurotoxicity [69]. The results of this paradox still need further verification. Later, two other water-soluble carboxylic acid $C_{60}$ derivatives ($C_{60}[C(COOH)_2]_3$) with C3 or D3 symmetry synthesized by Duan et al. were suggested to have similar neuroprotective functions [46]. These derivatives were also found to effectively alleviate the serum-deprivation or $A\beta_{1\text{-}42}$-mediated cell apoptosis and the NMDA- or AMPA-mediated cell necrosis without affecting the related receptors. Furthermore, compared with $D_3$ derivatives, $C_3$ derivatives exhibit stronger neuroprotective effects due to its excellent ability to cross cell membranes. Moreover, chronic infusion of $C_3$ derivatives delayed the deterioration of familial amyotrophic lateral sclerosis (FALS) in a transgenic mouse model carrying the human mutant (G93A) SOD gene [46].

As we mentioned in Sect. 6.3.1.2, the neuroprotective effect of $C_3$ derivatives is mainly attributable to their ROS-scavenging ability based on the SOD-like activity [49]. Besides, Xu et al. found that $C_{60}$-methionine derivatives (FMD) could reduce Pb-induced oxidative damage in human neuroblastoma SH-SY5Y cells, showed by increasing GSH level, reducing malondialdehyde content, and reducing DNA damage without obvious toxicity [70]. They also found that the protective effect of FMD was mainly based on its ability to remove ROS and that the protective effect of FMD was stronger than that of b-alanine $C_{60}$ derivatives and cystine $C_{60}$ derivatives.

In addition to fullerenes, recently, another carbon-based nanozyme has been reported by Zheng et al. to be used for neuroprotection [71]. Compared with large graphene oxide (GO) nanosheets, GO quantum dots (GOQDs) as catalase-like nanozymes can effectively reduce ROS in PC12 cells induced by 1-methyl-4-phenylpyridine ion ($MPP^+$). GOQDs can also effectively reduce ROS, apoptosis, and mitochondrial damage in zebrafish treated with $MPP^+$. Meanwhile, zebrafish sprayed with GOQD showed increased motor activity and Nissl bodies in the brain compared with GO nanotablets, confirming that GOQD alleviated $MPP^+$-induced neurotoxicity (Fig. 6.12. GOQDs can be translocated into zebrafish brain and mimic the activity of catalase to resist the oxidation of intracellular environment. GOQD improves neurotoxicity by increasing amino acid metabolism, reducing cycling activity of tricarboxylic acid, and reducing steroid biosynthesis, fatty acid biosynthesis, and galactose metabolic pathway activity related to antioxidation and nerve transmission [71].

The conversion between $Ce^{3+}$ and $Ce^{4+}$ provides nanoceria with an efficient ROS-scavenging capability, which has been found to protect rat nerves from oxidative damage caused by $H_2O_2$ in a serum-free cell culture model of adult rat spinal cord [72]. Also, nanoceria showed good biocompatibility in this model. Compared with the control group, single dose of 10 nM of nanoceria at the time of cell plating can effectively improve cell survival within 30 days, regardless of whether the cells were stimulated by $H_2O_2$ or not (Fig. 6.13) [72]. Nanoceria could also protect HT22 cells from oxidative damage induced by glutamic acid as reported in another study [73].

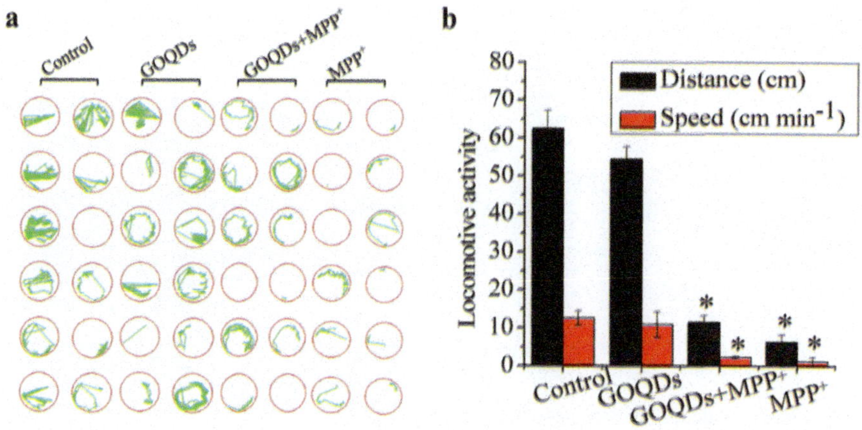


**Fig. 6.12** Effects of GOQDs on behavioral disturbances of zebrafish induced by $MPP^+$ [71]. (**a**) Spontaneous movement trajectories of larvae in 96-well plates treated with $MPP^+$ with or without preincubation with GOQDs. The green curves represent the movement trajectories of the larvae. (**b**) Distances and speeds of the larvae. $*p < 0.05$, compared with the control. (Copyright 2018 The Author(s))

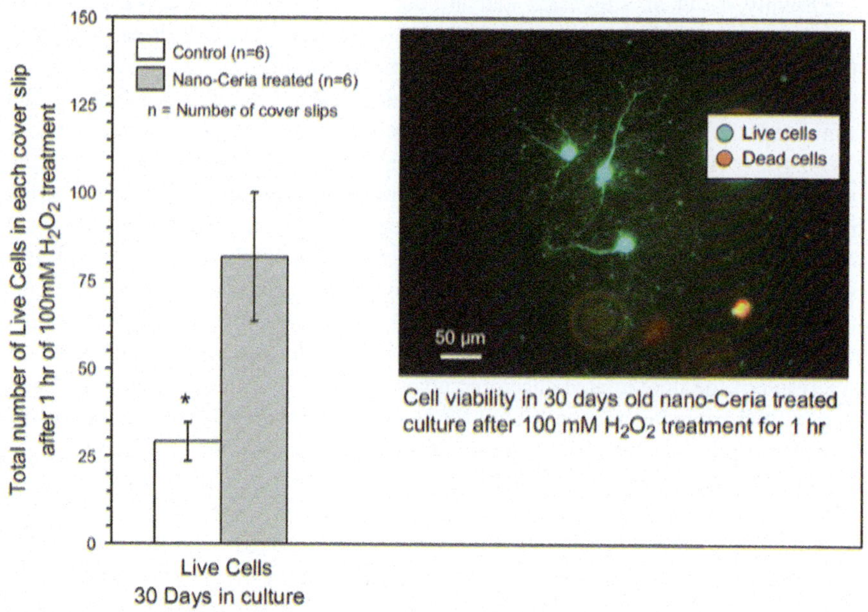


**Fig. 6.13** Results after $H_2O_2$-induced oxidative injury in control and treated cultures of adult rat spinal cord at day 30 [72]. Live–dead cell assay after $H_2O_2$ treatment indicates that nanoceria-treated cultures had a significantly higher number of surviving cells as compared to the control. (Copyright 2007 Elsevier)

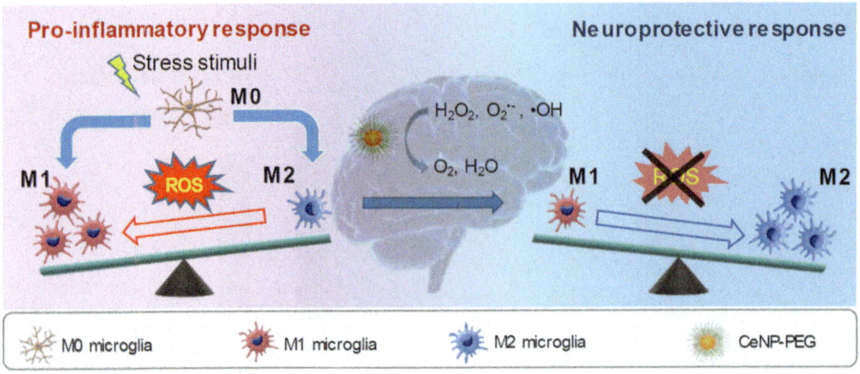


**Fig. 6.14** Remodeling the immuno-environment in brain disorders [74]. A bespoke ceria nanoparticle that can simultaneously eradicate diverse ROS with high efficiency and velocity was developed. It shifted microglial polarization from a pro-inflammatory M1 phenotype to an anti-inflammatory M2 phenotype after stress stimuli, which remodels the immuno-environment from detrimental to protective for the neuronal cells by blocking the pro-inflammatory signaling pathway. (Copyright 2018 John Wiley and Sons)

Recently, Li et al. suggested that the removal of ROS by nanoceria not only directly protects neurons from oxidative damage but also indirectly protects neurons by regulating the phenotypic polarization of microglia in the brain [74]. Microglia are resident immune cells in the brain, whose morphology is highly plasticity. When stimulated, the resting microglia (M0) will be polarized into the pro-inflammatory phenotype (M1) or the anti-inflammatory phenotype (M2). In most cases, microglia with M2 phenotype are recruited in large numbers because they can reconstruct tissue homeostasis and heal wounds [75]. However, in many CNS diseases, these anti-inflammatory M2 phenotype cells may shift to pro-inflammatory M1 phenotype, which aggravates the diseases [76]. Although the mechanism of M2 shift to M1 has not been fully studied, many studies have shown that ROS plays an important role in this process, and the cellular level of ROS in M1 is also higher than that in M2 as a result of the upregulated anaerobic glycolysis [77–79]. In the work done by Li et al., a PEG-coated nanoceria (CeNP-PEG) was synthesized and used to reverse the transformation from M2 to M1 in BV-2 cells and reconstruct cell homeostasis (Fig. 6.14) [74]. After hypoxia and glucose deprivation treatment, ROS in BV-2 cells increased significantly, while pretreatment with CeNP-PEG effectively inhibited the ROS level. Additionally, immunofluorescence tests using both CD16/32 as a M1 marker and CD206 as a M2 marker showed that the BV-2 cells pretreated with CeNP-PEG significantly upregulated CD206 and inhibited CD16/32 expression. Moreover, RT-PCR assay showed that the expression of pro-inflammatory factors (e.g., IL-1β, IL-6, and TNF-α) was downregulated in cells pretreated with CeNP-PEG, while the expression of anti-inflammatory factors (IL-4, IL-10, and TGF-β) was upregulated. Further study showed that the mechanism of CeNP-PEG regulating microglia phenotype transition is mainly due to its ability to block the ROS-induced inflammatory NF-κB signaling pathway, whose activation triggers a series of inflammatory activities [74, 80, 81].

### 6.4.2 Application of Nanozymes in Alzheimer's Disease

Oxidative stress events may exacerbate inflammatory-related pathological changes in Alzheimer's disease (AD) [82, 83]. Recent reports have shown that oxidative stress-mediated binding of advanced glycation end products (AGE) to their receptors (RAGE) can exacerbate the inflammatory response and cytokine release of microglia, resulting in the loss of AD neurons [84]. In AD patients, activation of both reactive astrocytes and microglia involves oxidative stress and inflammation mediated by β-amyloid protein (Aβ), as well as immune response to brain damage [85, 86]. Aβ can induce the production of NOX2-derived ROS and become the initiator of inflammatory cascade reaction [87]. Toll-like receptors (TLR2 and TLR4) as biomarkers to characterize the pathophysiology of AD play an important role in AD and have been fully studied [88]. The toxicity of Aβ or the increased lipid peroxidation product, 4-hydroxynonenal (HNE), significantly increased the expression of TLR4 in neurons, suggesting that oxidative stress did promote the loss of AD neurons [89].

Some water-soluble $C_{60}$ derivatives have been reported to be helpful in the treatment of AD in cell or animal models. As introduced in Sect. 6.3.1.2, the tris-malonic acid $C_{60}$ derivative ($C_3$) can alleviate $A\beta_{1-42}$-mediated nerve damage by clearing ROS [46]. Along with inhibiting ROS produced by Aβ aggregation, some $C_{60}$ derivatives have also been found to directly inhibit Aβ aggregation. Lee et al. have suggested that a kind of $C_{60}$ derivatives (1,2-dimethoxymethano-$C_{60}$) could inhibit the aggregation of Aβ by binding to the central hydrophobic motif (KLVFF) of Aβ peptide (Fig. 6.15) [90]. Compared with melatonin, a reported Aβ inhibitor, the inhibit constant ($IC_{50}$) value of this $C_{60}$ derivative is at least 4 times higher for $A\beta_{1-40}$ and 15 times higher for $A\beta_{11-25}$ [90].

Wei et al. used replication-exchange molecular dynamics (REMD) simulation to show that water-miscible $C_{60}$ can significantly prevent the formation of β-sheet of $A\beta_{16-22}$ peptides (with a molar ratio of $C_{60}$ to peptides greater than 1:8) [91]. The inhibitory effect of $C_{60}$ on $A\beta_{16-22}$ fibrillation was also confirmed by atomic force microscopy (AFM) experiments. Besides, using REMD simulation, these researchers found that fullerene $C_{180}$, which has the same number of carbon atoms as three $C_{60}$ molecules ($3C_{60}$) but smaller surface area than $3C_{60}$, exhibits a surprisingly stronger inhibitory effect on the β-sheet formation of $A\beta_{16-22}$ peptides (Fig. 6.16). Further analysis of the fullerene–peptide interaction shows that the strong interaction between fullerene and $A\beta_{16-22}$ peptide remarkably blocks the $A\beta_{16-22}$ peptide–peptide interaction, which is significant for the formation of

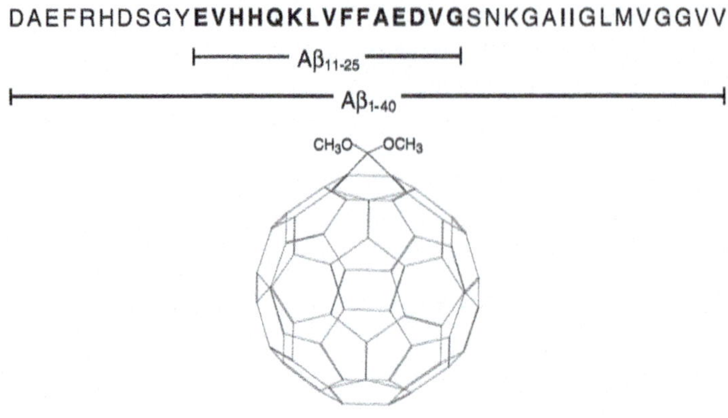


**1,2-(dimethoxymethano)fullerene**

**Fig. 6.15** The amyloid peptides and the molecular structure of 1,2-(dimethoxymethano) fullerene [90]. (Copyright 2003 Elsevier)

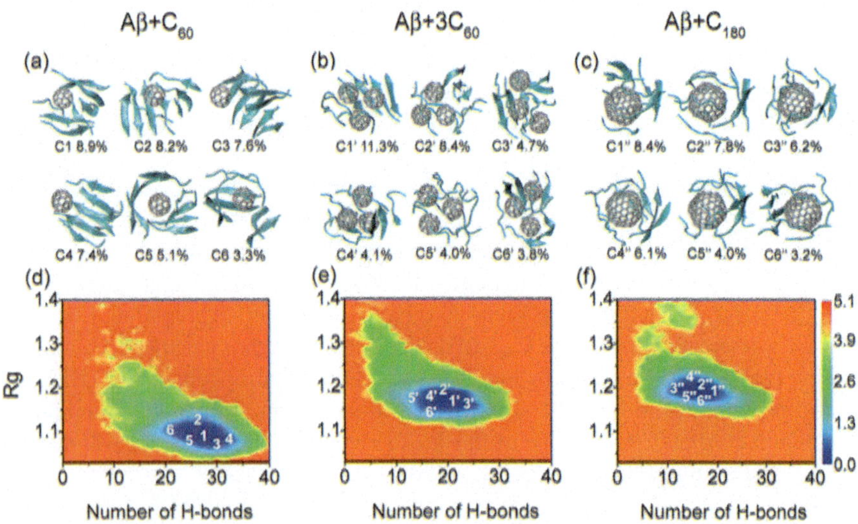


**Fig. 6.16** Structural analysis and potential of mean force (PMF) for $A\beta_{16-22}$ octamers in aqueous solution in the presence of $C_{60}$, $3C_{60}$, and $C_{180}$ nanoparticles [91]. Representative structures for the top six most-populated clusters of $A\beta_{16-22}$ octamers in Aβ+ $C_{60}$ (**a**), Aβ + $3C_{60}$ (**b**), and Aβ + $C_{180}$ (**c**) systems. The PMF (in kcal $mol^{-1}$) of $A\beta_{16-22}$ octamers plotted as a function of the number of intermolecular H-bonds and Rg of the $A\beta_{16-22}$ octamers in Aβ + $C_{60}$ (**d**), Aβ + $3C_{60}$ (**e**), and Aβ + $C_{180}$ (**f**) systems. (Copyright 2014 RSC Pub)

β-sheet, thus delaying the fibrillation of $A\beta_{16-22}$. As for the stronger inhibition of $C_{180}$ on β-sheet formation than $3C_{60}$, the stronger hydrophobic and aromatic stacking interaction between $C_{180}$'s hexagonal rings and the Phe rings as compared to the pentagonal rings of $C_{60}$ is proposed to be the main reason [91].

Because of its reliable ability to remove ROS based on their SOD-like or catalase-like activity, nanoceria has been widely studied for its potential in the treatment of AD. Cimini et al. treated human neuroblastoma cell line SH-SY5Y stimulated by aggregated $A\beta_{25-35}$ fibers with nanoceria and studied in detail the mechanism of nanoceria in protecting nerve cells [92]. They showed that nanoceria not only reduced the upregulation of ROS induced by $A\beta_{25-35}$ but also positively regulated neurodifferentiation markers (e.g., β-tubulin III, GAP-43), antioxidant enzymes (e.g., SOD, catalase), and signal pathways (e.g., PPARβ, BDNF, TrkB) related to neuro-survival to alleviate the neurological damage caused by $A\beta_{25-35}$. Unmodified nanoceria has no specific recognition ability for Aβ and poor biocompatibility in vivo. To directly target nanoceria to brain Aβ-aggregated pathological tissues in vivo, Cimini et al. further coupled nanoceria with antibodies targeting Aβ, and modified the surface of nanoceria with PEG, which can reduce the phagocytosis of immune system and increase the biocompatibility of nanoceria [93]. The protective effects of this polyethylene glycol (PEG)-coated and antibody-conjugated nanoceria on the human AD cell models were all similar to that of the unmodified nanoceria used in the previous-mentioned study [92]. The activation of BDNF/TrkB/ERK5, a neuronal survival pathway, was observed to be an essential factor in the process of nanoceria inhibiting Aβ-induced cell damage. Also, due to the coupling of PEG, nanoceria showed little toxicity to cells and exhibited more lasting neuroprotective effects. Antibody coupling enables nanoceria to target sites containing Aβ aggregates, but this study is only a proof of concept study in vitro, and there is no enough evidence to support this concept in vivo as yet [93]. These works imply that antioxidative nanozymes may interact directly or indirectly with many cellular components while scavenging ROS, thus exerting positive or negative effects on cells [92, 93].

In addition to affecting intracellular signaling pathways, the localization of antioxidative nanozymes in cells is also important. By modifying the surface of nanoceria with triphenylphosphonium (TPP), Jung et al. synthesized an antioxidative nanozyme (TPP-ceria) targeting to mitochondria (Fig. 6.17) [94]. Mitochondria is a major source of ROS production, and ROS-mediated mitochondrial dysfunction has been reported to accelerate the buildup of Aβ plaques in the brain [95]. Therefore, specific elimination of ROS in mitochondria may improve the

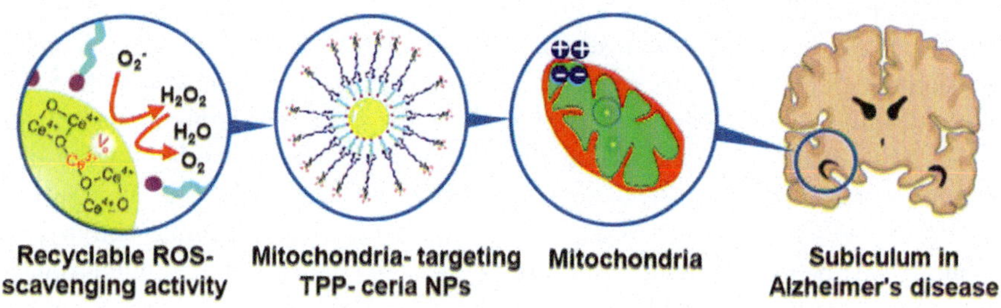


**Fig. 6.17** Mitochondria-targeting ceria nanoparticles as antioxidants for Alzheimer's disease [94]. (Copyright 2016 American Chemical Society)

therapeutic effect of AD. Generally, unmodified nanoceria mostly enters lysosomes in cells through endocytosis. However, in the research of Jung et al., because of the strong interaction between lipophilic cationic TPP and mitochondrial membrane, TPP modification makes it possible to localize nanoceria into mitochondria [94]. As well, by comparing different TPP-ceria, they found that the TPP-ceria with the highest efficiency of entering mitochondria should have the following properties: small core size (3 nm), small hydrodynamic diameter (22 nm), positive ζ-potential (+45 mv), good colloidal stability, and hydrophobicity. The antioxidative ability of TPP-ceria based on the SOD-like or catalase-like activities of nanoceria was also verified in SH-SY5Y cell model. Afterward, the researchers injected FITC-conjugated TPP-ceria into the subicula of 5XFAD mice, a recognized AD model with neuronal cell death and aggregation of Aβ plaque in subiculum, and found that TPP-ceria could effectively inhibit the activation of inflammatory-related microglia and the death of neurons. However, TPP-ceria showed no significant effects on the aggregation of Aβ, indicating that the neuroprotective effect of TPP-ceria on AD mice was mainly based on the ability of nanoceria to scavenge ROS [94].

Although nanoceria cannot directly block the aggregation of Aβ, it can combine with another nanozyme, polyoxometalates (POMs) with the proteinase-like activity for the hydrolysis of Aβ, to achieve this goal. Qu et al. synthesized a hybrid of nanoceria and POMs (CeONP@POMs) and found that this hybrid could protect nerve cells by scavenging ROS and degrading Aβ at the same time [29]. Moreover, it is noteworthy that this nanozyme hybrid has also been demonstrated to traverse BBB in vitro and in vivo models. Although the mechanism is still unclear, the researchers believe that these materials may be transported through BBB by macrophages [29]. POMs are a class of metal–oxygen clusters formed by oxygen bonding of pre-transition metal ions mainly including V, Nb, Ta, Mo, and W, which have the ability to form metal–oxygen cluster anions [96, 97]. POMs have been studied in AD for its interesting proteinase-like ability to hydrolyze Aβ, even if it is not an antioxidative nanozyme. Qu et al. have designed a kind of POMs with a Wells–Dawson structure (POM-Dawson) and found that it could effectively inhibit the peroxidase activity the aggregation of Aβ [98]. Then, in order to improve the specific affinity of POM-Dawson with Aβ, they synthesized a series of POM-Dawson derivatives (POMds) with defined histidine-chelating metal (e.g., Ni, Co, Cu, Fe, and Mn) binding site by high-throughput screening method and found that these POMds specifically bound to the histidine$_{13}$–histidine–glutamine–lysine$_{16}$ (HHQK) cationic cluster of Aβ [99]. Interestingly, these POMds are negatively charged under physiological conditions and can attack peptide bonds in a similar way like the electronegative hydroxyl groups of Ser amino acids in natural serine proteases.

By using gold nanoparticles as scaffolds, Qu et al. combined this protease-like POMds with a His-rich hepta-peptide (N-His-Sar-His-Sar-His-Sar-His, Sar = sarcosine), which have been reported to possess excellent SOD-like activity, to obtain a bifunctional nanozyme (AuNPs@POMD-8pep) capable of simultaneously degrading Aβ and eliminating ROS (Fig. 6.18a) [100]. With N-α-benzoyl-DL-arginine-4-nitroanilide (BAPNA) as substrate, the protease activity of AuNPs@POMD-8pep ($k_{cat}/K_M = (8.26 \pm 0.63) \times 10^5\ L \cdot g^{-1} \cdot min^{-1}$) has been proved

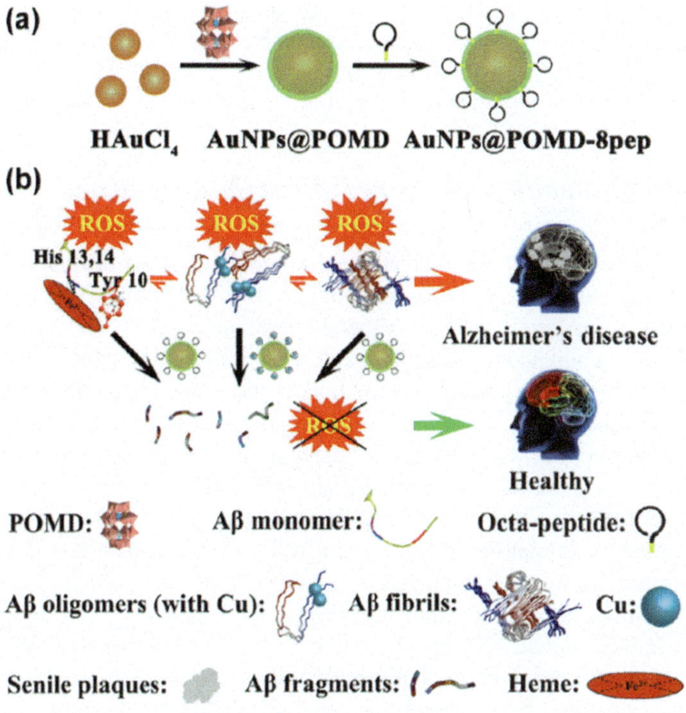


**Fig. 6.18** Aβ pathways influenced by AuNPs@POMD-8pep [100]. (**a**) Synthetic route of the nanozyme. (**b**) AuNPs@POMD-8pep acted as a multifunctional nanozyme to modulate multiple facets of Alzheimer's disease. (Copyright 2016 Springer Nature)

to be higher than that of commercialized trypsin ($k_{cat}/K_M = (2.86 \pm 0.31) \times 10^3$ L·g$^{-1}$ min$^{-1}$). With pyrogallol as substrate, the SOD-like activity of AuNPs@POMD-8pep was also confirmed, and its SOD-like activity was found to be enhanced in the presence of copper ions. Copper ions in cells can form Aβ/Cu complex with Aβ to induce cytotoxicity [101], while AuNPs@POMD-8pep was found to release Cu ions in Aβ/Cu complex and chelate with free Cu ions, and the chelated Cu can further enhance the SOD-like activity of AuNPs@POMD-8pep (Fig. 6.18b). Therefore, AuNPs@POMD-8pep could alleviate the symptoms of AD not only through degrading Aβ and eliminating ROS but also through regulating the metal homeostasis in nerves [100].

$Fe_3O_4$ nanoparticles exhibit pH-dependent enzymatic activity, peroxidase-like activity in acidic conditions (pH = 4.8), and catalase activity in neutral conditions (pH = 7.4) (Fig. 6.7) [62]. The peroxidase-like activity of $Fe_3O_4$ nanozyme has been widely used in antimicrobial and cancer treatment, but its catalase-like activity has not received much attention as yet. Song et al. have tested the protective effect of $Fe_3O_4$ nanozyme on a *Drosophila* AD model [102]. They found that dietary $Fe_3O_4$ nanozyme prolonged the life of aged *Drosophila* and inhibited the neurodegeneration of a *Drosophila* AD model. Although this study showed that

$Fe_3O_4$ retains catalase-like activity in vivo, the safety of $Fe_3O_4$ in vivo needs careful consideration because it may produce destructive free radicals catalyzed by its peroxidase-like activity under acidic conditions.

### *6.4.3 Application of Nanozymes in Cerebral Ischemia*

In cerebral ischemia (CI), the activation of microglia and astrocytes triggered by oxidative stress can significantly increase the levels of inflammatory mediators such as cytokines, chemokines, and matrix metalloproteinases [103]. Then, upregulation of cell adhesion molecules leads to neutrophil infiltration, which destroys the integrity of brain endothelial cells. In addition, it may cause secondary inflammation and ultimately permanent brain cell damage. The increase of free radicals after cerebral ischemia can also lead to the upregulation of several NF-κB-mediated pro-inflammatory genes [103]. Besides, although innate immunity activated by Toll-like receptors seems to promote regeneration in CI [104], ROS-induced TLR-mediated inflammatory response could still lead to neuronal loss [105].

Erlichman et al. found that nanoceria reduced ischemic cell death by about 50% in a mouse hippocampal brain slice model of cerebral ischemia [106]. In this study, nanoceria reduced about 15% of $O_2^{\bullet-}$ and NO in ischemic models. In addition, nanoceria was also observed to significantly reduce the level of 3-nitrotyrosine by about 70%. Because 3-nitrotyrosine is a stable metabolite produced by the nitration of protein tyrosine residues induced by $ONOO^-$, researchers believed that the effective removal of $ONOO^-$ by nanoceria may be the main reason for nanoceria's neuroprotective effect [106]. Later, Hyeon et al. studied the therapeutic effect of nanoceria on ischemic stroke in vivo [107]. They synthesized a kind of PEG-coated nanoceria (PEG-ceria) and confirmed the ROS-scavenging ability of PEG-ceria in vitro. Then, using an induced rat model of ischemic stroke, they found that intravenously injected PEG-ceria effectively crossed BBB to the location of brain lesions, while PEG-ceria could hardly cross BBB in normal rats. Moreover, 0.5 and 0.7 mg $kg^{-1}$ of PEG-ceria could effectively reduce the infarct volume by 50% compared with the control group, while lower (0.1 and 0.3 mg $kg^{-1}$) or higher doses (1.0 and 1.5 mg $kg^{-1}$) of PEG-ceria showed little therapeutic effect (Fig. 6.19) [107].

As a new type of antioxidant with SOD-like or catalase-like activity, platinum nanozyme has also been shown to alleviate the symptoms of ischemic stroke. Abe et al. studied the therapeutic effect of Pt nanoparticles (nPt) on cerebral infarction in a mice model which subjected to transient middle cerebral artery occlusion (tMCAO) for 60 min [108]. They found that nPt significantly reduced infarct size, protected the components of neuro-endothelial vessels (collagen IV), and inhibited the activation of pro-inflammatory factors (MMP-9). In a follow-up study, Abe et al. used thrombolytic tissue plasminogen activator (tPA) as a negative control, which deteriorated motor function and cerebral infarction, further confirming the therapeutic effect of nPt on ischemic stroke by exerting SOD-like activity [109].

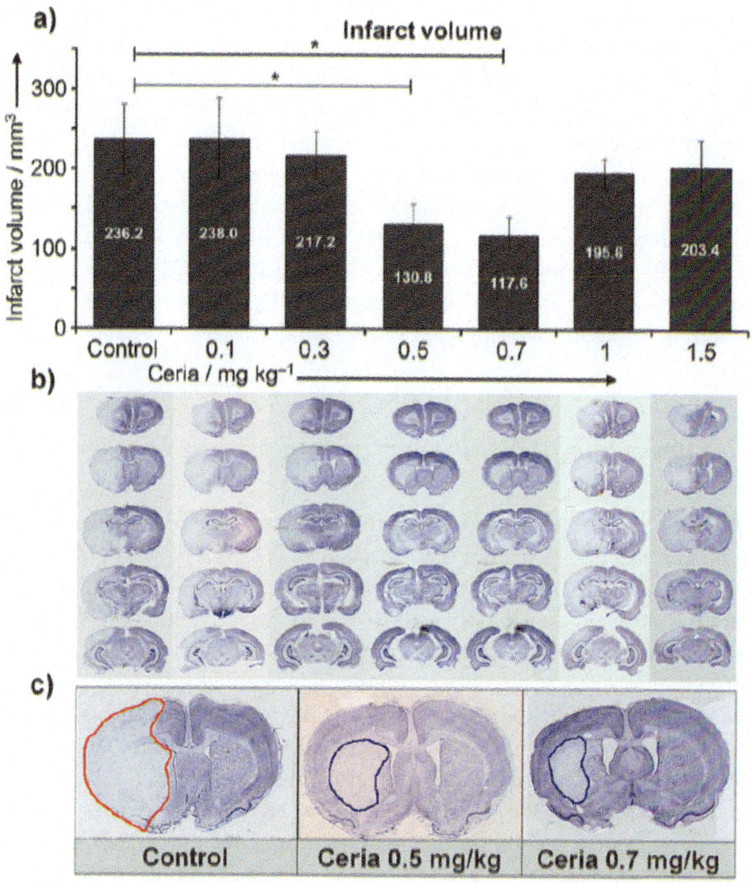


**Fig. 6.19** Infarct volume in vivo [107]. (**a**) Low-dose ceria nanoparticles (0.1 and 0.3 mg $kg^{-1}$) do not decrease infarct volumes, whereas 0.5 and 0.7 mg $kg^{-1}$ ceria nanoparticles considerably reduce infarct volumes, to as little as 50% of those of the control group (∗, $p<0.05$). Higher doses of ceria nanoparticles (1.0 and 1.5 mg $kg^{-1}$) do not exhibit protective effects against stroke (n = 12 for each group, except 0.1 and 1.5 mg $kg^{-1}$, where n = 6). (**b**) Brain slices from anterior (top) to posterior (bottom), with intervals of 2 mm. On Nissl-stained brains, infarcts are shown as pale blue-colored lesions, while undamaged region are stained as deep blue. Infarct areas maximally decreased at 0.5 and 0.7 mg $kg^{-1}$ ceria nanoparticles. (**c**) Representative slices, clearly showing that 0.5 and 0.7 mg $kg^{-1}$ ceria nanoparticles can significantly reduce infarct volumes. (Copyright 2012 John Wiley and Sons)

One third of stroke patients show hyperglycemia, which is closely related to ROS. Based on the SOD-like of PEG-conjugated carbon clusters (PEG-HCCs), Kent et al. tested the therapeutic potential of PEG-HCCs in a severe model of reversible middle cerebral artery stroke in acutely hyperglycemic rats [110]. In vitro experiments showed that PEG-HCCs could effectively relieve the damage of $H_2O_2$ in the neuro-endothelial cell line bEnd.3, which is an important target of hyperglycemia in stroke. After that, in a rat model of transient middle cerebral artery occlusion (tMCAO) complicated by acute hyperglycemia, PEG-HCCs alleviated the symptoms of reperfusion injury, such as infarct volume, hemorrhage conversion rate, and hemispheric swelling. Because vascular dysfunction caused by hyperglycemia has an important impact on stroke, based on experimental evidence in vitro and in vivo, researchers believed that the improvement of worsened vascular outcomes may be the main way for PEG-HCCs to play a therapeutic role in stroke [110]. Besides, some studies have also shown that PEG-HCCs alleviated cerebrovascular dysfunction in traumatic brain injury [111, 112]. Although PEG-HCCs have shown therapeutic potential in animal models of stroke, its ability to cross BBB, mechanism of action, and its safety in vivo still need further study.

### *6.4.4 Application of Nanozymes in Parkinson's Disease*

In Parkinson's disease (PD), there is considerable evidence of degeneration of both dopaminergic and non-dopaminergic cells [113, 114]. Loss of key dopaminergic neurons involves oxidative stress and neuroinflammation. Increased levels of inducible nitric oxide synthase (iNOS) lead to microglia activation, T-cell infiltration, astrocyte proliferation, and ultimately accumulation of $O_2^{\bullet-}$ and NO free radicals [114]. The higher levels of myeloperoxidase produced by astrocyte reaction can also increase the levels of •OH and $NO_2^-$ free radicals, which may subsequently lead to the loss of neurons in PD [114]. In addition, overexpression of cyclooxygenase-2 (COX-2) can cause the loss of dopaminergic neurons through oxidative stress-mediated inflammation [114]. Similarly, in PD, NADPH oxidase is upregulated, which is associated with ROS production and inflammation [115]. It also activates microglia and subsequently causes the loss of dopaminergic neurons. Because neuroglial cell activation is an important event in the pathogenesis of PD, altered levels of cytokines and ROS in neuroglial cells and accumulation of ROS derived from NADPH can cause neurotoxicity to PD [116, 117].

As mentioned above, PD is closely related to ROS. In addition, studies have shown that iron is enriched in the substantia nigra of Parkinson's patients [118, 119], and iron may produce a large number of ROS through the Fenton reaction, thus aggravating Parkinson's disease. Therefore, the effective removal of iron-induced ROS may be helpful for the treatment of PD. Ho et al. studied the protective effect of carboxyfullerenes (a water-soluble carboxylic acid derivative of fullerene, C regioisomer) on iron-induced neuro-oxidative damage in vitro and in vivo [120]. They found that carboxyfullerenes inhibited the autoxidation of brain homogenate (excluding iron) and the increase of iron-induced lipid peroxidation in a dose-dependent manner in vitro. Intracerebral perfusion of specific carboxyfullerenes did not increase lipid peroxidation in substantia nigra or depletion of dopamine in striatum. Co-infusion of carboxyfullerenes prevented oxidative damage induced by iron, which is manifested by reduced lipid peroxidation in substantia nigra, inhibition of K1-induced dopamine spillover, and maintenance of striatal dopamine content in vivo [120]. These studies suggest that carboxyfullerenes may be used as a potential antioxidant in the treatment of PD.

Mugesh et al. found that flower-shaped $Mn_3O_4$ nanozymes (Mnf) exhibited SOD-like, catalase-like, and GPx-like activities and that the redox regulation function of Mnf could effectively protect Parkinson's disease-related cell models (SHSY-5Y) from $MPP^+$-induced cytotoxicity [65]. Therefore, Mnf may be used as a potential therapeutic drug for Parkinson's disease caused by oxidative damage, but there is no evidence in vivo and further research is needed.

### 6.4.5 *Application of Nanozymes in Multiple Sclerosis*

In multiple sclerosis (MS), demyelination and axonal injury caused by oxidative stress and inflammation are widely reported [121]. NOX2-derived ROS in microglia is essential for the phagocytosis of myelin sheath cells, but overexpression of ROS can also damage myelin sheath cells [122]. Subsequent peroxynitrite molecules can cause serious damage to brain cells, especially neurons. Recently, ROS compounds have been considered as early pathogenic factors of MS-related inflammation and subsequently lead to a large loss of oligodendrocytes and axons [123]. In addition, oxidative stress-induced early mitochondrial damage plays an important role in subsequent inflammatory response and neuron loss in MS [124].

Erlichman et al. reported the in vivo properties of nanoceria nanozyme in a mice model with MS, experimental autoimmune encephalomyelitis (EAE), induced by ROS-mediated oxidative damage [125]. They synthesized a kind of nanoceria with a diameter of 2.9 nm and coated with citrate/EDTA, which remained monodispersed in high-ionic-strength saline for a long time. The half-life of nanoceria in vivo was about 4.0 h, and when injected intravenously into mice, nanoceria was well tolerated by mice and absorbed by liver and spleen. In addition, nanoceria can penetrate the brain, reduce the level of ROS, and alleviate the clinical symptoms and motor deficits of mice. In conclusion, this kind of nanoceria may help to alleviate MS caused by free radical accumulation in biological system.

## 6.5 Challenges and Perspectives

This chapter introduces a new type of antioxidants, inorganic nanozymes with intrinsic enzyme-like activity, and their research progress in the treatment of brain diseases. Abnormally elevated ROS plays an important role in the occurrence and deterioration of many brain diseases. Specific elimination of excessive ROS in the lesions and restoring homeostasis to normal levels are considered as the key to the treatment of brain diseases related to oxidative stress. At present, many antioxidants have been tested in the treatment of brain diseases. Although some antioxidants have been clinically approved, their efficacy is far from satisfactory. With the increasing incidence of brain-related diseases, it is urgent to develop new and effective antioxidant drugs. Traditional antioxidants are mostly organic natural molecules, including antioxidant enzymes and small organic molecules. Most of these antioxidants are inherent components of the body's antioxidant system, so they can effectively regulate the level of ROS. However, because of the poor stability of the biogenic molecules, the high requirement of reaction environment, and the short time to maintain curative effect, it is difficult to achieve ideal therapeutic effect in the practical application of treating brain diseases. Nanozymes are a new type of nanomaterials reported in recent years which have enzymatic activities such as SOD, catalase, peroxidase, oxidase, and so on. Because of its stable structure, long-term preservation of enzymatic activity, and easy preparation in large quantities, these materials have the advantages that bioorganic molecules do not possess, which attracts researchers to select nanozymes with antioxidant capacity for the treatment of neurological diseases. As mentioned earlier, nanozymes used in neurological diseases mainly contain SOD-like or catalase-like activities, which can effectively remove ROS accumulated in the body.

Although many antioxidative nanozymes have been studied for the treatment of neurological diseases, there is still a long way to go before they can be used in clinical treatment. The challenges of nanozymes in brain diseases are mainly as follows:

1. The catalytic efficiency of nanozymes is far from the level of natural enzymes, and the research on their substrate selectivity is also very limited. The substrate selectivity of nanozymes is so deficient that there are few one-substrate nanozymes being reported, which, on the one hand, may limit their catalytic efficiency and, on the other hand, make the in vivo behavior of nanozymes more complex. At present, most of the nanozymes contain more than one enzyme-like activities, which enable them to catalyze a variety of ROS at the same time, but nanozymes containing both SOD-like and catalase-like activities under the same conditions are rarely reported. In addition, although nanozymes have obvious catalytic activity in vitro, they may be affected by various biological components in vivo, which may affect their catalytic activity. Most of the nanozymes enter the liver, kidney, and other metabolic organs once injected in vivo, thus being a residual in the organs or being excreted out of the body. Besides, nanozymes may also bind to various cell components, such as proteins, lipids, ATP, and so on, thus blocking their active sites. Moreover, different localization of nanozymes in cells may also have an important impact on their enzymatic activity. How to use appropriate methods to avoid the negative effects of the in vivo environment on the activity of nanozymes and how to accurately measure the catalytic behavior of nanozymes in vivo is one of the biggest challenges in the future research of nanozymes.
2. The safety of nanozymes in vivo should not be neglected. Although nanozymes with stable structure may maintain a sustained catalytic efficiency in vivo compared with natural enzymes, they may also bring long-term chronic toxicity. The nanozymes currently used as antioxidants may become prooxidants under certain

conditions. For example, $Fe_3O_4$ has catalase-like activity in neutral condition but can catalyze the production of •OH once in acidic condition. Besides, as exogenous inorganic substances with poor biocompatibility, nanozymes can easily activate the immune system, which may lead to inflammation and brain damage. Because it is difficult to excrete nanozymes through physiological pathways once they are retained in brain tissues, the long-term effects of nanozymes on the body need to be further studied.

3. The penetration efficiency of nanozymes to BBB has not been deeply studied. BBB is a difficult obstacle in the treatment of brain diseases, because most foreign molecules cannot effectively cross BBB. Although nanozymes have been reported to alleviate symptoms of brain diseases in some animal models, their ability and mechanism to penetrate BBB has rarely been thoroughly studied. Some studies used macrophages as carriers to deliver antioxidants to the brain, but this strategy has not been tried on nanozymes. Nevertheless, because nanozymes are ideal nanocarriers, their BBB traversal ability may be achieved by coupling a variety of functional molecules.

Therefore, exploring novel nanozymes with high efficiency and good biosafety and applying them to brain diseases related to oxidative stress will be a hot research topic in the future. With the efforts of researchers, more and more new nanozyme antioxidants have been found and have shown good performance in the treatment of brain diseases. It is expected that nanozymes can be used from bench to bedside in the treatment of brain diseases in the future.

**Acknowledgments** This work was financially supported by the National Natural Science Foundation of China (No. 31530026, 31871005, 31900981), Chinese Academy of Sciences under Grant No. YJKYYQ20180048, Youth Innovation Promotion Association CAS (2019093), Strategic Priority Research Program (No. XDB29040101), Key Research Program of Frontier Sciences (No. QYZDY-SSW-SMC013), Chinese Academy of Sciences, National Key Research and Development Program of China (No. 2017YFA0205501).

## References

1. Hsieh HL, Yang CM. Role of redox signaling in neuroinflammation and neurodegenerative diseases. Biomed Res Int. 2013;2013:484–613.
2. Patel M. Targeting oxidative stress in central nervous system disorders. Trends Pharmacol Sci. 2016;37(9):768–78.
3. Jones DP. Redefining oxidative stress. Antioxid Redox Sign. 2006;8(9–10):1865–79.
4. Chovatiya R, Medzhitov R. Stress, inflammation, and defense of homeostasis. Mol Cell. 2014;54(2):281–8.
5. Jan AT, Azam M, Siddiqui K, Ali A, Choi I, Haq QMR. Heavy metals and human health: mechanistic insight into toxicity and counter defense system of antioxidants. Int J Mol Sci. 2015;16(12):29592–630.
6. Murphy MP. How mitochondria produce reactive oxygen species. Biochem J. 2009;417 (1):1–13.

7. Gao L, Zhuang J, Nie L, Zhang J, Zhang Y, Gu N, Wang T, Feng J, Yang D, Perrett S. Intrinsic peroxidase-like activity of ferromagnetic nanoparticles. Nat Nanotechnol. 2007;2(9):577–83.
8. Wei H, Wang E. Nanomaterials with enzyme-like characteristics (nanozymes): next-generation artificial enzymes. Chem Soc Rev. 2013;42(14):6060–93.
9. Day BJ. Catalytic antioxidants: a radical approach to new therapeutics. Drug Discov Today. 2004;9(13):557–66.
10. Shin HY, Park TJ, Kim MI. Recent research trends and future prospects in nanozymes. J Nanomater. 2015;2015:1–11.
11. Ragg R, Tahir MN, Tremel W. Solids go bio: inorganic nanoparticles as enzyme mimics. Eur J Inorg Chem. 2016;2016(13–14):1906–15.
12. Loschen G, Flohe L, Chance B. Respiratory chain linked $H_2O_2$ production in pigeon heart mitochondria. FEBS Lett. 1971;18(2):261–4.
13. Loschen G, Azzi A, Richter C, Flohé L. Superoxide radicals as precursors of mitochondrial hydrogen peroxide. FEBS Lett. 1974;42(1):68–72.
14. Weisiger RA, Fridovich I. Superoxide dismutase. Organelle specificity. J Biol Chem. 1973;248(10):3582–92.
15. Turrens JF. Mitochondrial formation of reactive oxygen species. J Physiol. 2010;552 (2):335–44.
16. Silvia S, Karl-Heinz K. NOX enzymes in the central nervous system: from signaling to disease. Antioxid Redox Sign. 2009;11(10):2481–504.
17. Bedard K, Krause KH. The NOX family of ROS-generating NADPH oxidases: physiology and pathophysiology. Physiol Rev. 2007;87(1):245–313.
18. B H. Reactive oxygen species and the central nervous system. J Neurochem. 1992;59 (5):1609–23.
19. Urrutia PJ, Mena NP, Nunez MT. The interplay between iron accumulation, mitochondrial dysfunction, and inflammation during the execution step of neurodegenerative disorders. Front Pharmacol. 2014;5(38):38.
20. Ortiz GG, Pacheco Moises FP, Mireles-Ramirez M, Flores-Alvarado LJ, Gonzalez-Usigli H, Sanchez-Gonzalez VJ, Sanchez-Lopez AL, Sanchez-Romero L, Diaz-Barba EI, Santoscoy-Gutierrez JF, Rivero-Moragrega P. Oxidative stress: love and hate history in central nervous system. Adv Protein Chem Struct Biol. 2017;108:1–31.
21. Batrakova EV, Li S, Reynolds AD, Mosley RL, Bronich TK, Kabanov AV, Gendelman HE. A macrophage-nanozyme delivery system for Parkinson's disease. Bioconjug Chem. 2007;18 (5):1498–506.
22. Brynskikh AM, Zhao Y, Mosley RL, Li S, Boska MD, Klyachko NL, Kabanov AV, Gendelman HE, Batrakova EV. Macrophage delivery of therapeutic nanozymes in a murine model of Parkinson's disease. Nanomedicine. 2010;5(3):379–96.
23. Rosenbaugh EG, Manickam DS, Batrakova EV, Kabanov AV, Zimmerman MC. Nanoformulated copper/zinc superoxide dismutase increases neuronal uptake via active endocytosis. Free Radical Bio Med. 2010;49:S195–6.
24. Haney MJ, Zhao Y, Li S, Higginbotham SM, Booth SL, Han H-Y, Vetro JA, Mosley RL, Kabanov AV, Gendelman HE, Batrakova EV. Cell-mediated transfer of catalase nanoparticles from macrophages to brain endothelial, glial and neuronal cells. Nanomedicine. 2011;6 (7):1215–30.
25. Zhao Y, Haney MJ, Mahajan V, Reiner BC, Dunaevsky A, Mosley RL, Kabanov AV, Gendelman HE, Batrakova EV. Active targeted macrophage-mediated delivery of catalase to affected brain regions in models of Parkinson's disease. J Nanomed Nanotechnol. 2011;S4 (1):003
26. Manickam DS, Brynskikh AM, Kopanic JL, Sorgen PL, Klyachko NL, Batrakova EV, Bronich TK, Kabanov AV. Well-defined cross-linked antioxidant nanozymes for treatment of ischemic brain injury. J Control Release. 2012;162(3):636–45.
27. Jiang Y, Brynskikh AM, S-Manickam D, Kabanov AV. SOD1 nanozyme salvages ischemic brain by locally protecting cerebral vasculature. J Control Release. 2015;213:36–44.

28. Lin Y, Ren J, Qu X. Catalytically active nanomaterials: a promising candidate for artificial enzymes. Acc Chem Res. 2014;47(4):1097–105.
29. Guan Y, Li M, Dong K, Gao N, Ren J, Zheng Y, Qu X. Ceria/POMs hybrid nanoparticles as a mimicking metallopeptidase for treatment of neurotoxicity of amyloid-beta peptide. Biomaterials. 2016;98:92–102.
30. Zelko IN, Mariani TJ, Folz RJ. Superoxide dismutase multigene family: a comparison of the CuZn-SOD (SOD1), Mn-SOD (SOD2), and EC-SOD (SOD3) gene structures, evolution, and expression. Free Radic Biol Med. 2002;33(3):337–49.
31. McCord JM, Fridovich I. Superoxide dismutase. An enzymic function for erythrocuprein (hemocuprein). J Biol Chem. 1969;244(22):6049–55.
32. Boaro M, Giordano F, Recchia S, Santo VD, Giona M, Trovarelli A. On the mechanism of fast oxygen storage and release in ceria-zirconia model catalysts. Appl Catal B Environ. 2004;52 (3):225–37.
33. Trovarelli A, Leitenburg CD, Boaro M, Dolcetti G. The utilization of ceria in industrial catalysis. Catal Today. 1999;50(2):353–67.
34. Korsvik C, Patil S, Seal S, Self WT. Superoxide dismutase mimetic properties exhibited by vacancy engineered ceria nanoparticles. Chem Commun (Camb). 2007;(10):1056–8.
35. Deshpande S, Patil S, Kuchibhatla SV, Seal S. Size dependency variation in lattice parameter and valency states in nanocrystalline cerium oxide. Appl Phys Lett. 2005;87(13):223–78.
36. Ying JY, Tschöpe A. Synthesis and characteristics of non-stoichiometric nanocrystalline cerium oxide-based catalysts. Chem Eng J Biochem Eng J. 1996;64(2):225–37.
37. Celardo I, Pedersen JZ, Traversa E, Ghibelli L. Pharmacological potential of cerium oxide nanoparticles. Nanoscale. 2011;3(4):1411.
38. Krusic PJ, Wasserman E, Keizer PN, Morton JR, Preston KF. Radical reactions of $C_{60}$. Science. 1991;254(5035):1183–5.
39. Lotharius J, Dugan LL, O'Malley KL. Distinct mechanisms underlie neurotoxin-mediated cell death in cultured dopaminergic neurons. J Neurosci. 1999;19(4):1284–93.
40. Lin AM, Chyi BY, Wang SD, Yu HH, Kanakamma PP, Luh TY, Chou CK, Ho LT. Carboxyfullerene prevents iron-induced oxidative stress in rat brain. J Neurochem. 2010;72(4):1634–40.
41. Dugan LL, Lovett EG, Quick KL, Lotharius J, Lin TT, O'Malley KL. Fullerene-based antioxidants and neurodegenerative disorders. Parkinsonism Relat Disord. 2001;7(3):243–6.
42. Dugan LL, Turetsky DM, Du C, Lobner D, Wheeler M, Almli CR, Shen KF, Luh TY, Choi DW, Lin TS. Carboxyfullerenes as neuroprotective agents. P Natl Acad Sci USA. 1997;94 (17):9434–9.
43. Huang SS, Tsai SK, Chih CL, Chiang L-Y, Hsieh HM, Teng CM, Tsai MC. Neuroprotective effect of hexasulfobutylated $C_{60}$ on rats subjected to focal cerebral ischemia. Free Radical Bio Med. 2001;30(6):643–9.
44. Kroto H, Allaf AW, Balm S. $C_{60}$: Buckminsterfullerene. Chem Rev. 1991;91(6):1213–35.
45. Dugan LL, Gabrielsen JK, Yu SP, Lin TS, Choi DW. Buckminsterfullerenol free radical scavengers reduce excitotoxic and apoptotic death of cultured cortical neurons. Neurobiol Dis. 1996;3(2):129–35.
46. Dugan LL, Turetsky DM, Du C, Lobner D, Wheeler M, Almli CR, Shen CK, Luh TY, Choi DW, Lin TS. Carboxyfullerenes as neuroprotective agents. P Natl Acad Sci USA. 1997;94 (17):9434–9.
47. Bensasson RV, Brettreich M, Frederiksen J, Göttinger H, Hirsch A, Land EJ, Leach S, Mcgarvey DJ, Schönberger H. Reactions of e−aq, $CO_2^{\bullet -}$, HO, $O_2^{\bullet -}$ and $O_2(1\delta g)$ with a dendro[60]fullerene and $C_{60}[C(COOH)_2]n$ (n = 2–6). Free Radical Bio Med. 2000;29 (1):26–33.
48. Okuda K, Hirota T, Hirobe M, Nagano T, Mochizuki M, Mashino T. Synthesis of various water-soluble $C_{60}$ derivatives and their superoxide-quenching activity. Fuller Sci Technol. 2000;8(1–2):89–104.

49. Ali SS, Hardt JI, Quick KL, Kim-Han JS, Erlanger BF, Huang TT, Epstein CJ, Dugan LL. A biologically effective fullerene ($C_{60}$) derivative with superoxide dismutase mimetic properties. Free Radic Biol Med. 2004;37(8):1191–202.
50. Ali SS, Hardt JI, Dugan LL. SOD activity of carboxyfullerenes predicts their neuroprotective efficacy: a structure-activity study. Nanomed Nanotechnol. 2008;4(4):283–94.
51. Belgorodsky B, Fadeev L, Ittah V, Benyamini H, Zelner S. Formation and characterization of stable human serum albumin-tris-malonic acid [$C_{60}$] fullerene complex. Bioconjug Chem. 2005;16(5):1058–62.
52. Errol LGS, Daniela CM, Vladimir B, Brittany RB, Gang W, Austin P, Roderic HF, Robia GP, Thomas AK, Ah-Lim T. Highly efficient conversion of superoxide to oxygen using hydrophilic carbon clusters. P Natl Acad Sci USA. 2015;112(8):2343–8.
53. Kajita M, Hikosaka K, Iitsuka M, Kanayama A, Toshima N, Miyamoto Y. Platinum nanoparticle is a useful scavenger of superoxide anion and hydrogen peroxide. Free Radic Res. 2007;41(6):615.
54. Hamasaki T, Kashiwagi T, Imada T, Nakamichi N, Aramaki S, Toh K, Morisawa S, Shimakoshi H, Hisaeda Y, Shirahata S. Kinetic analysis of superoxide anion radical-scavenging and hydroxyl radical-scavenging activities of platinum nanoparticles. Langmuir. 2008;24(14):7354–64.
55. Lianbing Z, Linda L, Wolfram M, Eckhard P, Ulrich GS, Matthias B, Mato K. Reducing stress on cells with apoferritin-encapsulated platinum nanoparticles. Nano Lett. 2010;10(1):219–23.
56. Clark A, Zhu A, Kai S, Petty HR. Cerium oxide and platinum nanoparticles protect cells from oxidant-mediated apoptosis. J Nanoparti Res. 2011;13(10):5547–55.
57. Deisseroth A, Dounce AL. Catalase: physical and chemical properties, mechanism of catalysis, and physiological role. Physiol Rev. 1970;50(3):319–75.
58. Sumner JB. The chemical nature of catalase. Adv Enzymol Relat Areas Mol Biol. Wiley. 2006, 163–176.
59. Pirmohamed T, Dowding JM, Singh S, Wasserman B, Heckert E, Karakoti AS, King JE, Seal S, Self WT. Nanoceria exhibit redox state-dependent catalase mimetic activity. Chem Commun (Camb). 2010;46(16):2736–8.
60. Ivanov VK, Shcherbakov AB, Usatenko AV. Structure-sensitive properties and biomedical applications of nanodispersed cerium dioxide. ChemInform. 2010;41(9)
61. Nelson BC, Johnson ME, Walker ML, Riley KR, Sims CM. Antioxidant cerium oxide nanoparticles in biology and medicine. Antioxidants (Basel). 2016;5(2):E5.
62. Chen Z, Yin JJ, Zhou YT, Zhang Y, Song L, Song M, Hu S, Gu N. Dual enzyme-like activities of iron oxide nanoparticles and their implication for diminishing cytotoxicity. ACS Nano. 2012;6(5):4001–12.
63. Jianshuai M, Yan W, Min Z, Li Z. Intrinsic peroxidase-like activity and catalase-like activity of $Co_3O_4$ nanoparticles. Chem Commun (Camb). 2012;48(19):2540–2.
64. Jia F, Jun-Jie Y, Bo N, Xiaochun W, Ye H, Mauro F, Anderson GJ, Jingyan W, Yuliang Z, Guangjun N. Direct evidence for catalase and peroxidase activities of ferritin-platinum nanoparticles. Biomaterials. 2011;32(6):1611–8.
65. Singh N, Savanur MA, Srivastava S, D'Silva P, Mugesh G. A redox modulatory $Mn_3O_4$ nanozyme with multi-enzyme activity provides efficient cytoprotection to human cells in a Parkinson's disease model. Angew Chem Int Edit. 2017;56(45):14267.
66. Gao L, Yan X. Nanozymes: an emerging field bridging nanotechnology and biology. Sci China Life Sci. 2016;59(4):400–2.
67. Wang X, Hu Y, Wei H. Nanozymes in bionanotechnology: from sensing to therapeutics and beyond. Inorg Chem Front. 2016;3(1):41–60.
68. Sabens Liedhegner EA, Gao XH, Mieyal JJ. Mechanisms of altered redox regulation in neurodegenerative diseases-focus on S-glutathionylation. Antioxid Redox Sign. 2012;16 (6):543–66.

69. Jin H, Chen WQ, Tang XW, Chiang LY, Yang CY, Schloss JV, Wu JY. Polyhydroxylated $C_{60}$, fullerenols, as glutamate receptor antagonists and neuroprotective agents. J Neurosci Res. 2015;62(4):600–7.
70. Chen T, Li YY, Zhang JL, Xu B, Lin Y, Wang CX, Guan WC, Wang YJ, Xu SQ. Protective effect of $C_{60}$ -methionine derivate on lead-exposed human SH-SY5Y neuroblastoma cells. J Appl Toxicol. 2011;31(3):255–61.
71. Ren C, Hu X, Zhou Q. Graphene oxide quantum dots reduce oxidative stress and inhibit neurotoxicity *in vitro* and *in vivo* through catalase-like activity and metabolic regulation. Adv Sci. 2018;5(5):1700595.
72. Das M, Patil S, Bhargava N, Kang JF, Riedel LM, Seal S, Hickman JJ. Auto-catalytic ceria nanoparticles offer neuroprotection to adult rat spinal cord neurons. Biomaterials. 2007;28 (10):1918–25.
73. Schubert D, Dargusch R, Raitano J, Chan SW. Cerium and yttrium oxide nanoparticles are neuroprotective. Biochem Biophys Res Commun. 2006;342(1):86–91.
74. Zeng F, Wu Y, Li X, Ge X, Guo Q, Lou X, Cao Z, Hu B, Long NJ, Mao Y, Li C. Custom-made ceria nanoparticles show a neuroprotective effect by modulating phenotypic polarization of the microglia. Angew Chem Int Edit. 2018;57(20):5808–12.
75. Ferreira R, Bernardino L. Dual role of microglia in health and disease: pushing the balance toward repair. Front Cell Neurosci. 2015;9:51.
76. Suenaga J, Hu X, Pu H, Shi Y, Hassan SH, Xu M, Leak RK, Stetler RA, Gao Y, Chen J. White matter injury and microglia/macrophage polarization are strongly linked with age-related long-term deficits in neurological function after stroke. Exp Neurol. 2015;272:109–19.
77. Palwinder KM, Aiste J, Guy CB. Microglia proliferation is regulated by hydrogen peroxide from NADPH oxidase. J Immunol. 2006;176(2):1046–52.
78. Zhang B, Bailey WM, Mcvicar AL, Gensel JC. Age increases reactive oxygen species production in macrophages and potentiates oxidative damage after spinal cord injury. Neurobiol Aging. 2016;47:157–67.
79. Orihuela R, Mcpherson CA, Harry GJ. Microglial M1/M2 polarization and metabolic states. Brit J Pharmacol. 2016;173(4):649–65.
80. Wang T, Zhang X, Li JJ. The role of NF-κB in the regulation of cell stress responses. Int Immunopharmacol. 2002;2(11):1509–20.
81. Zhang J, Johnston G, Stebler B, Keller ET. Hydrogen peroxide activates NF-κB and the interleukin-6 promoter through NF-κB-inducing kinase. Antioxid Redox Sign. 2001;3(3):493.
82. Leuner K, Reichert AS. From mitochondrial dysfunction to amyloid beta formation: novel insights into the pathogenesis of Alzheimer's disease. Mol Neurobiol. 2012;46(1):186–93.
83. Verri M, Pastoris O, Dossena M, Aquilani R, Guerriero F, Cuzzoni G, Venturini L, Ricevuti G, Bongiorno AI. Mitochondrial alterations, oxidative stress and neuroinflammation in Alzheimer's disease. Int J Immunopathol Pharmacol. 2012;25(2):345–53.
84. Parisa Y, Ali Y. Advanced glycation end-products and their receptor-mediated roles: inflammation and oxidative stress. Iran J Basic Med Sci. 2011;36(3):154–66.
85. Bornemann KD, Wiederhold KH, Pauli C, Ermini F, Stalder M, Schnell L, Sommer B, Jucker M, Staufenbiel M. A beta-induced inflammatory processes in microglia cells of APP23 transgenic mice. Am J Pathol. 2001;158(1):63–73.
86. Kitazawa M, Yamasaki TR, LaFerla FM. Microglia as a potential bridge between the amyloid beta-peptide and tau. Ann N Y Acad Sci. 2004;1035:85–103.
87. Chay KO, Koong KYN, Hwang S, Kim JK, Bae CS. NADPH oxidase mediates β-amyloid peptide-induced neuronal death in mouse cortical cultures. Chonnam Med J. 2017;53 (3):196–202.
88. Wei Z, Wang LZ, Yu JT, Chi ZF, Lan T. Increased expressions of TLR2 and TLR4 on peripheral blood mononuclear cells from patients with Alzheimer's disease. J Neurol Sci. 2012;315(1–2):67–71.
89. Sung-Chun T, Lathia JD, Selvaraj PK, Dong-Gyu J, Mughal MR, Aiwu C, Siler DA, Markesbery WR, Arumugam TV, Mattson MP. Toll-like receptor-4 mediates neuronal

apoptosis induced by amyloid beta-peptide and the membrane lipid peroxidation product 4-hydroxynonenal. Exp Neurol. 2008;213(1):114–21.
90. Jeong Eun K, Minyung L. Fullerene inhibits beta-amyloid peptide aggregation. Biochem Bioph Res Commun. 2003;303(2):576–9.
91. Luogang X, Yin L, Dongdong L, Wenhui X, Xinju Y, Guanghong W. The molecular mechanism of fullerene-inhibited aggregation of Alzheimer's β-amyloid peptide fragment. Nanoscale. 2014;6(16):9752–62.
92. D'Angelo B, Santucci S, Benedetti E, Loreto SD, Phani RA, Falone S, Amicarelli F, Cerù MP, Cimini A. Cerium oxide nanoparticles trigger neuronal survival in a human Alzheimer disease model by modulating BDNF pathway. Curr Nanosci. 2009;5(2):167–76.
93. Cimini A, D'Angelo B, Das S, Gentile R, Benedetti E, Singh V, Monaco AM, Santucci S, Seal S. Antibody-conjugated PEGylated cerium oxide nanoparticles for specific targeting of Aβ aggregates modulate neuronal survival pathways. Acta Biomater. 2012;8(6):2056–67.
94. Kwon HJ, Cha MY, Kim D, Kim DK, Soh M, Shin K, Hyeon T, Mook-Jung I. Mitochondria-targeting ceria nanoparticles as antioxidants for Alzheimer's disease. ACS Nano. 2016;10(2):2860–70.
95. Du H, Guo L, Yan S, Sosunov AA, McKhann GM, Yan SS. Early deficits in synaptic mitochondria in an Alzheimer's disease mouse model. P Natl Acad Sci USA. 2010;107(43):18670–5.
96. Misono M. Unique acid catalysis of heteropoly compounds (heteropolyoxometalates) in the solid state. Chem Commun (Camb). 2001;13(13):1141–52.
97. Izarova NV, Pope MT, Kortz U. Noble metals in polyoxometalates. Angew Chem Int Edit. 2012;51(38):9492–510.
98. Jie G, Meng L, Jinsong R, Enbo W, Xiaogang Q. Polyoxometalates as inhibitors of the aggregation of amyloid β peptides associated with Alzheimer's disease. Angew Chem Int Edit. 2011;50(18):4184–8.
99. Gao N, Sun H, Dong K, Ren J, Duan T, Xu C, Qu X. Transition-metal-substituted polyoxometalate derivatives as functional anti-amyloid agents for Alzheimer's disease. Nat Commun. 2014;5:3422.
100. Dong A, Zhao H, Ying W, Jinsong X. Polyoxometalate-based nanozyme: design of a multifunctional enzyme for multi-faceted treatment of Alzheimer's disease. Nano Res. 2016;9(4):1079–90.
101. Geng J, Li M, Wu L, Ren J, Qu X. Liberation of copper from amyloid plaques: making a risk factor useful for Alzheimer's disease treatment. J Med Chem. 2012;55(21):9146.
102. Zhang Y, Wang Z, Li X, Wang L, Yin M, Wang L, Chen N, Fan C, Song H. Dietary iron oxide nanoparticles delay aging and ameliorate neurodegeneration in Drosophila. Adv Mater. 2016;28(7):1387–93.
103. Lakhan SE, Kirchgessner A, Hofer M. Inflammatory mechanisms in ischemic stroke: therapeutic approaches. J Transl Med. 2009;7(1):97.
104. Pradillo JM, Fernández-López D, García-Yébenes I, Sobrado M, Hurtado O, Moro MA, Lizasoain I. Toll-like receptor 4 is involved in neuroprotection afforded by ischemic preconditioning. J Neurochem. 2010;109(1):287–94.
105. Javier RC, Jesús MP, Olivia H, Pedro L, María AM, Ignacio L. Toll-like receptor 4 is involved in brain damage and inflammation after experimental stroke. Circulation. 2007;115(12):1599–608.
106. Estevez AY, Pritchard S, Harper K, Aston JW, Lynch A, Lucky JJ, Ludington JS, Chatani P, Mosenthal WP, Leiter JC, Andreescu S, Erlichman JS. Neuroprotective mechanisms of cerium oxide nanoparticles in a mouse hippocampal brain slice model of ischemia. Free Radic Biol Med. 2011;51(6):1155–63.
107. Chi Kyung K, Taeho K, In-Young C, Min S, Dohoung K, Young-Ju K, Hyunduk J, Hye-Sung Y, Jun Yup K, Hong-Kyun P. Ceria nanoparticles that can protect against ischemic stroke. Angew Chem Int Edit. 2012;124(44):11334.

108. Motonori T, Yusei M, Toru Y, Kentaro D, Yasuyuki O, Yoshio I, Tohru M, Koji A. Neurological and pathological improvements of cerebral infarction in mice with platinum nanoparticles. J Neurosci Res. 2011;89(7):1125–33.
109. Takamiya M, Miyamoto Y, Yamashita T, Deguchi K, Ohta Y, Abe K. Strong neuroprotection with a novel platinum nanoparticle against ischemic stroke- and tissue plasminogen activator-related brain damages in mice. Neuroscience. 2012;221:47–55.
110. Fabian RH, Derry PJ, Rea HC, Dalmeida WV, Nilewski LG, Sikkema WKA, Mandava P, Tsai AL, Mendoza K, Berka V. Efficacy of novel carbon nanoparticle antioxidant therapy in a severe model of reversible middle cerebral artery stroke in acutely hyperglycemic rats. Front Neurol. 2018;9:199.
111. Bitner BR, Marcano DC, Berlin JM, Fabian RH, Leela C, Culver JC, Dickinson ME, Robertson CS, Pautler RG, Kent TA. Antioxidant carbon particles improve cerebrovascular dysfunction following traumatic brain injury. ACS Nano. 2012;6(9):8007–14.
112. Marcano DC, Bitner BR, Berlin JM, Jane J, Lee JM, Aakash J, Fabian RH, Kent TA, Tour JM. Design of poly(ethylene glycol)-functionalized hydrophilic carbon clusters for targeted therapy of cerebrovascular dysfunction in mild traumatic brain injury. J Neurotrauma. 2013;30 (9):789–96.
113. Olanow C. The pathogenesis of cell death in Parkinson's disease–2007. Neurology. 2010;22 (S17):S335–42.
114. Etienne CH, Stéphane H. Neuroinflammation in Parkinson's disease: a target for neuroprotection? Lancet Neurol. 2009;8(4):382–97.
115. Wu DC, Teismann P, Tieu K, Vila M, Jackson-Lewis V, Ischiropoulos H, Przedborski S. NADPH oxidase mediates oxidative stress in the 1-methyl-4-phenyl-1,2,3,6-tetrahydropyridine model of Parkinson's disease. P Natl Acad Sci USA. 2003;100 (10):6145–50.
116. Pabon MM, Bachstetter AD, Hudson CE, Gemma C, Bickford PC. CX3CL1 reduces neurotoxicity and microglial activation in a rat model of Parkinson's disease. J Neuroinflammation. 2011;8(1):1–7.
117. Joglar B, Rey P, Guerra M, Labandeira-Garcia J. The inflammatory response in the MPTP model of Parkinson's disease is mediated by brain angiotensin: relevance to progression of the disease. J Neurochem. 2010;109(2):656–69.
118. Dexter DT, Wells FR, Lees AJ, Agid F, Agid Y, Jenner P, Marsden CD. Increased nigral iron content and alterations in other metal ions occurring in brain in Parkinson's disease. J Neurochem. 1989;52(6):1830–6.
119. Sofic E, Paulus W, Jellinger K, Riederer P, Youdim MB. Selective increase of iron in substantia nigra zona compacta of parkinsonian brains. J Neurochem. 1991;56(3):978–82.
120. Lin AMY, Chyi BY, Wang SD, Yu HH, Kanakamma PP, Luh TY, Chou CK, Ho LT. Carboxyfullerene prevents iron-induced oxidative stress in rat brain. J Neurochem. 2010;72(4):1634–40.
121. Drechsel DA, Estévez AG, Barbeito L, Beckman JS. Nitric oxide-mediated oxidative damage and the progressive demise of motor neurons in ALS. Neurotox Res. 2012;22(4):251–64.
122. Gonsette RE. Neurodegeneration in multiple sclerosis: the role of oxidative stress and excitotoxicity. J Neurol Sci. 2008;274(1):48–53.
123. Thomas Z, Alfonse P, Andreas Johann S, Christine S, Wolfgang B, Nicole SW. Molecular changes in white matter adjacent to an active demyelinating lesion in early multiple sclerosis. Brain Pathol. 2010;19(3):459–66.
124. Fischer MT, Rakhi S, Lim JL, Lukas H, Frischer JM, Joost D, Don M, Monika B, Jack VH, Hans L. NADPH oxidase expression in active multiple sclerosis lesions in relation to oxidative tissue damage and mitochondrial injury. Brain. 2012;135(3):886–99.
125. Heckman KL, Decoteau W, Estevez A, Reed KJ, Costanzo W, Sanford D, Leiter JC, Clauss J, Knapp K, Gomez C. Custom cerium oxide nanoparticles protect against a free radical mediated autoimmune degenerative disease in the brain. ACS Nano. 2013;7(12):10582–96.

# Chapter 7
# The Advances of Biomacromolecule-based Nanomedicine in Brain Disease

**Yuhua Weng and Yuanyu Huang**

**Abstract** Brain diseases affect a sizable portion of people in the world. Various treatment modalities have been pursued to control, alleviate, or cure these disorders. Biomacromolecules, e.g., antibody, peptide, enzyme, cytokine, nucleic acid, etc., are one kind of important and promising therapeutic regimens that have forced researchers to make great efforts to realize their clinical applications. However, effective and safe systemic delivery of biomacromolecules into the brain faces diverse challenges such as insufficient drug administration, degradation in the blood, first pass clearance, physical brain barriers, off-target accumulation, immune response, and toxicity to normal tissues. Nanotechnology offers advanced strategies to address these problems through rational design and fabrication of biomacromolecule-loaded nanomedicine. In this chapter, we summarized the administration strategies to the brain and design concepts of various biomacromolecular nanomedicines, highlighted their recent advances in preclinical and clinical studies, and discussed the existing challenges and our perspectives on this field.

**Keywords** Biomacromolecule · Nanomedicine · Antibody · Peptide · Enzyme · Nucleic acid

## 7.1 Introduction

The brain is the master organ in the body. It controls most of physical and psychological behaviors of human beings. Although there are physical barriers in the brain, such as the blood-brain barrier (BBB), it is also susceptible to a variety of infections and other disorders of varying intensity, such as brain cancer, tumors, Alzheimer's disease, alcoholism, amnesia, altitude sickness, autism, epilepsy, etc. The incidence and mortality of global brain diseases, in particular the brain tumors and

Y. Weng · Y. Huang (✉)
School of Life Science, Advanced Research Institute of Multidisciplinary Science, and Institute of Engineering Medicine, Beijing Institute of Technology, Beijing, China
e-mail: yyhuang@bit.edu.cn



X. Xue (ed.), *Nanomedicine in Brain Diseases*,
https://doi.org/10.1007/978-981-13-8731-9_7

neurodegenerative diseases, are growing at an alarming rate. In future, brain diseases will become one of the most deadly and expensive medical expenses in the world [1]. Unfortunately, most of current treatments do not yield satisfactory results. Many biomacromolecules such as antibodies, polypeptides, enzymes, cytokines, nucleic acids, other proteins, etc. are effective drugs for brain disease therapy. However, due to the unique structure of brain, drugs always face diverse obstacles to develop efficiency. These obstacles include drug administration route, degradation in the blood, physical brain barriers, first pass clearance, off-target accumulation, immune response, and toxicity to normal tissues. Biomacromolecules generally have large molecular weights and bioactive properties. They are less likely to overcome these barriers, let alone enter into the brain compared with other small molecule drugs, making them more challenging in treating brain diseases.

Nanomedicine, which is explained as the medical application of nanotechnology, is an important field in basic research and pharmaceutical industry. It has become a hot spot in brain-targeting therapy and imaging with its unique characteristics. Studies have shown that nanocarriers can carry drugs through the blood-brain barrier (BBB) and produce therapeutic effects [2, 3]. Moreover, the degradation, clearance, and limited concentration of drugs in the brain can be significantly improved after rational design and fabrication of nanocarrier. Promising nanocarriers applied in the brain are mainly including liposomes, [4] micelles, [5] dendrimers, [6] nanogels, [7] polymeric, [8] inorganic, [9] and organic-inorganic hybrid nanoparticles [10]. These excellent nanocarriers have provided advanced strategies for biomacromolecules delivery in the brain.

In this chapter, we focused on the problems around biomacromolecules delivery in the brain and outlined the delivery strategies of biomacromolecules. We summarized the various types and design concepts of nanotechnology-assisted biomacromolecule-based nanomedicines, highlighted their recent advances in preclinical and clinical research, and discussed the existing challenges and our prospective on this exciting field toward more effective brain disease therapy.

## 7.2 Key Constraint Factors for Biomacromolecule Delivery into the Brain

### *7.2.1 Blood-Brain Barrier (BBB)*

The consistent delivery of drugs to the central nervous system (CNS) is a challenge in the biopharmaceutical industry due to the multiple barriers in the brain, especially when delivering larger molecular weight biomacromolecules. The BBB mainly consists of brain microvascular endothelial cells, astrocyte foot ends, pericytes, the tight junctions between these cells, and the basement membranes. It is reported that the BBB could prevent access of almost all the large molecule drugs and most of small molecule drugs from systemic circulation. Only small molecules such as lipids with molecular weight < 400 Daltons (Da) can cross the normal BBB [11, 12]. Few peptides and peptide analogues can transport across BBB according to limited reports [13]. In addition to cellular factors, various influx/efflux transport proteins, ion channels, and receptors that expressed on the BBB also display as a metabolic barrier that regulates the permeability of pharmacological agents [11, 12].

### 7.2.2 *Blood Cerebrospinal Fluid Barrier (BCFB) and Blood Tumor Barrier (BTB)*

The BCFB also plays a significant role in maintaining the homeostasis in the brain [14]. It is composed of choroid plexuses and outer arachnoid membrane epithelial cells. The monolayer of epithelial cells is joined together by tight junctional proteins, restricting permeability of nutrients and drugs. The main function of these epithelial cells is to secrete and maintain the homeostatic composition of the cerebrospinal fluid (CSF). The CSF is a transparent liquid filled in the subarachnoid space and ventricles, supporting and surrounding the entire brain and spinal cord, changing four to five times a day. It provides the brain with a drainage system called the sink effect, where metabolic products and inflammatory exudates are diluted and subsequently removed [15–17]. The large molecular weight and hydrophilic substances have a greater sink effect [18].

For the treatment of malignant brain tumors such as glioblastoma, a major obstacle for drug delivery to the tumor is the blood tumor barrier (BTB). BTB mainly consists of endothelial cells and their tight junctions, basement membranes and glioma cells foot ends, etc. [19] Penetrating across the BTB is needed to open the tight junctions in the paracellular route or increase transcellular transportation; [12, 20] both tasks are very challenging.

### 7.2.3 *Drug Diffusion in the Brain*

Even if biomacromolecules overcome all the barriers along with the brain and enter into the CNS, they still need to diffuse long distance through the narrow extracellular spaces of the brain microenvironment to produce therapeutic effects. Many factors associated with brain microenvironment affect the diffusion and distribution of biomacromolecules, such as the geometry, volume fraction, viscosity of extracellular space, interactions with cell surfaces, components of interstitial fluid, as well as extracellular matrix [21]. Therefore, rational modification of biomacromolecule for brain application is particularly important. There are several developmental ways to achieve this goal. One commonly used method is to reduce the size of drug molecule [22]. It is known that small size particles can penetrate and distribute easily in the brain compared with larger particles. In terms of antibody, for example, selection of smaller fragments of antigen-binding domain will not only ensure the targeting characteristic of antibody but also reduce its size [23, 24]. It is reported that this method may increase the permeability and diffusion ability of some antibody drugs in the brain [25]. However, in most cases, reducing the size of drug molecule is difficult and inefficient for many drugs. Other choices for enhancing drug distribution and transportation in the brain include conjugating a helper molecule to the drug and using an efficient drug carrier, both of which have been studied extensively. For example, cell-penetrating peptides and adeno-associated virus (AAV) vectors have yielded exciting drug delivery efficiencies in various animal models [26–29]. Another point to note is that some growth factor-based biomacromolecule drugs tend to bind to heparin in the extracellular matrix, which limits drug distribution in the CNS. This problem can be solved by co-injection of exogenous heparin and growth factors such as the glial cell-derived neurotrophic factor (GDNF) [30, 31]. The exogenous heparin will block the heparin-binding site in the extracellular matrix, resulting in significantly enhanced distribution of GDNF in brain tissue.

Efflux transporter is another key restriction for distribution of therapeutics in the brain. Inhibition of efflux transporter such as P-glycoprotein (P-gp) and breast cancer resistance protein (BCRP) has proven to be an efficient method for increasing therapeutic doses in the brain and achieving a better therapy [32–35]. Transporter inhibitors, by combining with nanocarriers, can effectively exert their inhibitory effects and reduce systemic toxicity and pharmacokinetic interactions. Applying this method, significant accumulation of multidrug resistance 1 (MDR1) substrates in the brain and positive outcomes toward to chemoresistant cancer treatment were achieved [36]. This strategy may provide some insights for the treatment of brain tumors.

## 7.3 Strategies for Biomacromolecular Nanomedicine Delivery into the Brain

### *7.3.1 Direct Injections*

At present, direct injection of drug intracranial, intracerebral implants, intraventricular infusion, as well as convection-enhanced delivery (CED) are common local administration strategies for treating CNS diseases. Upon direct injection, the BBB, BCFB, BTB, efflux pumps, and receptors that present to block the action of drugs are all bypassed. Drugs can reach the lesion directly, avoiding the toxicity caused by the first-pass effect and off-target effect [37]. However, these methods are extremely invasive and are more likely to cause permanent brain damage, especially when repeated injections are needed [21, 37–40].

### *7.3.2 Convection-Enhanced Delivery (CED)*

Among various local delivery methods, CED is a widely applied invasive method that directly delivered drugs into intracerebral tissues. It allows for constant drug delivery through fenestrated catheters placed at the time of surgery and is driven by a hydrostatic pressure gradient from a prolonged slow infusion in the brain parenchyma, which prompt drugs to distribute into intracerebral tissues via convective flow [41, 42]. The CED has been extensively applied for delivery of biomacromolecules such as proteins, oligonucleotides, liposomes, nanoparticles, as well as viral carriers [30, 43, 44]. Some fusion proteins and antisense oligonucleotides being delivered by CED have even entered into clinical status but at last failed, for example, IL4-PE (PRX32), IL13-PE38QQR, and trabedersen (AP-12009), etc. Drawbacks such as high dependence on technical expertise, limited volume of drugs distributed in the brain, etc. restrict the efficacy of CED [45].

### *7.3.3 Intranasal Administration*

One of the noninvasive and convenient brain delivery routes is intranasal administration. After intranasal administration, drug molecules will reach the cerebrospinal fluid through the neural pathway of the olfactory mucosa, thus bypassing the brain physical barriers. Due to the large surface area of the nasal mucosa and abundant blood capillaries, the drug molecules are rapidly absorbed into the systemic circulation during intranasal administration. Compared with oral drugs, intranasal drugs avoid degradation in gastrointestinal fluids and liver first-pass effects [41]. Generally, the intranasal bioavailability of small molecule drugs is close to intravenous injection, while intranasal biomacromolecular drugs considerably have less bioavailability than intravenous injection due to enzymatic degradation, mucociliary clearance, and poor mucosal permeability [46, 47]. This situation can be improved after the introduction of nanocarriers to promote large molecule penetration [48]. The leucine$^5$-enkephalin-loaded chitosan nanoparticles and neurotoxin II peptide-loaded, polysorbate 80-coated nanoparticles are two successful examples of intranasal delivery, both of which have resulted in enhanced antinociceptive effect compared with administration of free drugs [49, 50]. Since intranasal administration has many superior advantages, this strategy has been applied widely. The biggest problem of intranasal administration is the toxicity of drugs and their excipient to nasal mucociliary and the poor penetration of large molecular drugs [51, 52]. Reducing or eliminating the toxicity and selecting potent absorption enhancers are important tasks for nanocarrier assisted intranasal delivery.

### 7.3.4 Oral Administration

Intravenous injection, oral administration, as well as BBB disruption are important systemic administration routes of brain delivery. Only large dosage of systemic administration drug can achieve a therapeutic concentration in the brain. There are many limiting factors in the blood stream that blocking the effect of nanomedicine upon systemic administration. For example, enzymes and proteins in the blood will combine with nanomedicine, causing drug degradation and clearance.

Oral administration is a noninvasive strategy for brain delivery but is less perused by researchers. For biomacromolecule drug delivery, oral absorption is limited by a high molecular weight (> 50 Da) and low gastrointestinal permeability, which restricts adequate drugs transporting across BBB to reach an adequate therapeutic concentration and thus results in a low bioavailability of biomacromolecule drug [21, 53, 54]. These problems may be overcome by nanotechnology-based strategies, which are reported to have the ability to protect oral drugs from the harsh environment of the gastrointestinal tract, enhance drug absorption by the gastrointestinal mucosa, and prolong the circulation half-life in the blood stream. As reported by several researchers, quaternary ammonium palmitoyl glycol chitosan (GCPQ) nanoparticles facilitated brain delivery of peptides such as leucine $^{5}$-enkephalin through oral administration [53, 54]. Although most of the GCPQ nanoparticles (85–90%) were not absorbed by the gastrointestinal tract, the residual dosage appeared to deliver their payloads to the systemic circulation, enabling the nanoparticle-stabilized peptides to reach the brain [55].

### 7.3.5 BBB Disruption

BBB disruption strategy represents a promising approach for effective and noninvasive brain disease therapy. High concentration osmotic agents, bradykinin analogues, as well as efflux pump inhibitors are classical regents used for temporary disruption of BBB [56–59]. The most commonly used hyperosmolar disruptor for BBB disruption purpose is mannitol. Mannitol can cause osmotic efflux of water from brain microvascular endothelial cells, resulting in shrinkage of cells and tight junction dysfunction [59]. The bradykinin analogues are known to have the ability to enhance BBB penetration. Combination administration of RMP-7 (lobradimil) with chemotherapeutic drug carboplatin has increased the BBB permeability for carboplatin in a bradykinin B2 receptor-mediated manner and has reached clinical trials [60]. However, the clinical trial was failed at last. The bradykinin receptors, which distributed uniformly in the brain and worked very transient, are capable of limiting drug distribution in the brain.

### 7.3.6 Ultrasound and Microbubble with Nanoparticles

It has been reported that focused ultrasound in combination with microbubbles can increase the BBB and BBTB permeability with no or minimal damage to the normal brain tissues [61–63]. The microbubbles consist of a gas core and a semirigid lipid or albumin shell with less than 50 μm in diameter. When exposed to focus ultrasound, the microbubbles will convert the acoustic energy into mechanical power, which disrupts the BBB temporarily and allow the drug permeating into the brain [64]. In recent years, magnetic resonance imaging (MRI)-guided focused ultrasound with microbubbles has been applied in delivery of chemotherapeutic drugs and antibodies into the brain and received promising results. Liu et al. reported that focused ultrasound combined with magnetic nanoparticles improved the delivery of anticancer drugs into the brain. A better delivery and therapeutic outcome of the combination therapy was achieved compared with focused ultrasound and magnetic nanoparticles alone [65]. In another study, focused ultrasound combined with intranasal administration of fluorescently labeled dextran showed higher intracranial concentration of dextran compared to intranasal administration alone [47, 66]. These results further expand the application feasibility of focused ultrasound in brain administration.

### 7.3.7 Targeted Brain Delivery

In general, nanomedicines can overcome the brain's physiological barriers to some extent, but this effect is not strong enough. In order to achieve effective and practical brain-targeted therapy, choosing an efficient and viable targeting strategy is very important. Several directions are being developed and straightforward, including (i) trans-BBB targeted delivery; (ii) trans-BBTB targeted delivery; (iii) combined trans-BBB/BBTB targeted delivery; (iv) enhanced permeability and retention (EPR)-based delivery; and (v) whole process targeted delivery. For options i–iii, the conjugation of reliable transport ligands to drug-loaded nanostructures is a viable strategy for delivery of biomacromolecules such as peptides, proteins, and oligonucleotides to the brain. Those ligands are specific for putative receptors or transporters expressed on the brain barrier and can be efficiently uptake by brain cells. It is worth noting that the endogenous expression level of ligands should also be considered, as the large amount of endogenous ligands would compete with the conjugated ligands for receptor occupancy [67, 68]. The most studied receptors and transporters in brain barriers are transferrin, leptin, insulin and diphtheria toxin receptors and low density lipoprotein (LDL) receptor related protein 1 and 2 [69, 70]. Another approach to improve the targeting ability is camouflaging the nanocomplexes with erythrocytes, leukocytes membranes, as well as exosomes [71, 72]. This biomimetic strategy greatly reduced the immune response caused by nanocomplexes and improved their safety.

The EPR effect is an excellent feature of nanoparticles for treatment of solid tumors. Brain tumors such as gliomas contain variable regions of BBB and BBTB disruption. The tumor vasculature is heterogeneous, and its permeability increases in most part of tumor. This microenvironment could be used by nanoparticles and nanoformulations which can increase the systemic circulation time and prolong tumor retention time of drugs [73]. The EPR effect is thought to relate to the size of nanoparticles. Some studies showed that the disrupted BBTB can accumulate particles around 100 nm upon the EPR effect, while others considered that only particles as small as 20 nm can penetrate the BBTB [72, 74]. Besides, it is feasible to develop a whole process brain-targeted nano-delivery system that takes full advantages of the physical properties of nanoparticles, such as their small size, surface coating and modification, and combination with appropriate administration route.

## 7.4 Preclinical and Clinical Advances of Biomacromolecular Nanomedicines for Brain Disease Treatment

### *7.4.1 Protein-Based Nanomedicine*

Many brain diseases are associated with loss of function in a protein. The efficacy of protein-based therapies for treating CNS diseases is limited by insufficient amounts of free therapeutics in the brain and their short half-lives. Moreover, proteins are unlikely to overcome blood circulation and cellular uptake while remaining active. These dilemmas have forced researchers to find ways for improving their delivery. One of the most classic attempts to increase protein delivery to the brain is to modify proteins with either hydrophobic fatty acid residues or amphiphilic block copolymers, such as N-(2-hydroxypropyl)-methacrylamide (HPMA), poly(2-oxazoline)s (POx), and poly(ethylene) glycol (PEG) [75–77]. This strategy is developed based on the fact that amphiphilic molecules can transport across the cell membrane more easily. Until now, modification with PEG may be one of the most successful approaches to improve protein bioavailability, known as "PEGylation." Several PEGylated proteins are approved for clinical use, such as pegfilgrastim, peginterferon α-2a, pegaptanib sodium, etc. PEGylated nanoparticles such as liposomes and albumin nanoparticles are also very attractive vehicles for brain delivery.

As an excellent biocompatible carrier, albumin has received extensive attention. A number of studies have been carried out to verify its effectiveness [75, 78–80]. Albumin is the most abundant plasma protein in the blood. Many tumor cells overexpress albumin-binding proteins for transportation of albumin as a nutrition and energy source for fast-growing cancer cells [78]. Since albumin has a long circulation time in vivo, it is an attractive attempt to conjugate anticancer molecules or cytokines with albumin to improve the pharmacokinetic profile of drugs. Clinically, methotrexate-albumin conjugate, doxorubicin derivate-albumin conjugate, and Levemir, a myristic acid derivative of insulin that binds to the fatty acid binding sites of albumin, have been tested [78]. Levemir has even been approved for the treatment of diabetes. In 2005, the first human serum albumin-bound paclitaxel (Abraxane®) was approved by the FDA for treatment of breast cancer. Then, the approved applications were expanded to include treatment of NSCLC and pancreatic cancer in 2012 and 2013, respectively. It has been reported that the new nanoparticle formulation improves solubility of paclitaxel and reduces toxicity compared to the conventional paclitaxel formulation [81]. For brain tumor treatment, albumin nanoparticles loaded with dual drugs or tagged with brain-targeting ligands facilitate BBB penetration of therapeutics and therefore improve the treatment. Based on the fact that glioma cells overexpress albumin-binding proteins, researchers constructed paclitaxel and fenretinide (4-HPR)-loaded albumin nanoparticles, with modification of cell-penetrating peptides on their surface. The nanoparticles effectively promoted BBB penetration of the drugs and significantly improved the anti-glioma effect [82]. Another polysorbate 80-coated albumin nanoparticles resulted in a threefold increase of drug concentration in the brain and a significant reduction in convulsion compared to the free drug [79]. Furthermore, pharmaceutical enterprises have developed the albumin fusion technology that could generate albumin protein conjugates via gene engineering. For example, the Human Genome Sciences Company has produced a fusion protein of albumin and interferon α-2b named Albuferon using this technology. Albuferon is used for chronic hepatitis B and chronic hepatitis C treatment. Several phase II stages of clinical trials have been completed or terminated around 2009. It is reported that conjugation to albumin prolongs the half-life of interferon α-2b ~ 6 days, which has exceeded several other interferon-α-regents [83, 84]. Meanwhile, the administration times and immunogenicity have been reduced drastically.

### *7.4.2 Enzyme-Based Nanomedicine*

The brain is a very complicated and special organ. Many brain diseases are resulted from a missing or deficient enzyme. The chemical and biological flexibility of nanoparticles make them suitable to load larger cargo like enzymes for brain delivery. Although there are abundant researches focusing on preparation, characterization, and therapeutic efficacy of enzyme-loaded nanoparticles, few have been designed specifically for brain application. In a proof of concept study, different nanoparticle delivery systems including liposome, poly(butyl cyanoacrylate) (PBCA), and PLGA nanoparticle were evaluated for superoxide dismutase (SOD) enzyme delivery to the brain. These SOD enzyme-loaded nanoparticles were tagged or untagged with specific antibodies. It is showed that the tagged nanoparticles protected primary neurons in vitro from oxygen-glucose deprivation and showed protection against ischemia and reperfusion injury in vivo [85]. Some other typical enzyme-based nanomedicines for brain delivery are showed in Table 7.1.

**Table 7.1** Protein- and enzyme-based nanomedicines

| Drug name | Formulation | Developmental stage | Therapeutic advantages | Reference |
|---|---|---|---|---|
| bFGF | Liposome | Preclinical | Intranasal administration of bFGF-loaded liposomes improved bFGF accumulation in the brain and showed effective neuroprotective effects in rats for 21 days | [86] |
| BSA, $DT_{390}$ | Lipidoid-telodendrimer hybrid nanoparticle | Preclinical | The 100 nm sized nanoparticles showed improved intratumoral biodistribution of proteins and effective glioblastoma inhibition activity in vivo | [87] |
| PTX and 4-HPR | Albumin nanoparticle | Preclinical | The dual drug-loaded albumin nanoparticles facilitated BBB penetration of drugs via interaction with albumin-binding proteins in glioma cells. Cell-penetrating peptide modification of albumin nanoparticles significantly improved treatment of glioma | [82] |
| Loperamide | Albumin nanoparticle | Preclinical | The albumin nanoparticle formulation produced obvious antinociceptive effect, while the loperamide solution did not | [88] |
| BSA, anti-VEGF antibody | $Fe_3O_4$nanoparticle | Preclinical | The BSA-coated and antibody-targeted nanoparticles promoted their binding to glioma cells in vitro and showed effective MRI visualization of glioma | [89] |

(continued)

**Table 7.1** (continued)

| Drug name | Formulation | Developmental stage | Therapeutic advantages | Reference |
|---|---|---|---|---|
| Gabapentin | Albumin nanoparticle coated with polysorbate 80 | Preclinical | Polysorbate 80-coated albumin nanoparticles significantly reduced convulsion and increased the concentration of gabapentin in the brain by threefold compared to free drug | [79] |
| α-Synuclein antibody, apolipoprotein B domain | Fusion protein | Preclinical | The fusion protein-degraded α-synuclein aggregates in a Lewy bodies/Parkinson's disease mouse model | [90] |
| Metalloproteinases-I | Polysorbate 80-coated PLGA nanoparticle | Preclinical | The coated nanoparticles showed BBB penetration both in vitro and in vivo | [91] |
| SOD enzyme | Polymer nanoparticle tagged with antibody | Preclinical | The targeted nanoparticles protected primary neurons in vitro and showed protection against ischemia and reperfusion injury in vivo | [85] |
| NAP | Lactoferrin-conjugated PEG-PCL nanoparticle | Preclinical | The nanomedicine exhibited neuroprotective and memory improvement effect in vivo at low dosage by intranasal administration | [92] |
| β-Galactosidase | Iron oxide nanoparticle | Preclinical | The nanomedicine increased β-galactosidase's activity by tenfold in tumor lesions compared to normal brain tissue in gliosarcoma-bearing rats | [93] |

*bFGF* fibroblast growth factor, *DT390* a truncated diphtheria toxin, *PTX* paclitaxel, *4-HPR* fenretinide, *SOD* superoxide dismutase, *NAP* a neuroprotective peptide

Although the versatility of nanoparticles increases the half-life, tissue targeting, and delivery efficiency of therapeutic proteins and enzymes in the brain, it still does not create a complete system in which all brain delivery characteristics are combined [40]. These shortcomings drastically limit the number of clinical trials and successful treatments that have entered the market.

### 7.4.3 Peptide-Based Nanomedicine

Peptides are excellent therapeutics that been developed over 40 years. According to database in www.peptidetherapeutics.org in 2018, there are 64 therapeutic peptides approved in the USA and Europe, and 163 peptides are now in active development. Several peptides are being investigated in treating brain disorders, such as brain tumors, neurodegeneration, stroke, psychiatric disorders, and pain. In addition to the excellent pharmacological properties, many peptides have poor physical and metabolic stability. Moreover, due to their hydrophilicity and relatively high molecular weight (>500 Da), some peptides have difficulty penetrating biological membranes, especially the physical barriers in the brain. The nanotechnologies have shown promise in delivering peptide to the brain. Nanoparticles exhibit specific features such as small size, controllable shape, easy to be decorated, etc., making them ideal carriers to deliver therapeutic drugs to the target tissue. Such evidences are abundant in preclinical cancer therapy. Those strategies have also been applied in delivering peptides across the BBB. For example, when a synthetic opioid peptide DAMGO was loaded into glutathione (GSH)-PEG liposomes and injected intravenously, the liposome formulation prolonged the half-life of DAMGO by 4.5-fold compared to free DAMGO [94]. Another interesting targeted delivery of peptide is macrophage-mediated delivery of catalase, an anti-inflammatory peptide, to the brain. Catalase was first packaged into a block copolymer poly(ethylenimine)-poly(ethylene glycol) (PEI-PEG) to form the "nanozyme" and was then loaded into bone marrow-derived macrophages (BMM). The BMM served as a "depot" for "nanozyme" and slowly released it in active form. This cell-mediated formulation increased the bioavailability of catalase in the brain significantly and attenuated neuroinflammation and produced neuro protection effect in rodent models of Parkinson's disease (PD) [95]. In addition to being a drug candidate, peptide is also widely used as targeting ligand or carrier for other biomacromolecules' delivery. The cell-penetrating peptide (CPP) is a typical representative targeting ligand. CPP is a short peptide of heterogeneous in size (10–27 amino acid residues) and sequence, possessing multiple positive charges at physiological pH. They can deliver biomacromolecules into cells by an adsorptive-mediated transcytosis pathway [96–98]. The most widely used CPP includes the transactivator of transcription (TAT), penetratin, and the SynB vectors [67]. There is also a class of peptide ligands work through the receptor-mediated transcytosis. Taking a viral coat peptide rabies virus glycoprotein (RVG) as an example, a 29-amino acid RVG peptide (RVG29) exhibits excellent brain-targeting ability. It can accumulate in the brain efficiently after intravenous administration in mice [99]. RVG29-modified nanomedicines are widely employed in preclinical stage for treating brain diseases such as PD, AD, and brain injury [100–102]. Demeule et al. reported a 19-amino acid peptide-based carrier called Angiopeps for brain delivery. It has e-targeting affinity to glioblastoma cells after brain entry and belongs to a family of Kunitz domain-derived peptides [103]. Evidences showed that the Angiopeps facilitated more transcytosis and parenchymal accumulation of drugs in the brain than other targeting moieties such as avidin and lactoferrin by interaction with the low-density lipoprotein receptor-related protein 1 (LRP-1) [103, 104]. Angiopep peptides developed by Angiochem Inc. are promising techniques, and one Angiopep conjugate ANG1005 (an Angiopep-2-paclitaxel conjugate) has entered into several clinical trials [105, 106]. Preclinical studies using Angiopep-2-conjugated dendrimer delivery system, as well as organic-inorganic hybrid nanocomposites, have shown positive results in brain delivery or imaging [107–109].

Another property of peptide is that it can arrange into various structures upon rational regulation of the molecular structure and alter the external environment. Numerous studies have reported that peptides could be self-assembled into spherical, membrane, hydrogel, and fibrous aggregates mediated by hydrogen bonding, hydrophobic and π-stacking interactions, etc. [110–112] Mazza et al. reported that an amphipathic derivative of peptide dalargin (an opioid receptor agonist) could form nanofibers in a self-assemble manner. The latter was able to deliver the peptide to the brain and elicit a pharmacological response, while the underivatized peptide was not detectable in the brain [113]. The possible mechanism may be that the peptide chains in the nanofibers were wrapped tightly around the nanofiber core, which could prevent degradation of peptides in vivo. This self-assemble concept further broadens the application of peptides in the pharmaceutical industry. Although several peptides have reached into clinical stages, peptide-based nanomedicines for the treatment of brain diseases are mostly in preclinical research. Some typical peptide-based nanomedicines for brain delivery are summarized in Table 7.2.

### *7.4.4 Antibody-Based Nanomedicine*

Similar to peptide, antibody is also commonly chosen as a targeting ligand of nanomedicine or can be directly loaded by the nanocarrier. Nanoparticles modified with antibodies are widely used in brain delivery. In recent decades, there have been numerous attempts to treat brain diseases by combining antibodies with chemotherapy drugs and biological macromolecules such as antisense oligonucleotides and siRNA. Antibodies against the receptors that have a high expression level in brain cells have been extensively applied in brain delivery. OX26, a rat monoclonal antibody against the transferrin receptor, has been widely used in conjugation with polymersomes and nanoparticles for peptide, plasmids, as well as chemotherapeutics penetrating the BBB [114–116]. However, OX26 and another anti-transferrin receptor antibody 8D3 [117] are rodent antibodies, which are not suitable for human CNS-targeted delivery.

**Table 7.2** Peptide- and antibody-based nanomedicines

| Drug name | Formulation | Developmental stage | Therapeutic advantages | Reference |
|---|---|---|---|---|
| DAMGO | GSH-PEG liposomes | Preclinical | The liposomes increased the half-life of DAMGO in the brain by 4.5-fold compared with free DAMGO | [94] |
| NA | g21 conjugated PLGA nanoparticle | Preclinical | After intravenous injection for 2 h, the PLGA nanoparticles could cross the BBB and reach the brain | [120] |
| Leucine $^{5}$-enkephalin | GCPQ nanoparticles | Preclinical | GCPQ nanoparticles facilitated both oral and intravenous administration of peptide and delivered peptide to the brain | [55] |
| DOX | Angiopep-2-modified gold nanoparticle | Preclinical | The nanomedicine displayed superior anti-glioma effect in mice compared with free DOX | [121] |
| NA | Angiopep-2-modified carbonaceous nanodots | Preclinical | The nanomedicine showed significant glioma cell-targeting effect in vitro and in vivo and displayed good biocompatibility | [122] |
| DTX | Angiopep-2-modified PLGA@Au nanoparticle | Preclinical | The targeted nanoparticles showed potent anti-glioma efficiency through chemo-photothermal therapy and exhibited potential for X-ray imaging in vivo | [107] |
| ANG1005 | NA | Phase III | ANG1005 (the conjugate between paclitaxel and Angiopep-2) is active against breast cancer metastasis both within and outside the CNS | [105, 123] |
| Dalargin | Nanofibers | Preclinical | The amphipathic derivative of dalargin peptide self-assembled into nanofibers and prolonged the antinociceptive time to 8 h | [113] |
| Dalargin | Polysorbate 80-coated PBCA nanoparticles | Preclinical | The coating nanoparticles facilitated dalargin transporting across the BBB and elicited a short-lived | [124] |

(continued)

**Table 7.2** (continued)

| Drug name | Formulation | Developmental stage | Therapeutic advantages | Reference |
|---|---|---|---|---|
| | | | antinociceptive effect after oral and intravenous administration | |
| ASV-30 | Iron oxide nanoparticles | Preclinical | Systemic administration of ASV-30-conjugated nanoparticles improved the bioavailability of ASV-30 in the brain and reduced anxiety-like behavior without affecting locomotion | [125] |
| Catalase | Bone marrow-derived macrophages | Preclinical | The macrophages loaded with nano-formulated catalase improved catalase's bioavailability and transportation across the BBB | [95] |
| Peptide, daunomycin, and plasmid | OX26 antibody-functionalized chitosan nanoparticle, liposome, polymersome | Preclinical | Functional nanomedicine enhanced therapeutics transporting across the BBB and delivering to the brain | [114–116] |
| Plasmid DNA | 8D3 antibody-tagged liposome | Preclinical | The targeted liposomes increased tissue-specific expression of exogenous gene in the brain after intravenous injection | [117] |
| BVZ antibody | Solid lipid nanoparticles | Preclinical | BVZ-loaded nanoparticles enhanced BVZ's activity and BBB permeation in vitro | [118] |
| Anti-A$\beta_{1-42}$ antibody | Nanoparticles | Preclinical | The anti-A$\beta_{1-42}$ nanoparticles could significantly reduce the soluble forms of A$\beta$-peptides and rescue memory in AD mice | [119] |
| IMC-C225 antibody | Liposome | Preclinical | The liposomes showed specific binding and internalization in glioblastoma cells and produced efficient anticancer activity in vivo | [126] |

*DAMGO* a synthetic opioid peptide, *GSH* glutathione, *g21* leptin fragment, *GCPQ* quaternary ammonium palmitoyl glycol chitosan nanoparticle, *DTX* docetaxel, *PBCA* poly(butyl cyanoacrylate), *ASV-30* antisauvagine-30, *BVZ* bevacizumab, *NA* not available.

In the aspect of glioblastoma (GBM) treatment, bevacizumab (BVZ) is an approved agent for treatment of recurrent GBM. In an experimental work, BVZ-loaded solid lipid nanoparticles showed increased activity of 100- to 200-fold compared with free BVZ on human umbilical vein endothelial cells (HUVECs). The nanoparticles could also enhance the permeation of BVZ in an in vitro BBB model [118]. In a latest study, anti-A$\beta_{1-42}$ antibody-decorated PEGylated nanoparticles were tested in AD-like transgenic mice. The authors found the anti-A$\beta_{1-42}$ nanoparticles could significantly reduce the soluble forms of Aβ-peptides and rescue memory in AD mice [119]. Different from other conventional antibody therapeutic mechanisms, the anti-A$\beta_{1-42}$ nanoparticles functioned through the interaction with soluble A$\beta_{1-42}$ in the blood and then eliminated them through a "sink effect." After intravenous injection of the anti-A$\beta_{1-42}$ nanoparticles, the soluble A$\beta_{1-42}$ will be absorbed by the nanoparticle and then eliminated with nanoparticles through the nanoparticle clearance mechanisms. The concentration of Aβ-peptides in the brain is reduced accordingly. This study presented the first successful example of using anti-Aβ1–42 antibody-modified nanoparticles to treat AD in an experimental model.

### *7.4.5 Nucleic Acid-Based Nanomedicine*

#### 7.4.5.1 Current Status of Nucleic Acid-Based Therapeutics

In recent years, several antisense oligonucleotide (ASO) drugs have been approved by FDA. Furthermore in August 2018, the first RNAi drug, patisiran, developed by Alnylam Pharmaceuticals was also approved by FDA for treatment of hereditary transthyretin-mediated (hATTR) amyloidosis with polyneuropathy. These breakthroughs have doubled the confidence of nucleic acid pharmaceuticals. Because of the distinguished efficiency and mature manufacturing techniques of nucleic acid drugs such as siRNA, ASO, mRNA, DNA, etc., they have attracted great interests from researchers and have been used to treat various brain diseases, ranging from motor neuron diseases and dementia to brain tumor and brain injuries. Some nucleic acids have even entered into the clinical status (Table 7.3). Nevertheless, challenges encountered in blood circulation, after brain entry, accessing targeted sites, cell endocytosis, as well as intracellular transportation are key factors that restrict the application of nucleic acid therapeutics. Naked nucleic acids such as siRNA and mRNA can be rapidly degraded by RNase during the blood circulation. Moreover, the physical barriers which exist in the brain are considered to be the biggest obstacles limiting the entry of nucleic acids. Even few nucleic acids crossing the BBB and arrived in the brain, issues such as nonspecific targeting are still intractable. Fortunately, the chemical modification of nucleic acid structural elements developed in recent years affords a solution to protecting nucleic acid from degradation to some extent. Meanwhile, the nanoplatform also offers intriguing potential to addressing these challenges for nucleic acid drugs in brain delivery.

**Table 7.3** Current clinical trials of nucleic acid therapeutics for brain disease treatment

| Therapeutic name | Indication | Nucleic acid types | Clinical status | Sponsor | NCT number |
|---|---|---|---|---|---|
| Inotersen | Polyneuropathy | ASO | Phase III-completed | Ionis | NCT01737398 |
| Trabedersen | Glioblastoma | ASO | Phase III-terminated | Isarna therapeutics | NCT00761280 |
| Imetelstat | Glioblastoma Brainstem tumors | ASO | Phase II-terminated | Geron | NCT01836549 |
| Ex vivo transfected DCs | Glioblastoma | mRNA | Phase I/II-recruiting | University Hospital, Antwerp | NCT02649582<br>NCT00965224 |
| Ex vivo transfected DCs | Glioblastoma | mRNA | Phase I/II-recruiting | Guangdong 999 Brain Hospital | NCT02709616<br>NCT02808364<br>NCT02808416 |

*DC* dendritic cell, *ASO* antisense oligonucleotides

As mentioned above, nanoparticles can be tailored for various purposes. They can be precisely functionalized with targeted ligands. Their size, morphology, charge, etc. also can be modulated in specific circumstances. Nonviral nanocarriers such as liposomes, polysomes, lipoplexes, and polyplexes are widely used for loading with nucleic acids. They are mainly comprised of cationic polymers and ionizable lipids such as protamine, PEI, 1,2-dilinoleyloxy-N,N-diethyl-3-aminopropane (DLinDMA), (6Z,9Z,28Z,31Z)-heptatriaconta-6,9,28,31-tetraen-19-yl-4-(dimethylamino)butanoate(DLin-MC3-DMA), etc. [127–129] The positively charged carriers can spontaneously form into nanocomplexes with negatively charged DNA, siRNA, and mRNA by electrostatic interaction. Not only can the nanocomplex condensate nucleic acid payloads, they also protect their cargo from rapidly degradation [130–132]. However, the positively charged nanocomplexes can also interact with anionic blood components such as proteins and thus be excluded by the reticuloendothelial system clearance [133]. Such issues could be avoided by coating the nanocomplex surface with either zwitterionic ligands or polyethylene glycol (PEG). These strategies are proved to further improving the circulation time of nanocomplexes and enhanced their targeting ability. Until now, the commercially available Lipofectamine™, MegaFectin™, and Transint™ lipid-based transfection reagents have been reported to successfully deliver mRNA and siRNA in vitro or in vivo [134–138]. Other lipid carriers such as stabilized nucleic acid lipid particles (SNALP) are under clinical stage for siRNA delivery [139].

#### 7.4.5.2 siRNA

To date, more than 30 RNAi clinical trials are under way or have been completed to treat orphan diseases that lack a large market. Many brain diseases such as AD,

Huntington's disease (HD), amyotrophic lateral sclerosis (ALS), etc. are orphan diseases or very similar to orphan diseases [133, 140]. Since siRNA drug candidates are easy to design and develop, they are particular suitable for treating brain orphan diseases. However, among the ongoing siRNA clinical trials, none of them are associated with brain diseases. There are large amount of preclinical studies focusing on modification or engineering a siRNA nanoformulation that is able to overcome existing brain barriers. The receptor-mediated targeting approach is mainly used for enhancing siRNA penetrating the BBB. For example, in an siRNA/docetaxel (DTX) co-loaded liposomal formulation, the modification of dual peptides Angiopep-2 and tLyP-1 promoted targeted therapy of gliomas [141]. Results showed that the combined dual peptide and dual therapeutics achieved a synergistic effect in tumor growth inhibition. The targeting ability of this siRNA liposome was achieved by Angiopep-2 which exhibits the receptor-mediated targeting through interaction with LRP-1. In addition, more BBB transportation approaches such as cell-mediated transport and glycemia-controlled GLUT1 receptor-mediated transport require further evaluation for siRNA delivery to the brain [133].

#### 7.4.5.3 ASO

Antisense oligonucleotides (ASOs) play a therapeutic role by interfering gene expression of disease-related target mRNAs. They are the earliest and most mature species in the field of nucleic acid drug research, with high specificity, low toxicity, and side effects. To date, at least 5 ASOs have been approved by FDA or EU, and more than 100 are undergoing clinical trials. Application of ASOs in brain diseases is mainly focused on the treatment of glioma, AD, and PD. It is worth noting that an ASO drug inotersen was recently approved in the EU for treatment of stage 1/2 polyneuropathy in patients with hATTR [142]. Inotersen was developed by Ionis Pharmaceuticals and Akcea Therapeutics. It is the inhibitor of mutant and wild-type human TTR. Mutation of TTR gene leads to accumulation of TTR protein as amyloid deposits throughout the organs, including the nervous system. In a phase III clinical trial, inotersen significantly reduced neurological progression and improved quality of life in patients. Now inotersen is under evaluation in the USA and Canada for a similar indication [143].

#### 7.4.5.4 DNA and mRNA

In contrast to siRNA and ASO, DNA and mRNA can be used as a complement to the missing proteins in the brain. The plasmid DNA needs to enter the nucleus and integrate with the host genome to be functional. After the transcription and translation process, a longtime expression of targeted proteins will be realized. For example, when plasmid DNA encoding for glial cell line-derived neurotrophic factor (GDNF) was packaged into polymer nanoparticles and used to treat dopamine-injured rats, researchers observed an overexpression of GDNF in the rat brain, which results in neuroprotective effect to the rats [144, 145]. Compared to DNA, mRNA functions in the cytoplasm without the need to insert into the genome, and only transiently expresses the encoded protein. mRNA therapy has been very hot in the last decade, with nearly a hundred ongoing clinical trials. The vast majority of these clinical trials are focusing on cancer immunotherapy, infectious diseases vaccines, and treatment of cardiovascular diseases [146]. Among them, personalized mRNA-based DC vaccines in particular the glioma vaccines are being widely studied. At first, the specific tumor antigen is screened from the patient-derived glioma or other cancer stem cells. After an in vitro transcribed process, tumor antigen-encoding mRNA is obtained and is used to transfect patients' own dendritic cells (DCs). The readministered DCs will further present tumor antigen-encoding mRNA to other antitumor immune cells, such as T cells and natural killer cells. So the tumor cells are attacked by the immune system for therapeutic purposes [147–149].

## 7.5 Concerns over Biomacromolecular Nanomedicines

One of the major challenges and disadvantages with nanomedicines is their potential neurotoxicity [150–152]. The neurotoxicities generated by nanocarriers include generating excess reactive oxygen species (ROS), inducing cell apoptosis, upregulation of inflammatory cytokines and inducing lipid peroxidation, protein damage and depleted glutathione reserve in the CNS, etc. [152–154] One study reported a case in which titanium dioxide (P25) produced ROS in brain microglia and resulted in damages to neurons in vitro [155, 156]. The lipid and polymer nanocarriers are thought to be less harmful to the brain. However, in-depth toxicity studies of nanocarrier loading with payloads are needed to be conducted to provide potential risks around brain therapy. Another challenge for nanomedicine is that it is difficult to draw similar conclusions from different studies. Because of the versatile elements of nanoparticles, ranging from the physical-chemical characteristics to functional strategies, each study offers different positive features for innovation. A complete system that combined with all elements is insufficient. These disadvantages have drastically limited successful nanomedicine reaching the market. Moreover, a better understanding of the mechanisms of nanoparticles crossing brain barriers and the fate of nanoparticles after entry of brain is very important to design a complete nanosystem and predict its efficiency. At last, establishment of reliable in vitro brain barrier model and homogenous animal model is also very crucial for brain therapeutics development. This requires a clear understanding of the molecular mechanisms underlying the pathogenesis of brain diseases.

## 7.6 Conclusion and Prospective

In summary, biotechnological therapeutics with high molecular weight, e.g., peptide, antibody, enzyme, protein, DNA, RNA, etc., have been perused by the pharmaceutical industry for a long time. Because of the existing barriers of the brain, these biomacromolecules are often unable to cross these barriers and reach a therapeutic concentration. However, they could substantially benefit from the use of nanocarriers. Until now, an increasing number of various biomacromolecule-loaded nanomedicines showed meaningful results in CNS disease therapy both in vitro and in vivo. But to be honest, this field is still in its infancy. Many considerations should be emphasized and resolved before CNS nanomedicine becomes applicable in clinical stage. Considering the complexity of the brain structures and CNS diseases, in-depth and comprehensive studies of nanomedicine's fate, mechanism, and toxicity are required.

For future research, increasing the drug-trafficking performance and specificity for brain tissues is especially important for CNS nanomedicine development. As a result, utilizing the respective advantages of different biomacromolecules and suitable administration route to realize a combined treatment is necessary. For example, combining intranasal delivery with nanoparticles; advancing modification or bioconjugation technologies to improve the stability, pharmacokinetics, and pharmacodynamics of biomacromolecule drug candidates; and enhancing the targeted delivery and BBB-penetrating ability are necessary. Only if all of these factors are combined into a complete system will it be possible to create a clinically relevant biomacromolecule nanomedicine for brain diseases therapy.

## References

1. Lindsley CW, Lindsley CW, Lindsley CW. New 2016 data and statistics for global pharmaceutical products and projections through 2017. ACS Chem Neurosci. 2017;8(8):1635–6.
2. Alyautdin R, Khalin I, Nafeeza MI, Haron MH, Kuznetsov D. Nanoscale drug delivery systems and the blood-brain barrier. Int J Nanomedicine. 2014;9:795–811.
3. Azad TD, James P, Connolly ID, Austin R, Wilson CM, Grant GA. Therapeutic strategies to improve drug delivery across the blood-brain barrier. Neurosurg Focus. 2015;38(3):E9.
4. Saito R, Bringas JR, McKnight TR, Wendland MF, Mamot C, Drummond DC, Kirpotin DB, Park JW, Berger MS, Bankiewicz KS. Distribution of liposomes into brain and rat brain tumor models by convection-enhanced delivery monitored with magnetic resonance imaging. Cancer Res. 2004;64(7):2572–9.
5. Tian Y, Mi G, Chen Q, Chaurasiya B, Li Y, Shi D, Zhang Y, Webster TJ, Sun C, Shen Y. Acid-induced activated cell penetrating peptide modified cholesterol-conjugated polyoxyethylene sorbitol oleate mixed micelles for pH-triggered drug release and efficient brain tumor targeting based on a charge reversal mechanism. ACS Appl Mater Interfaces. 2018;10:43411.
6. Dwivedi N, Shah J, Mishra V, Mohd Amin MC, Iyer AK, Tekade RK, Kesharwani P. Dendrimer-mediated approaches for the treatment of brain tumor. J Biomater Sci Polym Ed. 2016;27(7):557–80.

7. Picone P, Sabatino MA, Ditta LA, Amato A, San Biagio PL, Mule F, Giacomazza D, Dispenza C, Di Carlo M. Nose-to-brain delivery of insulin enhanced by a nanogel carrier. J Control Release. 2018;270:23–36.
8. Elzoghby AO, Abd-Elwakil MM, Abd-Elsalam K, Elsayed MT, Hashem Y, Mohamed O. Natural polymeric nanoparticles for brain-targeting: implications on drug and gene delivery. Curr Pharm Des. 2016;22(22):3305–23.
9. Nigro A, Pellegrino M, Greco M, Comande A, Sisci D, Pasqua L, Leggio A, Morelli C. Dealing with skin and blood-brain barriers: the unconventional challenges of mesoporous silica nanoparticles. Pharmaceutics. 2018;10(4):E250.
10. Tomitaka A, Arami H, Huang Z, Raymond A, Rodriguez E, Cai Y, Febo M, Takemura Y, Nair M. Hybrid magneto-plasmonic liposomes for multimodal image-guided and brain-targeted HIV treatment. Nanoscale. 2017;10(1):184–94.
11. Banks WA. From blood-brain barrier to blood-brain interface: new opportunities for CNS drug delivery. Nat Rev Drug Discov. 2016;15(4):275–92.
12. Keaney J, Campbell M. The dynamic blood-brain barrier. FEBS J. 2015;282(21):4067–79.
13. Banks WA. Characteristics of compounds that cross the blood-brain barrier. BMC Neurol. 2009;9(Suppl 1):S3.
14. Hladky SB, Barrand MA. Fluid and ion transfer across the blood-brain and blood-cerebrospinal fluid barriers; a comparative account of mechanisms and roles. Fluids and barriers of the CNS. 2016;13(1):19.
15. Segal MB. The choroid plexuses and the barriers between the blood and the cerebrospinal fluid. Cell Mol Neurobiol. 2000;20(2):183–96.
16. Davson H, Hollingsworth G, Segal MB. The mechanism of drainage of the cerebrospinal fluid. Brain J Neurol. 1970;93(4):665–78.
17. Saunders N, Habgood M. Understanding barrier mechanisms in the developing brain to aid therapy for the dysfunctional brain. Future Neurol. 2011;6(2):187–99.
18. Spector R, Johanson CE. The mammalian choroid plexus. Sci Am. 1989;261(5):68–74.
19. Wang Z, Cai XJ, Qin J, Xie FJ, Han N, Lu HY. The role of histamine in opening blood-tumor barrier. Oncotarget. 2016;7(21):31299–310.
20. Zhou W, Chen C, Shi Y, Wu Q, Gimple RC, Fang X, Huang Z, Zhai K, Ke SQ, Ping YF, Feng H, Rich JN, Yu JS, Bao S, Bian XW. Targeting glioma stem cell-derived pericytes disrupts the blood-tumor barrier and improves chemotherapeutic efficacy. Cell Stem Cell. 2017;21(5):591–603.
21. Wolak DJ, Thorne RG. Diffusion of macromolecules in the brain: implications for drug delivery. Mol Pharm. 2013;10(5):1492–504.
22. Saltzman WM, Radomsky ML. Drugs released from polymers: diffusion and elimination in brain tissue. Chem Eng Sci. 1991;46(10):2429–44.
23. Nelson AL. Antibody fragments: hope and hype. MAbs. 2010;2(1):77–83.
24. T. Yokota, ., D.E. Milenic, M. Whitlow, ., J. Schlom, . Rapid tumor penetration of a single-chain Fv and comparison with other immunoglobulin forms, Cancer Res 52(12) (1992) 3402–3408.
25. Jain RK. Physiological barriers to delivery of monoclonal antibodies and other macromolecules in tumors. Cancer Res. 1990;50(3 Suppl):814s–9s.
26. Lim ST, Airavaara M, Harvey BK. Viral vectors for neurotrophic factor delivery: a gene therapy approach for neurodegenerative diseases of the CNS. Pharm Res. 2010;61(1):14–26.
27. Kamei N. Nose-to-brain delivery of peptide drugs enhanced by co-administration of cell-penetrating peptides: therapeutic potential for dementia. Yakugaku Zasshi : J Pharm Soc Jpn. 2017;137(10):1247–53.
28. Joshi S, Cooke JRN, Ellis JA, Emala CW, Bruce JN. Targeting brain tumors by intra-arterial delivery of cell-penetrating peptides: a novel approach for primary and metastatic brain malignancy. J Neuro-Oncol. 2017;135(3):497–506.

29. Mastakov MY, Baer K, Kotin RM, During MJ. Recombinant adeno-associated virus serotypes 2- and 5-mediated gene transfer in the mammalian brain: quantitative analysis of heparin co-infusion. Mol Ther. 2002;5(4):371–80.
30. Hamilton JF, Morrison PF, Chen MY, Harvey-White J, Pernaute RS, Phillips H, Oldfield E, Bankiewicz KS. Heparin coinfusion during convection-enhanced delivery (CED) increases the distribution of the glial-derived neurotrophic factor (GDNF) ligand family in rat striatum and enhances the pharmacological activity of neurturin. Exp Neurol. 2001;168(1):155–61.
31. Yu YJ, Zhang Y, Kenrick M, Hoyte K, Luk W, Lu Y, Atwal J, Elliott JM, Prabhu S, Watts RJ, Dennis MS. Boosting brain uptake of a therapeutic antibody by reducing its affinity for a transcytosis target. Sci Transl Med. 2011;3(84):84ra44.
32. Carcaboso AM, Elmeliegy MJ, Juel SJ, Zhang ZM, Calabrese C, Tracey L, Waters CM, Stewart CF. Tyrosine kinase inhibitor gefitinib enhances topotecan penetration of gliomas. Cancer Res. 2010;70(11):4499–508.
33. Doyle L, Ross DD, et al. Oncogene. 2003;22(47):7340.
34. Enokizono J, Kusuhara H, Ose A, Schinkel AH, Sugiyama Y. Quantitative investigation of the role of breast cancer resistance protein (Bcrp/Abcg2) in limiting brain and testis penetration of xenobiotic compounds. Drug Metab Dispos. 2008;36(6):995.
35. Salama NN, Kelly EJ, Bui T, Ho RJ. The impact of pharmacologic and genetic knockout of P-glycoprotein on nelfinavir levels in the brain and other tissues in mice. J Pharm Sci. 2005;94 (6):1216–25.
36. Wong HL, Bendayan R, Rauth AM, Wu XY. Simultaneous delivery of doxorubicin and GG918 (Elacridar) by new polymer-lipid hybrid nanoparticles (PLN) for enhanced treatment of multidrug-resistant breast cancer. J Control Release. 2006;116(3):275–84.
37. Glascock JJ, Osman EY, Coady TH, Rose FF, Shababi M, Lorson CL. Delivery of therapeutic agents through intracerebroventricular (ICV) and intravenous (IV) injection in mice. J Vis Exp Jove. 2010;56:e2968.
38. Tosi G, Musumeci T, Ruozi B, Carbone C, Belletti D, Pignatello R, Vandelli MA, Puglisi G. The “fate” of polymeric and lipid nanoparticles for brain delivery and targeting: strategies and mechanism of blood–brain barrier crossing and trafficking into the central nervous system. J Drug Delivery Sci Technol. 2016;32:66–76.
39. Roland N, Fritz SR, Helmut E. Penetration of drugs through the blood-cerebrospinal fluid/ blood-brain barrier for treatment of central nervous system infections. Clin Microbiol Rev. 2010;23(4):858–83.
40. Duskey JT, Belletti D, Pederzoli F, Vandelli MA, Forni F, Ruozi B, Tosi G. Current strategies for the delivery of therapeutic proteins and enzymes to treat brain disorders. Int Rev Neurobiol. 2017;137:1–28.
41. Garg T, Bhandari S, Rath G, Goyal AK. Current strategies for targeted delivery of bio-active drug molecules in the treatment of brain tumor. J Drug Target. 2015;23(10):865–87.
42. Parrish KE, Sarkaria JN, Elmquist WF. Improving drug delivery to primary and metastatic brain tumors: strategies to overcome the blood-brain barrier. Clin Pharmacol Ther. 2015;97 (4):336–46.
43. R.H. Bobo, D.W. Laske, A. Akbasak, ., P.F. Morrison, R.L. Dedrick, E.H. Oldfield, Convection-enhanced delivery of macromolecules in the brain, Proc Natl Acad Sci U S A 91(6) (1994) 2076–2080.
44. Raghu R, Brady ML, Andreas H, Christoph P, Sampson JH. Convection-enhanced delivery of therapeutics for brain disease, and its optimization. Neurosurg Focus. 2006;20(4):E12.
45. Barua NU, Gill SS, Seth L. Convection-enhanced drug delivery to the brain: therapeutic potential and neuropathological considerations. Brain Pathol. 2014;24(2):117–27.
46. Betbeder D, Spérandio S, Latapie JP, Nadaí JD, Etienne A, Zajac JM, Francés B. Biovector™ nanoparticles improve Antinociceptive efficacy of nasal morphine. Pharm Res. 2000;17 (6):743–8.
47. Illum L. Is nose-to-brain transport of drugs in man a reality? J Pharm Pharmacol. 2010;56 (1):3–17.

48. Xiaoling G, Weixing T, Wei L, Qizhi Z, Yan Z, Xinguo J, Shoukuan F. Lectin-conjugated PEG-PLA nanoparticles: preparation and brain delivery after intranasal administration. Biomaterials. 2006;27(18):3482–90.
49. Ruan Y, Yao L, Zhang B, Zhang S, Guo J. Nanoparticle-mediated delivery of neurotoxin-II to the brain with intranasal administration: an effective strategy to improve antinociceptive activity of neurotoxin. Drug Dev Ind Pharm. 2012;38(1):123–8.
50. Kumar M, Pandey RS, Patra KC, Jain SK, Soni ML, Dangi JS, Madan J. Evaluation of neuropeptide loaded trimethyl chitosan nanoparticles for nose to brain delivery. Int J Biol Macromol. 2013;61(10):189–95.
51. Alsarra IA, Hamed AY, Alanazi FK, Maghraby GME. Vesicular Systems for Intranasal Drug Delivery. NeuroMethods. 2010;45:175–203.
52. Hongbing W, Kaili H, Xinguo J. From nose to brain: understanding transport capacity and transport rate of drugs. Expert Opin Drug Deliv. 2008;5(10):1159–68.
53. Siew A, Le H, Thiovolet M, Gellert P, Schatzlein A, Uchegbu I. Enhanced oral absorption of hydrophobic and hydrophilic drugs using quaternary ammonium palmitoyl glycol chitosan nanoparticles. Mol Pharm. 2012;9(1):14–28.
54. Chooi KW, Carlos MIS, Soundararajan R, Gaisford S, Arifin N, Schätzlein AG, Uchegbu IF. Physical characterisation and Long-term stability studies on quaternary ammonium Palmitoyl glycol chitosan (GCPQ)—a new drug delivery polymer. J Pharm Sci. 2014;103 (8):2296–306.
55. Lalatsa A, Garrett NL, Ferrarelli T, Moger J, Schatzlein AG, Uchegbu IF. Delivery of peptides to the blood and brain after oral uptake of quaternary ammonium palmitoyl glycol chitosan nanoparticles. Mol Pharm. 2012;9(6):1764–74.
56. S. Nakano, ., K. Matsukado, ., K.L. Black, Increased brain tumor microvessel permeability after intracarotid bradykinin infusion is mediated by nitric oxide, Cancer Res 56(17) (1996) 4027–4031.
57. E.A. Neuwelt, M. Glasberg, ., J. Diehl, ., E.P. Frenkel, P. Barnett, . Osmotic blood-brain barrier disruption in the posterior fossa of the dog, J Neurosurg 55(5) (1981) 742–748.
58. John BA, Apostolos TJ, Christoph HP, Ilhami K, Sherese F, Kartik K, Stephen SM, Susan PC, Theodore SH, Philip S. Safety and maximum tolerated dose of superselective intraarterial cerebral infusion of bevacizumab after osmotic blood-brain barrier disruption for recurrent malignant glioma. Clinical article. J Neurosurg. 2011;114(3):624.
59. G. Nilaver, ., L.L. Muldoon, R.A. Kroll, M.A. Pagel, X.O. Breakefield, B.L. Davidson, E.A. Neuwelt, Delivery of herpesvirus and adenovirus to nude rat intracerebral tumors after osmotic blood-brain barrier disruption, Proc Natl Acad Sci U S A 92(21) (1995) 9829–9833.
60. K. Matsukado, ., T. Inamura, ., S. Nakano, ., M. Fukui, ., R.T. Bartus, K.L. Black, Enhanced tumor uptake of carboplatin and survival in glioma-bearing rats by intracarotid infusion of bradykinin analog, RMP-7, Neurosurgery 39(1) (1996) 125–133.
61. Nickolai S, Nathan MD, Shipra S, Kullervo H. Effect of focused ultrasound applied with an ultrasound contrast agent on the tight junctional integrity of the brain microvascular endothelium. Ultrasound Med Biol. 2008;34(7):1093–104.
62. Park J, Aryal M, Vykhodtseva N, Zhang YZ, Mcdannold N. Evaluation of permeability, doxorubicin delivery, and drug retention in a rat brain tumor model after ultrasound-induced blood-tumor barrier disruption. J Control Release. 2017;250:77–85.
63. Alison B, Kullervo H. Noninvasive and targeted drug delivery to the brain using focused ultrasound. ACS Chem Neurosci. 2013;4(4):519.
64. Yao L, Song Q, Bai W, Zhang J, Miao D, Jiang M, Wang Y, Shen Z, Hu Q, Gu X. Facilitated brain delivery of poly (ethylene glycol)-poly (lactic acid) nanoparticles by microbubble-enhanced unfocused ultrasound. Biomaterials. 2014;35(10):3384–95.
65. Hao-Li L, Mu-Yi H, Hung-Wei Y, Chiung-Yin H, Po-Chun C, Jia-Shin W, I-Chou T, Jiun-Jie W, Tzu-Chen Y, Pin-Yuan C. Magnetic resonance monitoring of focused ultrasound/magnetic nanoparticle targeting delivery of therapeutic agents to the brain. Proc Natl Acad Sci U S A. 2010;107(34):15205–10.

66. Jeffrey LJ, Robert TG. Intranasal delivery of biologics to the central nervous system. Adv Drug Deliv Rev. 2012;64(7):614–28.
67. Lalatsa A, Schatzlein AG, Uchegbu IF. Strategies to deliver peptide drugs to the brain. Mol Pharm. 2014;11(4):1081–93.
68. Visser CC, Voorwinden LH, Crommelin DJ, Danhof M, de Boer AG. Characterization and modulation of the transferrin receptor on brain capillary endothelial cells. Pharm Res. 2004;21 (5):761–9.
69. Pardridge WM, Boado RJ. Reengineering biopharmaceuticals for targeted delivery across the blood-brain barrier. Methods Enzymol. 2012;503:269–92.
70. Pardridge WM. Drug transport across the blood-brain barrier. J Cereb Blood Flow Metab. 2012;32(11):1959–72.
71. Sun H, Su J, Meng Q, Yin Q, Chen L, Gu W, Zhang P, Zhang Z, Yu H, Wang S. Cancer-cell-biomimetic nanoparticles for targeted therapy of homotypic tumors. Adv Mater. 2016;28 (43):9581–8.
72. Luk BT, Jiang Y, Copp JA, Hu CJ, Krishnan N, Gao W, Li S, Fang RH, Zhang L. Biomimetic targeting of nanoparticles to immune cell subsets via cognate antigen interactions. Mol Pharm. 2018;15:3723.
73. Taurin S, Nehoff H, Aswegen TV, Greish K. Tumor vasculature, EPR effect, and anticancer nanomedicine: connecting the dots. In: Cancer targeted drug delivery: an elusive dream. New York: Springer; 2013.
74. Maedaa H, Daruwalla J. Polymeric drugs for efficient tumor-targeted drug delivery based on EPR-effect. Eur J Pharm Biopharm. 2009;71(3):409–19.
75. Yi X, Kabanov AV. Brain delivery of proteins via their fatty acid and block copolymer modifications. J Drug Target. 2013;21(10):940–55.
76. Kinstler O, Molineux G, Treuheit M, Ladd D, Gegg C. Mono-N-terminal poly(ethylene glycol)–protein conjugates. Adv Drug Deliv Rev. 2002;54(4):477–85.
77. Lei T, Jingquan L, Davis TP. Branched polymer-protein conjugates made from mid-chain-functional P(HPMA). Biomacromolecules. 2009;10(10):2847–51.
78. Kratz F. Albumin as a drug carrier: design of prodrugs, drug conjugates and nanoparticles. J Control Release. 2008;132(3):171–83.
79. Wilson B, Lavanya Y, Priyadarshini SR, Ramasamy M, Jenita JL. Albumin nanoparticles for the delivery of gabapentin: preparation, characterization and pharmacodynamic studies. Int J Pharm. 2014;473(1–2):73–9.
80. Tarun G, Animesh K, Goutam R, Goyal AK. Gastroretentive drug delivery systems for therapeutic management of peptic ulcer. Crit Rev Ther Drug Carrier Syst. 2014;31(6):531–57.
81. Green MR, Manikhas GS, Afanasyev B, Makhson AM, Bhar P, Hawkins MJ. Abraxane, a novel Cremophor-free, albumin-bound particle form of paclitaxel for the treatment of advanced non-small-cell lung cancer. Ann Oncol. 2006;17(8):1263–8.
82. Lin T, Zhao P, Jiang Y, Tang Y, Jin H, Pan Z, He H, Yang VC, Huang Y. Blood–brain-barrier-penetrating albumin nanoparticles for biomimetic drug delivery via albumin-binding protein pathways for antiglioma therapy. ACS Nano. 2016;10(11):9999–10012.
83. Blaire OL, Henrik OS, Bernardetta N, James MH, Joe ZXH, Andrew G, Gordon M, Liubov ZS, Cynthia S. Pharmacokinetic and pharmacodynamic studies of a human serum albumin-interferon-alpha fusion protein in cynomolgus monkeys. J Pharmacol Exp Ther. 2002;303 (2):540–8.
84. Balan V, Sulkowski M, Nelson D, Everson G, Bambury T, Recta J, Zhong J, Mesghali H, Murray J, Osborn B. 313 A 1/2phase study to evaluate the pharmacokinetics, safety, tolerability, immunogenicity, and pharmacodynamics of Albuferon™ in the treatment experienced subjects with chronic hepatitis C. Hepatology. 2003;38(5):307.
85. Yun X, Maximov VD, Yu J, Zhu G, Vertegel AA, Kindy MS. Nanoparticles for targeted delivery of antioxidant enzymes to the brain after cerebral ischemia and reperfusion injury. J Cereb Blood Flow Metab. 2013;33(4):583–92.

86. Zhao YZ, Lin M, Lin Q, Yang W, Yu XC, Tian FR, Mao KL, Yang JJ, Lu CT, Wong HL. Intranasal delivery of bFGF with nanoliposomes enhances in vivo neuroprotection and neural injury recovery in a rodent stroke model. J Control Release. 2016;224:165–75.
87. Wang X, Bodman A, Shi C, Guo D, Wang L, Luo J, Hall WA. Tunable Lipidoid-Telodendrimer hybrid nanoparticles for intracellular protein delivery in brain tumor treatment. Small. 2016;12(31):4185–92.
88. K. Michaelis, ., M.M. Hoffmann, S. Dreis, ., E. Herbert, ., R.N. Alyautdin, M. Michaelis, ., J. Kreuter, ., K. Langer, . Covalent linkage of apolipoprotein e to albumin nanoparticles strongly enhances drug transport into the brain, J Pharmacol Exp Ther 317(3) (2006) 1246.
89. Abakumov MA, Nukolova NV, Sokolsky-Papkov M, Shein SA, Sandalova TO, Vishwasrao HM, Grinenko NF, Gubsky IL, Abakumov AM, Kabanov AV, Chekhonin VP. VEGF-targeted magnetic nanoparticles for MRI visualization of brain tumor. Nanomedicine. 2015;11(4):825–33.
90. Spencer B, Emadi S, Desplats P, Eleuteri S, Michael S, Kosberg K, Shen J, Rockenstein E, Patrick C, Adame A, Gonzalez T, Sierks M, Masliah E. ESCRT-mediated uptake and degradation of brain-targeted alpha-synuclein single chain antibody attenuates neuronal degeneration in vivo. Mol Ther. 2014;22(10):1753–67.
91. Chaturvedi M, Molino Y, Sreedhar B, Khrestchatisky M, Kaczmarek L. Tissue inhibitor of matrix metalloproteinases-1 loaded poly(lactic-co-glycolic acid) nanoparticles for delivery across the blood-brain barrier. Int J Nanomedicine. 2014;9:575–88.
92. Liu Z, Jiang M, Kang T, Miao D, Gu G, Song Q, Yao L, Hu Q, Tu Y, Pang Z, Chen H, Jiang X, Gao X, Chen J. Lactoferrin-modified PEG-co-PCL nanoparticles for enhanced brain delivery of NAP peptide following intranasal administration. Biomaterials. 2013;34(15):3870–81.
93. Chertok B, David AE, Yang VC. Delivery of functional proteins to brain tumor using MRI-monitored, magnetically-targeted nanoparticles. J Control Release. 2008;132(3):e61–2.
94. Lindqvist A, Rip J, Gaillard PJ, Bjorkman S, Hammarlund-Udenaes M. Enhanced brain delivery of the opioid peptide DAMGO in glutathione pegylated liposomes: a microdialysis study. Mol Pharm. 2013;10(5):1533–41.
95. Zhao Y, Haney MJ, Mahajan V, Reiner BC, Dunaevsky A, Mosley RL, Kabanov AV, Gendelman HE, Batrakova EV. Active targeted macrophage-mediated delivery of catalase to affected brain regions in models of Parkinson's disease. J Nanosci Nanotechnol. 2011;01: S4.
96. Morris MC, Deshayes S, Heitz F, Divita G. Cell-penetrating peptides: from molecular mechanisms to therapeutics. Biol Cell. 2008;100(4):201–17.
97. Eiriksdottir E, Konate K, Langel U, Divita G, Deshayes S. Secondary structure of cell-penetrating peptides controls membrane interaction and insertion. Biochim Biophys Acta. 2010;1798(6):1119–28.
98. Herve F, Ghinea N, Scherrmann JM. CNS delivery via adsorptive transcytosis. AAPS J. 2008;10(3):455–72.
99. Kumar P, Wu H, McBride JL, Jung KE, Kim MH, Davidson BL, Lee SK, Shankar P, Manjunath N. Transvascular delivery of small interfering RNA to the central nervous system. Nature. 2007;448(7149):39–43.
100. Park TE, Singh B, Li H, Lee JY, Kang SK, Choi YJ, Cho CS. Enhanced BBB permeability of osmotically active poly(mannitol-co-PEI) modified with rabies virus glycoprotein via selective stimulation of caveolar endocytosis for RNAi therapeutics in Alzheimer's disease. Biomaterials. 2015;38:61–71.
101. Haroon MM, Dar GH, Jeyalakshmi D, Venkatraman U, Saba K, Rangaraj N, Patel AB, Gopal V. A designed recombinant fusion protein for targeted delivery of siRNA to the mouse brain. J Control Release. 2016;228:120–31.
102. Kim JY, Choi WI, Kim YH, Tae G. Brain-targeted delivery of protein using chitosan- and RVG peptide-conjugated, pluronic-based nano-carrier. Biomaterials. 2013;34(4):1170–8.

103. Demeule M, Currie JC, Bertrand Y, Che C, Nguyen T, Regina A, Gabathuler R, Castaigne JP, Beliveau R. Involvement of the low-density lipoprotein receptor-related protein in the transcytosis of the brain delivery vector angiopep-2. J Neurochem. 2008;106(4):1534–44.
104. Bertrand Y, Currie JC, Demeule M, Regina A, Che C, Abulrob A, Fatehi D, Sartelet H, Gabathuler R, Castaigne JP, Stanimirovic D, Beliveau R. Transport characteristics of a novel peptide platform for CNS therapeutics. J Cell Mol Med. 2010;14(12):2827–39.
105. Thomas FC, Kunal T, Vinay R, Satyanarayana G, Thorsheim HR, Gaasch JA, Mittapalli RK, Diane P, Steeg PS, Lockman PR. Uptake of ANG1005, a novel paclitaxel derivative, through the blood-brain barrier into brain and experimental brain metastases of breast cancer. Pharm Res. 2009;26(11):2486.
106. Regina A, Demeule M, Che C, Lavallee I, Poirier J, Gabathuler R, Beliveau R, Castaigne JP. Antitumour activity of ANG1005, a conjugate between paclitaxel and the new brain delivery vector Angiopep-2. Br J Pharmacol. 2008;155(2):185–97.
107. Hao Y, Zhang B, Zheng C, Ji R, Ren X, Guo F, Sun S, Shi J, Zhang H, Zhang Z, Wang L, Zhang Y. The tumor-targeting core-shell structured DTX-loaded PLGA@Au nanoparticles for chemo-photothermal therapy and X-ray imaging. J Control Release. 2015;220(Pt A):545–55.
108. Huang R, Ma H, Guo Y, Liu S, Kuang Y, Shao K, Li J, Liu Y, Han L, Huang S, An S, Ye L, Lou J, Jiang C. Angiopep-conjugated nanoparticles for targeted long-term gene therapy of Parkinson's disease. Pharm Res. 2013;30(10):2549–59.
109. Yan H, Wang L, Wang J, Weng X, Lei H, Wang X, Jiang L, Zhu J, Lu W, Wei X, Li C. Two-order targeted brain tumor imaging by using an optical/paramagnetic nanoprobe across the blood brain barrier. ACS Nano. 2012;6(1):410–20.
110. S. Zhang, ., T. Holmes, ., C. Lockshin, ., A. Rich, . Spontaneous assembly of a self-complementary oligopeptide to form a stable macroscopic membrane, Proc Natl Acad Sci U S A 90(8) (1993) 3334–3338.
111. Ulijn RV, Smith AM. Designing peptide based nanomaterials. Chem Soc Rev. 2008;37(4):664–75.
112. Zhang S, Marini DM, Hwang W, Santoso S. Design of nanostructured biological materials through self-assembly of peptides and proteins. Curr Opin Chem Biol. 2002;6(6):865–71.
113. Mazza M, Notman R, Anwar J, Rodger A, Hicks M, Parkinson G, McCarthy D, Daviter T, Moger J, Garrett N, Mead T, Briggs M, Schatzlein AG, Uchegbu IF. Nanofiber-based delivery of therapeutic peptides to the brain. ACS Nano. 2013;7(2):1016–26.
114. Pang Z, Lu W, Gao H, Hu K, Chen J, Zhang C, Gao X, Jiang X, Zhu C. Preparation and brain delivery property of biodegradable polymersomes conjugated with OX26. J Control Release. 2008;128(2):120–7.
115. Aktas Y, Yemisci M, Andrieux K, Gursoy RN, Alonso MJ, Fernandez-Megia E, Novoa-Carballal R, Quinoa E, Riguera R, Sargon MF, Celik HH, Demir AS, Hincal AA, Dalkara T, Capan Y, Couvreur P. Development and brain delivery of chitosan-PEG nanoparticles functionalized with the monoclonal antibody OX26. Bioconjug Chem. 2005;16(6):1503–11.
116. Yue PJ, He L, Qiu SW, Li Y, Liao YJ, Li XP, Xie D, Peng Y. OX26/CTX-conjugated PEGylated liposome as a dual-targeting gene delivery system for brain glioma. Mol Cancer. 2014;13(1):191.
117. Shi N, Zhang Y, Zhu C, Boado RJ, Pardridge WM. Brain-specific expression of an exogenous gene after i.v. administration. Proc Natl Acad Sci U S A. 2001;98(22):12754–9.
118. Battaglia L, Gallarate M, Peira E, Chirio D, Solazzi I, Giordano SM, Gigliotti CL, Riganti C, Dianzani C. Bevacizumab loaded solid lipid nanoparticles prepared by the coacervation technique: preliminary in vitro studies. Nanotechnology. 2015;26(25):255102.
119. Carradori D, Balducci C, Re F, Brambilla D, Le Droumaguet B, Flores O, Gaudin A, Mura S, Forloni G, Ordonez-Gutierrez L, Wandosell F, Masserini M, Couvreur P, Nicolas J, Andrieux K. Antibody-functionalized polymer nanoparticle leading to memory recovery in Alzheimer's disease-like transgenic mouse model. Nanomedicine. 2018;14(2):609–18.

120. Giovanni T, Luca B, Barbara R, Anna Valeria V, Lucia B, Anna F, Francesco R, Flavio F, Maria Angela V. Can leptin-derived sequence-modified nanoparticles be suitable tools for brain delivery? Nanomedicine. 2012;7(3):365–82.
121. Ruan S, Yuan M, Zhang L, Hu G, Chen J, Cun X, Zhang Q, Yang Y, He Q, Gao H. Tumor microenvironment sensitive doxorubicin delivery and release to glioma using angiopep-2 decorated gold nanoparticles. Biomaterials. 2015;37:425–35.
122. Ruan S, Qian J, Shen S, Chen J, Zhu J, Jiang X, He Q, Yang W, Gao H. Fluorescent carbonaceous nanodots for noninvasive glioma imaging after angiopep-2 decoration. Bioconjug Chem. 2014;25(12):2252–9.
123. Tang SC, Bates S, Kesari S, Brenner AJ, Anders CK, Garcia A, Ibrahim NK, Tkaczuk K, Kumthekar P. Abstract P6-17-04: A phase II, open-label, multi-center study of ANG1005, a novel brain-penetrant taxane derivative, in breast cancer patients with recurrent CNS metastases. Cancer Res. 2016;76(4 Supplement):P6-17-04.
124. Yong L, Pan Y, Shi Y, Huang X, Jia N, Jiang J. Delivery of large molecules via poly(butyl cyanoacrylate) nanoparticles into the injured rat brain. Nanotechnology. 2012;23(23):165101.
125. Vinzant N, Scholl JL, Wu CM, Kindle T, Koodali R, Forster GL. Iron oxide nanoparticle delivery of peptides to the brain: reversal of anxiety during drug withdrawal. Front Neurosci. 2017;11:608.
126. Mamot C, Drummond DC, Noble CO, Kallab V, Guo Z, Hong K, Kirpotin DB, Park JW. Epidermal growth factor receptor-targeted immunoliposomes significantly enhance the efficacy of multiple anticancer drugs in vivo. Cancer Res. 2005;65(24):11631–8.
127. Muthusamy J, Ansell SM, Mui BL, Tam KY, Jianxin C, Xinyao D, David B, Laxman E, Shigeo M, Narayanannair JK. Maximizing the potency of siRNA lipid nanoparticles for hepatic gene silencing in vivo. Angew Chem. 2012;51(34):8529–33.
128. Wei L, Guo XY, Yang T, Yu MZ, Chen DW, Wang JC. Brain tumor-targeted therapy by systemic delivery of siRNA with transferrin receptor-mediated core-shell nanoparticles. Int J Pharm. 2016;510(1):394–405.
129. Rungta RL, Choi HB, Lin PJ, Ko RW, Ashby D, Nair J, Manoharan M, Cullis PR, Macvicar BA. Lipid nanoparticle delivery of siRNA to silence neuronal gene expression in the brain. Mol Ther–Nucleic Acids. 2013;2(12):e136.
130. Dwarki VJ, Malone RW, Verma IM, Wu R. Cationic liposome-mediated RNA transfection. Proc Natl Acad Sci U S A. 1989;86(16):6077–81.
131. Kwok A. The challenges and current advances in delivering RNAi as therapeutics. Berlin/Heidelberg: Springer; 2013.
132. Jiehua Z, Shum KT, Burnett JC, Rossi JJ. Nanoparticle-based delivery of RNAi therapeutics: Progress and challenges. Pharmaceuticals. 2013;6(1):85–107.
133. Zheng M, Tao W, Zou Y, Farokhzad OC, Shi B. Nanotechnology-based strategies for siRNA brain delivery for disease therapy. Trends Biotechnol. 2018;36(5):562–75.
134. Dalby B, Cates S, Harris A, Ohki EC, Tilkins ML, Price PJ, Ciccarone VC. Advanced transfection with Lipofectamine 2000 reagent: primary neurons, siRNA, and high-throughput applications. Methods. 2004;33(2):95–103.
135. Julia B, Meng D, Sebastian T, Kamilla P, Elke K, Godehard F, Ulrich S, Claus-Michael L, Ulrich K, Mürdter TE. Efficient telomerase inhibition in human non-small cell lung cancer cells by liposomal delivery of 2′-O-methyl-RNA. J Pharm Sci. 2010;98(5):1765–74.
136. Xingfang S, Jennifer F, Kavanagh DG, Irvine DJ. In vitro and in vivo mRNA delivery using lipid-enveloped pH-responsive polymer nanoparticles. Mol Pharm. 2011;8(3):774–87.
137. Kormann MSD, Günther H, Aneja MK, Gabriela N, Flemmer AW, Susanne HJ, Marceline H, Mays LE, Marta I, Andrea S. Expression of therapeutic proteins after delivery of chemically modified mRNA in mice. Nat Biotechnol. 2011;29(2):154–7.
138. Thess A, Grund S, Mui BL, Hope MJ, Baumhof P, Fotin-Mleczek M, Schlake T. Sequence-engineered mRNA without chemical nucleoside modifications enables an effective protein therapy in large animals. Mol Ther. 2015;23(9):1456–64.

139. Gomesdasilva LC, Fonseca NA, Moura V, Mc PDL, Simões S, Moreira JN. Lipid-based nanoparticles for siRNA delivery in cancer therapy: paradigms and challenges. Acc Chem Res. 2012;45(7):1163.
140. Joppi R, Bertele V, Garattini S. Orphan drugs, orphan diseases. The first decade of orphan drug legislation in the EU. Eur J Clin Pharmacol. 2013;69(4):1009–24.
141. Yang ZZ, Li JQ, Wang ZZ, Dong DW, Qi XR. Tumor-targeting dual peptides-modified cationic liposomes for delivery of siRNA and docetaxel to gliomas. Biomaterials. 2014;35 (19):5226–39.
142. Keam SJ. Inotersen: first global approval. Drugs. 2018;78:1–6.
143. Benson MD, Waddington-Cruz M, Berk JL, Polydefkis M, Dyck PJ, Wang AK, Plante-Bordeneuve V, Barroso FA, Merlini G, Obici L, Scheinberg M, Brannagan TH 3rd, Litchy WJ, Whelan C, Drachman BM, Adams D, Heitner SB, Conceicao I, Schmidt HH, Vita G, Campistol JM, Gamez J, Gorevic PD, Gane E, Shah AM, Solomon SD, Monia BP, Hughes SG, Kwoh TJ, McEvoy BW, Jung SW, Baker BF, Ackermann EJ, Gertz MA, Coelho T. Inotersen treatment for patients with hereditary transthyretin amyloidosis. N Engl J Med. 2018;379(1):22–31.
144. Yurek DM, Flectcher AM, Kowalczyk TH, Padegimas L, Cooper MJ. Compacted DNA nanoparticle gene transfer of GDNF to the rat striatum enhances the survival of grafted fetal dopamine neurons. Cell Transplant. 2009;18(10):1183–96.
145. Yurek DM, Hasselrot U, Cass WA, Sesenoglu-Laird O, Padegimas L, Cooper MJ. Age and lesion-induced increases of GDNF transgene expression in brain following intracerebral injections of DNA nanoparticles. Neuroscience. 2015;284:500–12.
146. Ugur S, Katalin K, Türeci Ö. mRNA-based therapeutics--developing a new class of drugs. Nat Rev Drug Discov. 2014;13(10):759–80.
147. Kreiter S, Diken M, Selmi A, Türeci Ö, Sahin U. Tumor vaccination using messenger RNA: prospects of a future therapy. Curr Opin Immunol. 2011;23(3):399–406.
148. Axel H, Doris C, Jens D, Donna Y, Maurice MA, Lallas CD, Philipp D, Donna N, Eli G, Johannes V. Autologous dendritic cells transfected with prostate-specific antigen RNA stimulate CTL responses against metastatic prostate tumors. J Clin Investig. 2002;109(3):409–17.
149. Geall AJ, Mandl CW, Ulmer JB. RNA: the new revolution in nucleic acid vaccines. Semin Immunol. 2013;25(2):152–9.
150. Wong HL, Wu XY, Bendayan R. Nanotechnological advances for the delivery of CNS therapeutics. Adv Drug Deliv Rev. 2012;64(7):686–700.
151. Hussain SM, Javorina AM, Schrand AK, Duhart HM, Ali SF, Schlager JJ. The interaction of manganese nanoparticles with PC-12 cells induces dopamine depletion. Toxicol Sci. 2006;92 (2):456–63.
152. Wang J, Rahman MF, Duhart HM, Newport GD, Patterson TA, Murdock RC, Hussain SM, Schlager JJ, Ali SF. Expression changes of dopaminergic system-related genes in PC12 cells induced by manganese, silver, or copper nanoparticles. Neurotoxicology. 2009;30(6):926–33.
153. Hu YL, Gao JQ. Potential neurotoxicity of nanoparticles. Int J Pharm. 2010;394(1):115–21.
154. Wu J, Wang C, Sun J, Xue Y. Neurotoxicity of silica nanoparticles: brain localization and dopaminergic neurons damage pathways. ACS Nano. 2011;5(6):4476–89.
155. Xiaoyong D, Qixia L, Wenting C, Yanli W, Minghong W, Haijiao Z, Zheng J. Nanosized zinc oxide particles induce neural stem cell apoptosis. Nanotechnology. 2009;20(11):115101.
156. Long TC, Saleh N, Tilton RD, Lowry GV, Veronesi B. Titanium dioxide (P25) produces reactive oxygen species in immortalized brain microglia (BV2): implications for nanoparticle neurotoxicity. Environ Sci Technol. 2006;40(14):4346–52.

# Chapter 8
# Carbon-Based Nanomedicine

**Peng Zhang, Ming Zhang, and Jia Geng**

## 8.1 Introduction

Carbon-based nanomaterials, such as graphene oxide (GO), carbon nanotubes (CNTs), and nanodiamonds (NDs), have been considered as excellent carriers for anti-cancer drugs because of their high drug-loading capability, nanoscale size, and high specific surface areas, enabling them to penetrate the mammalian cell membrane. Therefore, it's meaningful to explore these carbon-based nanomaterials as versatile cancer drug carriers [1]. This chapter reviews the recent advances in carbon-based nanomedicine, including application, pharmacodynamics and metabolism, diagnosis and treatment, as well as biodistribution of carbon nanomaterials.

P. Zhang
Department of Laboratory Medicine, State Key Laboratory of Biotherapy, West China Hospital, Sichuan University and Collaborative Innovation Center for Biotherapy, Chengdu, China

School of Pharmaceutical Sciences, Guangzhou Higher Education Mega Center, Sun Yat-sen University, Guangzhou, China

M. Zhang · J. Geng (✉)
Department of Laboratory Medicine, State Key Laboratory of Biotherapy, West China Hospital, Sichuan University and Collaborative Innovation Center for Biotherapy, Chengdu, China
e-mail: geng.jia@scu.edu.cn



X. Xue (ed.), *Nanomedicine in Brain Diseases*,
https://doi.org/10.1007/978-981-13-8731-9_8

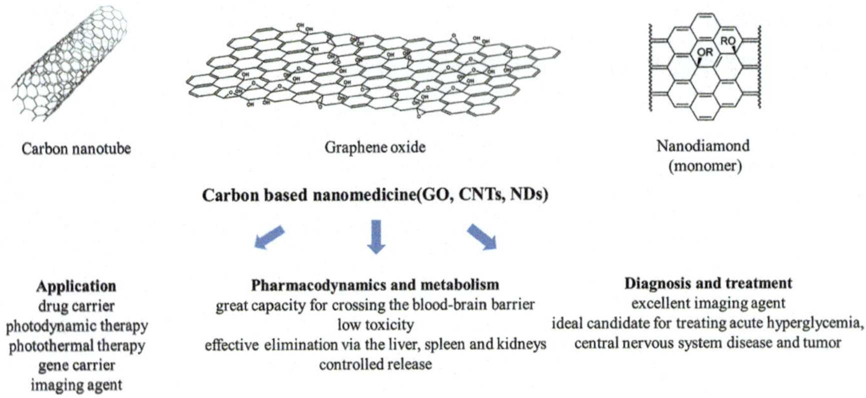


## 8.2 Application

### *8.2.1 Drug Carrier*

The ideal criterion of drug carrier is having best bioavailability at specific sites in patients at the necessary time. The usage of suitable carriers can improve the drug loading and promote therapeutic effects. Among a number of drug carrier systems, carbon nanomaterials offer a potential and efficient alternative for delivering therapeutic agents. The usage of carbon nanomaterials could lead transport drugs to cancerous tissues specifically without affecting the normal tissue. Tripisciano et al. encapsulated cisplatin, a platinum-based chemotherapeutic drug, into single-walled carbon nanotubes (SWCNTs) with a diameter of 1.3–1.6 nm. The efficiency of the developed delivery system was confirmed by inhibiting prostate cancer cells (PC3 and DU145) viability using tubes encapsulating cisplatin [2]. Naderi et al. developed a SWCNTs functionalized by an octa-ammonium polyhedral oligomeric silsesquioxanes (octa-ammonium-POSS) for delivering PTX to colon cancer cell (HT-29) and MCF-7 cell [3]. Pan et al. established a GO-based drug delivery carrier for treatment of the liver tumor cell line SMMC-7721 in vitro. In this nanocarrier, GO was firstly functionalized with fluorescein isothiocyanate (FIC), carboxymethyl chitosan (CMC), and lactobionic acid (LA), then it transported with DOX by physical adsorption. This nanosystem exhibited high drug loading (>96%). In addition, the functionalized GO showed good biocompatibility with the SMMC-7721 cell and induced cell death after incubation for 24 h [4].

Meanwhile, the binding between antibodies or ligands and the surface receptors of cancer cell promotes site-specific accumulation of lesion; the therapeutic effect of the drug could also be enhanced. Zhou et al. developed a GO nanocarrier (GO-Abs/PEI/PAH-Cit/DOX) functionalized with pH-responsive charge-reversal polyelectrolyte and integrin $\alpha_V\beta_3$ mono-antibody. They used these functionalized nanocarriers for targeted transportation and controllable release of DOX into tumor cells U87 MG cells. DOX was proved to be highly loaded and efficiently released using the nanocarrier. They also demonstrated that U87 MG cells could efficiently absorb the nanocarriers, and release DOX intracellularly, followed by delivery of DOX into the nucleus, leaving the nanocarrier in the cytoplasm [5]. Recently, Zhang et al. fabricated a hyaluronic acid (HA) functionalized MWCNTs (HA-MWCNTs) cancer-targeted nanosystem, acting as a multifunctional drug carrier. Transferrin (Tf) was used as a targeting ligand, and artemisinin (ART) as a model drug to conjugate with the HA-MWCNTs, thus developing the HA-MWCNTs/Tf@ART nanosystem, enabling it to treat MCF-7 cells in vitro. Meanwhile, a rat model carrying tumors was applied in in vivo studies. Compared with free ART, the drug carrier had significantly improved anti-cancer effect in vitro and in vivo [6]. Zhang et al. applied functional nanoscale graphene oxide (NGO) in order to control loading and targeting delivery of mixed anti-tumor drugs in MCF-7 cells. They found that FA-bound NGO (FA-NGO) carrying both anti-cancer drugs, camptothecin (CPT) and doxorubicin (DOX), suggested specific targeting to MCF-7 cells and significantly high cytotoxicity compared to that of NGO which is loaded with either DOX or CPT only. This indicated that the combined usage of two or more drugs, a widely applied clinical practice, often showed much better therapeutic efficiency than that with a single dosage form [7] (Figs. 8.1 and 8.2).

### *8.2.2 Gene Carrier*

The treatment of diseases by introducing gene therapy has earned significant attention. The advancement of nontoxic, biocompatible, and nonviral carriers for gene therapy is urgently needed for their applications for delivering drugs. Carbon nanovectors offer a promising strategy because of their advantages including the easier access into tumor cells, a significant enhanced solubility, bioavailability, and residence time of nucleic acids which protect the loaded nucleic acids from enzyme digestion and degradation efficiently.

Several measures have been developed to modify the surface of carbon-based nanomaterials for improving delivery of nucleic acids, including pDNA [8–11], siRNA [12, 13], and miRNA [14]. Feng et al. synthesized a physiologically stable dual-polymer-functionalized GO conjugate (NGO-PEG-PEI). In their study, polyethylene glycol (PEG) and PEI were covalently conjugated to GO through amide bonds. The product had excellent gene transfection effects without serum interference, and reduced cytotoxicity. Using the NIR optical absorbance of NGO, the cellular uptake of NGO-PEG-PEI was proved to be enhanced with a low-power NIR laser irradiation. With NGO-PEG-PEI as the light-responsive gene transportation, the transfection effects of plasmid DNA induced by the NIR laser were significantly enhanced. Being controlled by NIR light, NGO-PEG-PEI also showed the ability to deliver siRNA into cells, leading to markable target gene, Polo-like kinase 1 (Plk1), downregulation under the induction of laser irradiation [15].

Dendritic grafts PEI were responsible for stabilizing complexes between DNA and MWCNTs and avoiding DNA degradation. The chitosan-folic acid

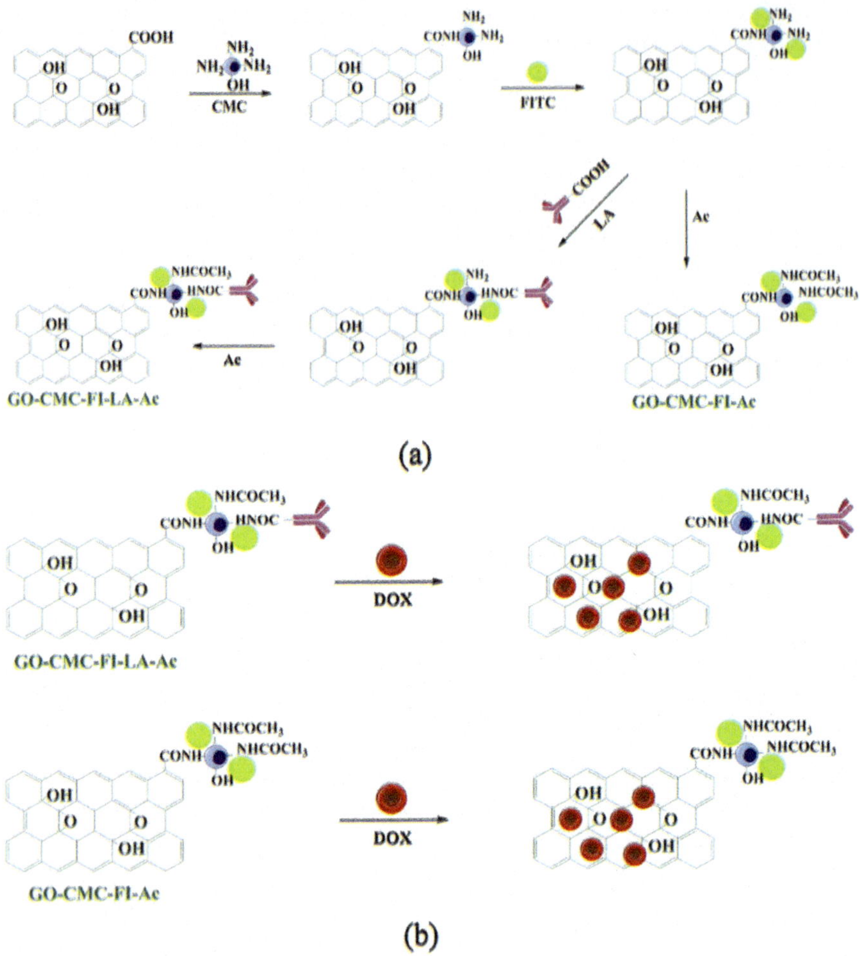


**Fig. 8.1** The synthetic strategy of drug delivery nanosystem: (**a**) the synthesis of modified GO carriers and (**b**) DOX loading [4]

nanoparticles were used to functionalize MWCNTs of different lengths and showed high efficiency to transport green fluorescent protein (GFP) plasmid DNA into HeLa and MCF-7 cells. The expression of GFP gene was identified by bright-field and fluorescence images of transfected cells [16]. A developed multi-walled carbon nanotubes (MWCNTs) functionalized by cationic polymers PEI, polyallylamine (PAA) or a mixture of both polymers possessed the ability to efficiently transfer plasmid DNA (pCMV-β-Galactosidase) to human lung carcinoma (A549) cell line [17]. Al-Jamal et al. studied whether silencing of Caspase-3 with carbon nanotube-mediated in vivo RNA interference (RNAi) could provide a therapeutic opportunity against stroke. They used functionalized carbon nanotubes (f-CNTs) to achieve the peri-lesional stereotactic administration of a Caspase-3 siRNA (siCas 3) in an

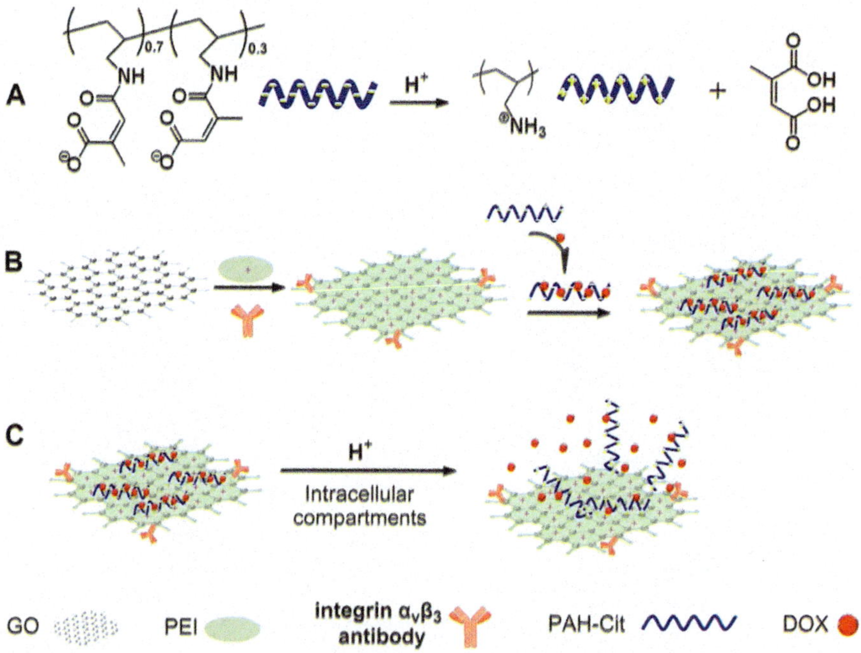


**Fig. 8.2** (**a**) The side chain of citracamide of the anion charge-inverting polyelectrolyte (PAH-Cit) was hydrolyzed in mild acidic conditions to obtain a cationic PAH (poly(allylamine)). (**b**) Development of the targeting charge-reversal nano delivery system (GO-Abs/PEI/PAH-Cit/DOX). (**c**) Controllable release of DOX in endosome or lysosome was stimulated by pH-dependent charge inversion of charge-reversal of polyelectrolytes on GO [5]

endothelin-1-induced stroke model. The results showed that the neurodegeneration was reduced and the functional preservation was enhanced before and after focal ischemic damage of the rodent motor cortex [13] (Fig. 8.3).

### 8.2.3 *Photodynamic Therapy (PDT) and Photothermal Therapy (PTT)*

It is necessary to increase temperature at tumor sites because the dielectric loss of tumors is higher than that of normal tissues. The carbon-based nanomaterials can transport radiofrequency (RF) energy to tumor cells, generating a remarkable amount of heat at specific sites and conducting hyperthermia [18]. Santos et al. combined CNTs and NIR radiation to induce hyperthermia treatment of primary brain tumors, glioma tumor cell lines (U251, U87, LN229 and T98G). Glioma cells internalized CNTs preferentially. Upon exposure to NIR radiation, the internalized CNTs produced heat, leading to death of necrotic cells. This continuous hyperthermia therapy was effective in the rodent tumor model in vivo, achieving tumor

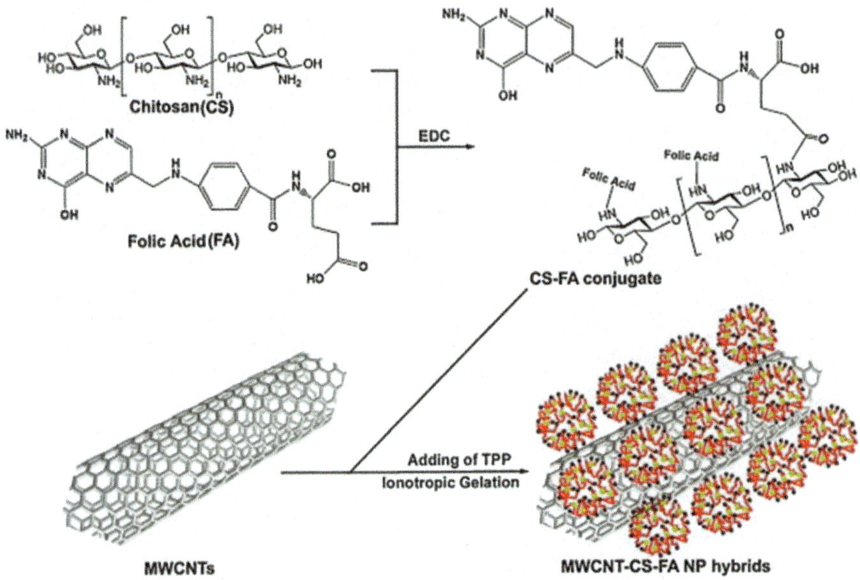


**Fig. 8.3** The scheme illustrating the construction of MWCNT-CS-FA NP hybrids [16]

reduction and no recurrence with only one treatment [19]. Zhang et al. fabricated a self-amplified drug delivery nanosystem (CMWCNTs-PEG) targeting fibrin for the photothermal therapy of tumors. The results showed that CMWCNTs-PEG could significantly increase the temperature in the tumor region than its counterpart 24 h after an initial NIR irradiation by an in vivo temperature monitoring test [20]. Zhou et al. reported an unmodified SWCNTs to develop a mitochondria-targeting photoacoustic model for anti-tumor in vitro and in vivo. Under pulsed laser irradiation (1064 nm), 79.4% of the tumor cells with intracellular SWCNTs were observed to be dead within 20 s, while 82.3% of the normal cells remained alive without using the SWCNTs [21]. Taratula et al. reported a new nano-theranostic platform by employing a new low-oxygen graphene nanosheet. The nanosheet was chemically functionalized with polypropylenimine dendrimers which was loaded with phthalocyanine (Pc) as a photosensitizer, for the photothermal therapy (PTT) and photodynamic therapy (PDT). This combination of PTT-PDT treatment showed an increased destruction of ovarian tumor cells, with a killing efficiency of 90–95% at a low Pc and low-oxygen graphene dose [22] (Fig. 8.4).

### 8.2.4 *Imaging Agent*

Among a variety of imaging techniques for cancer cells, fluorescence imaging mediated by carbon-based nanomaterials has aroused the most attention. If they

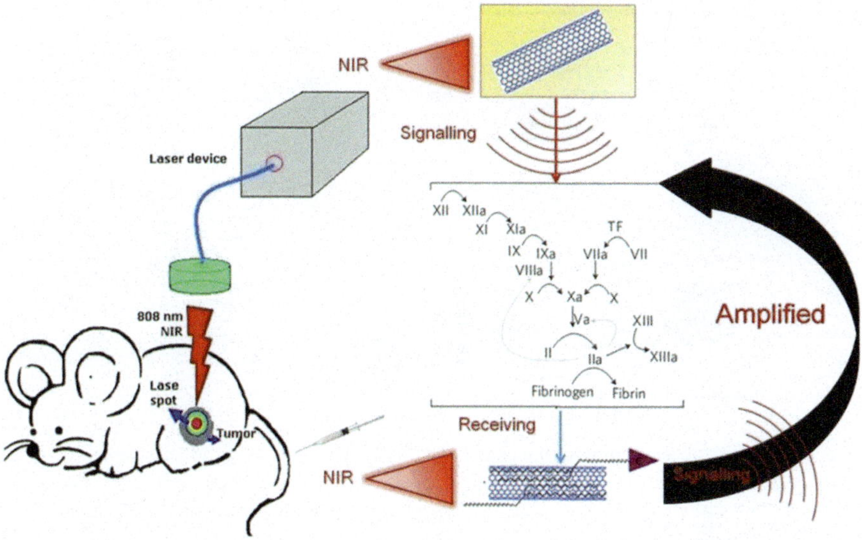


**Fig. 8.4** The scheme of the strategy of self-amplified targeted system. CMWCNTs-PEG which was targeted to fibrin was transported into A549 xenografts-bearing mouse models by intravenous injection. A NIR illumination could cause the photo-thermal effect, which damaged tumor vessels and triggered coagulation reaction. The fibrin produced in the vessels could attract a great deal of CMWCNTs-PEG to the tumor location, and additional local clotting can be induced when treated with NIR again, hence generating new fibrin for more CMWCNTs-PEG entrapment. Finally, targeting amplification of tumor can be achieved through the positive feedback mechanism of coagulation reaction [20]

are modified properly, they have potential excellent advantages such as high light stability, low cytotoxicity, non-quenching and photobleaching in cells, and enough penetration depth in tissues. Battogtokh et al. developed a GO nanocomplex functionalized by a bovine serum albumin (BSA)-cis-aconityl pheophorbide-a (c-PheoA), which was bound to folic acid (FA), thus producing the tumor-targeting FA-BSA-c-PheoA-GO nanocomplex. The nanocomplex was used to image the B16F10 cancer cell line (murine melanoma) and the MCF-7 cell line [23].

SWCNT-based nanosystems have been used as multimodality nanoprobes for cancer imaging [24]. In addition, the large difference of inter-band between the excitation and emission wavelengths of SWCNTs contributes to high signal-to-noise ratios which improve the detection limit for cancer cells. To apply SWCNTs as a biological probe, appropriate modification is required to enhance their specific binding capacities and reduce their toxic side effects. Welsher et al. applied SWCNTs as bright and biocompatible imaging agents for near-infrared imaging of a rat model. SWCNTs were sonicated with sodium cholate followed by surfactant exchange to fabricate phospholipid-polyethylene glycol functionalized nanotubes, obtaining a high quantum yield. These SWCNT conjugations represented the

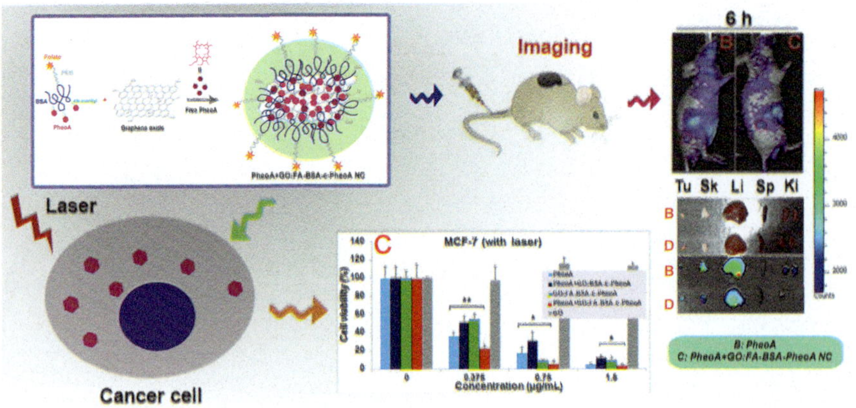


**Fig. 8.5** Schematic illustration of the generation of photosensitizer which loaded graphene oxide and FA-BSA-c-PheoA nanocomplex [23]

capability to obtain high contrast images at a relatively low dose (17 mg/L) for whole rats imaging [25]. Hong et al. used SWCNTs in NIR-II region to evaluate the blood flow in small vessels in mice hind limbs by applying an in vivo real time epifluorescence imaging technology. A high resolution was achieved for small vessel imaging in both spatial (~30 μm) and temporal (<200 ms per frame) aspects, imaging 1–3 mm deep in the hind limbs. This work proved the potential advantages of SWCNTs used as fluorophores in NIR-II imaging, thus overcoming the challenges of traditional imaging modes [26]. Yi et al. investigated the M13 phage-modified SWCNT probe for NIR-II imaging of targeted tumors. The results revealed that even a low dose of the M13-SWCNT probe (2 μg/mL) could be detected in deep tissues and up to 2.5 cm in tissue-like phantoms [27]. Ghosh et al. reported a targeted M13 virus-functionalized SWCNT probe for detection of human ovarian tumor in a rat model. Compared to visible and near-infrared (NIR1) dyes, the M13-stabilized SWCNT nanoprobe showed a higher signal-to-noise ratio, enabling this probe to be utilized to diagnose tumors even in the submillimeter range [28]. Xie et al. developed an Evans Blue (EB) modified SWCNT nanosystem (SWCNTs/EB), followed by loading with Ce6 (photosensitizer) encapsulated albumin to produce a SWCNTs/EB/albumin/Ce6 as a multifunctional nanosystem. Owing to the high optical absorption of SWCNTs and the fluorescence characteristics of Ce6, SWCNTs/EB/albumin/Ce6 promoted both the fluorescence and photoacoustic imaging of tumors. In addition, the comprehensive characteristics of this nanosystem contributed to image-guided photodynamic therapy (PDT) and photothermal therapy (PTT) [29] (Figs. 8.5 and 8.6).

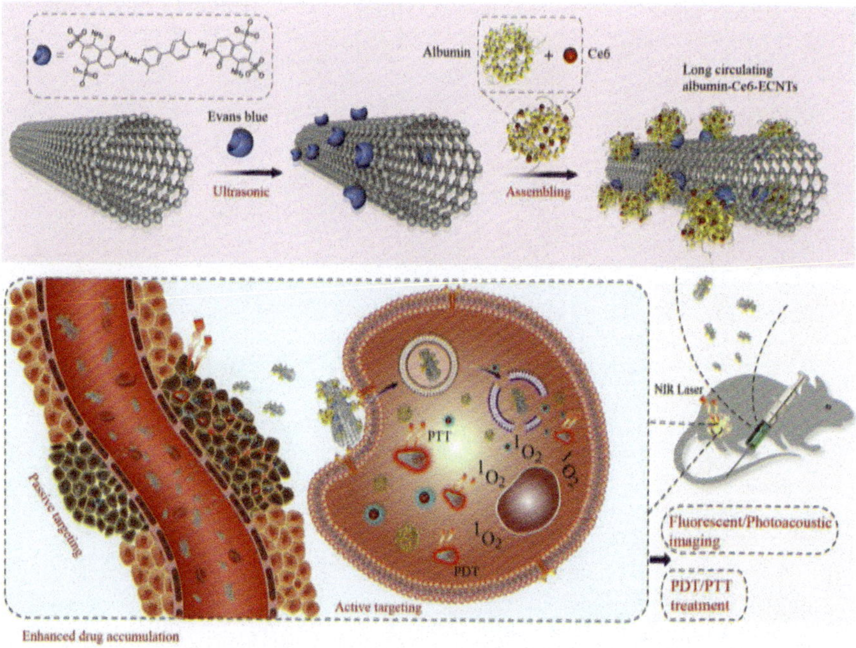


**Fig. 8.6** Fabrication and application of a novel transport system which is based on albumin/Ce6 engineered EB/carbon nanotube (ACEC) [29]

# 8.3 Pharmacodynamics and Metabolism

## *8.3.1 Delivery to the Brain*

As most drugs are difficult to cross the blood-brain barrier (BBB), nanoparticles are considered to be promising tools for delivering drugs to the brain. Among them, carbon-based nanomaterials have intrinsic capability to infiltrate BBB in vitro and in vivo. Kafa et al. analyzed the ability of Angiopep-2 (ANG)-targeting chemically functionalized multi-walled carbon nanotubes (f-MWCNTs) to penetrate the BBB in vitro and in vivo. Conjugation of ANG to f-MWCNTs improved BBB transportation of w- and t-MWCNTs-ANG compared to nontargeting equivalents of the co-cultured BBB models in vitro. Furthermore, after intravenous injection of w-MWCNTs-ANG, uptake of this drug showed remarkably higher than that of the nontargeted w-MWCNT into whole brain in vivo, achieving about 2% of injected dose per gram of the brain (%ID/g) within 1 h postinjection. Moreover, the uptake of w-MWCNT-ANG in glioma brain was higher than in normal brain at 24 h after injection applying a syngeneic/homologous glioma model [30]. Zhao et al. analyzed CNT as a CpG delivery carrier in vitro and in intracranial GL261 gliomas-bearing mice. The results revealed that CNT-CpG was nontoxic and the uptake of CpG was enhanced both in vitro and intracranial gliomas. CNT-mediated CpG delivery also enhanced production of proinflammatory cytokine by primary monocytes.

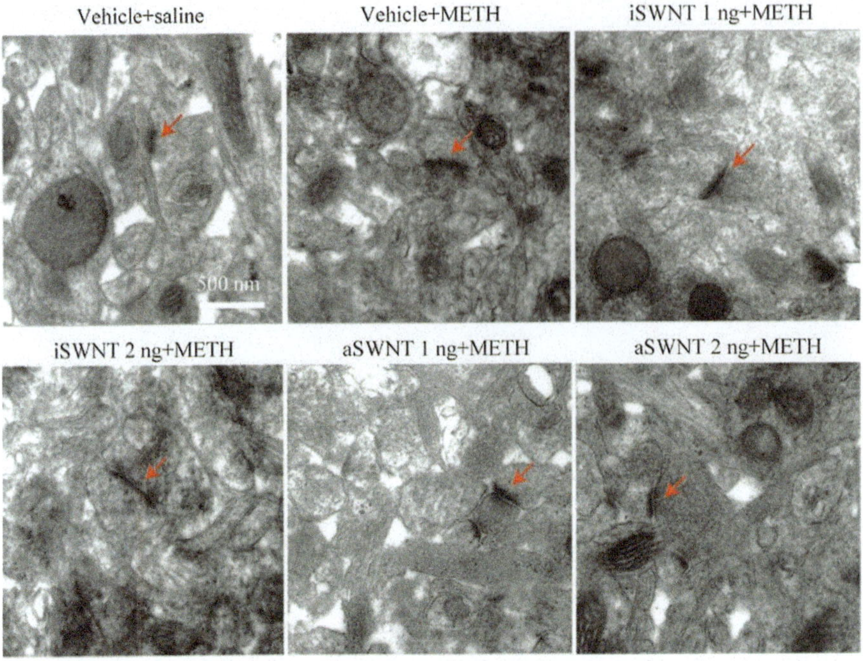


**Fig. 8.7** Representative images of synapses (marked with red arrows) in the striatum between the different treatment groups by transmission electron microscope (TEM) [32]

Furthermore, surviving rats showed prolonged tumor-free remission (>3 months), and were protected against rechallenge of intracranial tumor, suggesting induction of long-term anti-tumor immunity [31]. Xue et al. revealed that aggregated single-walled carbon nanotubes (aSWCNTs) significantly inhibited self-administration of methamphetamine (METH), METH-induced conditioned location preference, and METH- or cue-induced relapse of drug-seeking behavior in rats. And they reported that aSWCNTs may treat METH addiction through oxidation of enhanced METH extracellular dopamine in the striatum [32]. Wang et al. determined the distribution of f-MWCNT after intravenous injection using a variety of qualitative and quantitative techniques. The results showed that in addition to brain endothelial cells, there existed f-MWCNT in rat brain parenchyma [33] (Fig. 8.7).

### 8.3.2 Toxicological Characteristics

Because of increasing concerns of toxicology regarding the usage of nanomaterials and the rising demand for safe materials to protect public health, the toxicological and bioavailable characteristics of nanomaterials are urgently needed to be analyzed. A carbon nanomaterials-based drug delivery system should exhibit low toxicity, sustained drug release, as well as continued circulation without aggregation. Yang et al. systematically studied the potential toxicity of graphene over time. The results revealed that PEGylated NGS could be gradually excreted, possibly through kidneys and feces. Besides, blood biochemistry, hematological analysis and histological examinations confirmed that no significant toxicity was observed in treated mice in a period of three months at tested dosage (20 mg/kg) [34].

The toxicological evaluation in terms of cell viability and cell morphological changes suggested that surface features played an important role in the biological response of functionalized CNTs [35]. Moore et al. fabricated a biocompatible block-co-polymer which was composed of poly(lactide)-poly(ethylene glycol) (PLA-PEG), followed by coating with a CNT. The coated CNTs could reduce short-term and long-term toxicity, consistently release drug paclitaxel (PTX), and prevent aggregation. The copolymer coating on the surface of CNTs remarkably reduced toxicity of human umbilical vein endothelial cells (HUVEC) and U-87 glioblastoma cells in vitro. Furthermore, the coating reduced inflammatory response in mouse lung epithelial cells in vitro. Compared to uncoated CNTs, in vivo analyses have revealed no long-term inflammatory response with CNT coated with PLA-PEG (CLP). In addition, in the presence of the surface coating, acute toxicity decreased significantly by doubling the maximum tolerable dosage in rats. And they showed that some of the CLP nanotubes may be removed by this mechanism; however, the majority of tubes were eliminated by the liver and spleen similar to other nanomaterial drug transport systems [36]. Some researches [37–41] had confirmed that dispersed SWCNTs were cleared through the kidneys because of the alignment of thin diameter tubes with the pores of the glomerulus, about 5–20 nm in diameter [38, 39].

### 8.3.3 *Drug Release*

The rapid and efficient release of encapsulated drugs is an important factor for evaluating the performance of carriers. In addition to efficient loading of drugs, carbon-based drug delivery vectors should be designed to ensure that the loaded drug is released only after the nanomaterials enter the cells. Moore et al. developed a CNT-based drug delivery system coated with a biocompatible block-co-polymer which was composed of poly (lactide)-poly(ethylene glycol) (CLP) for sustaining drug release of PTX and preventing aggregation. Using polymer coatings, this nanosystem could encapsulate PTX and release over 7 days to improve the therapeutic effects compared to free drugs [36].

Because the pH of microenvironment of tumor cells is different from that of normal cells, some targeted drug delivery systems that are triggered by changing pH are developed. Zhang et al. constructed a targeted drug delivery system based on a carboxylate groups derivatized SWCNTs which were coated with polysaccharide materials. This system could be loaded with the anti-cancer agent doxorubicin (DOX). The drug bound at the physiological pH (pH 7.4) and was only released at a lower pH, for instance, lysosomal pH and the pH characteristic of some tumor environments [42]. Mo et al. developed carboxyl SWCNTs to load with DOX.

The results suggested that at pH 5.5, this system could achieve a high loading of DOX with 107.73 ± 0.67%. Moreover, depending on the pH, this nanosystem had the capability to perform the controlled and sustained release of DOX. It was shown that at pH 7.4, this nanosystem released less than 10% of the loaded DOX post 72 h, whereas at pH 5.5, ~85% was released [43]. Tan et al. formulated drug delivery nanosystems for the transportation of levodopa (LD) via non-covalent functionalization of carboxylated SWCNTs. In vitro drug release tests performed in simulated human environment at pH 7.4 indicated that the drug release curve of LD was consistent with the pseudo-second-order kinetic model. Nevertheless, the drug LD release curve conducted at pH 4.8 could not be extended to a specific model. The cumulative release of LD at pH 7.4 was higher than that at pH 4.8, revealing that the release rate was triggered by pH [44] (Fig. 8.8).

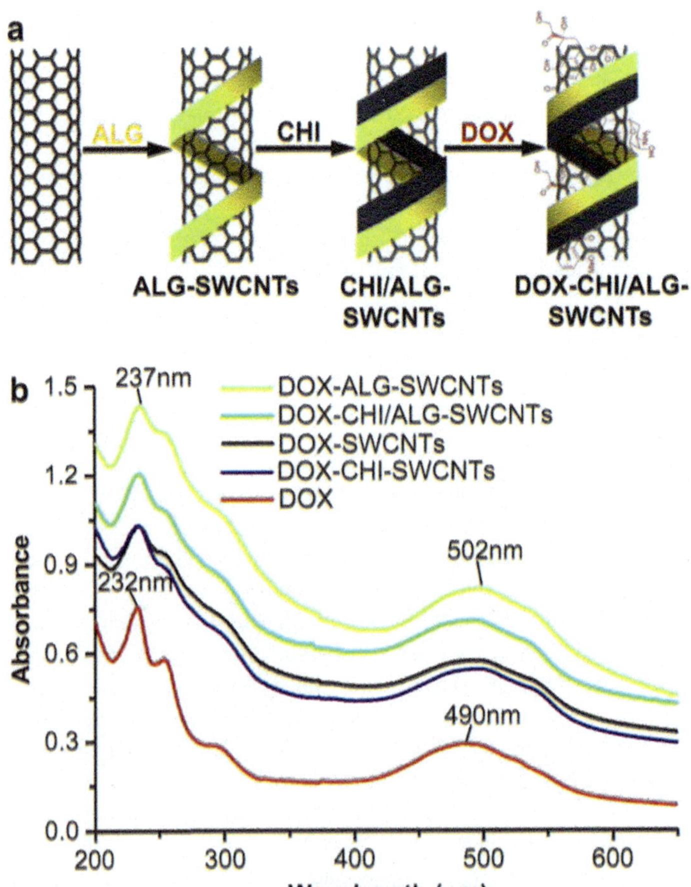


**Fig. 8.8** Preparation of functionalized SWCNTs. (**a**) SWCNTs (derivatized with -$CO_2H$ groups) were modified with ALG, CHI and DOX. (**b**) UV-vis absorption spectra of DOX and SWCNTs which loaded DOX [42]

## 8.4 Diagnosis and Treatment

### *8.4.1 Neurodegenerative Disease*

With the aging of the population, the main consequence of this phenomenon is that the number of patients with neurodegenerative diseases, including Alzheimer's disease (AD) and Parkinson's disease (PD), has increased. These deadly clinical diseases are considered to be social, economic, and medical challenges and have developed into a public health problem. One area where nanomedicine may provide better performances and efficiency compared to current measures is the diagnosis and treatment of neurodegenerative diseases. Tan et al. developed four drug delivery systems via non-covalent modification of carboxylated SWCNTs applying biocompatible polymers as coating agent to transport levodopa, a drug applied for treatment of PD [44]. Yang et al. combined acetylcholine (Ach) with SWCNTs to deliver Ach to the brain of mice with AD to relieve symptoms through gastrogavage. 5 mg/kg Ach-loaded SWCNTs could improve the learning and memory ability of mouse model with Alzheimer's disease and the safe dosage of SWCNTs was 12 mg/kg [45].

### *8.4.2 Tumor*

Nanomaterials have broad prospects of applications in the field of electronics, the environment, energy, biomedicine, and bioengineering. Their excellent performances make them widely applicable as cancer prognostic, diagnostic, and therapeutic drugs. Zhang et al. reported a modified nano graphene oxide (NGO) which was bound to folic acid (FA). It was a new nanocarrier for loading and targeted delivering anti-cancer agents, DOX and CPT. Its properties enabled it to specifically target MCF-7 cells with FA receptors. Besides, its cytotoxicity to target cells was much higher than that loaded with either drug [7]. Bhirde et al. fabricated a drug-SWCNT bioconjugation for killing cancer cells in vivo. Cisplatin, a first-line anti-cancer drug, and epidermal growth factor (EGF) were bound to SWCNTs to specifically target the squamous tumor. Most notably, targeted SWCNT-cisplatin-EGF had a rapid regression on tumor growth in rats, suggesting that its efficiency was superior to that of nontargeted SWCNT-cisplatin [46]. Xie et al. developed a SWCNTs/EB/albumin/Ce6 nanosystem for image-guided photodynamic therapy (PDT) and photothermal therapy (PTT) which showed a synergistic effect to eliminate SCC-7 cancer cells in a rat model in vivo. The results revealed that SWCNTs/EB/albumin/Ce6 nanosystem had the potential for the simultaneous image-guided diagnosis and treatment of cancer [29]. Naderi et al. constructed a nanoconjugate system based on SWCNTs to deliver paclitaxel (PTX), and hence the death of MCF-7 and colon cancer cells (HT-29) was observed [3]. Nurunnabi et al. developed carboxyl functionalized graphene nanodots (cGdots) acting as a multifunctional

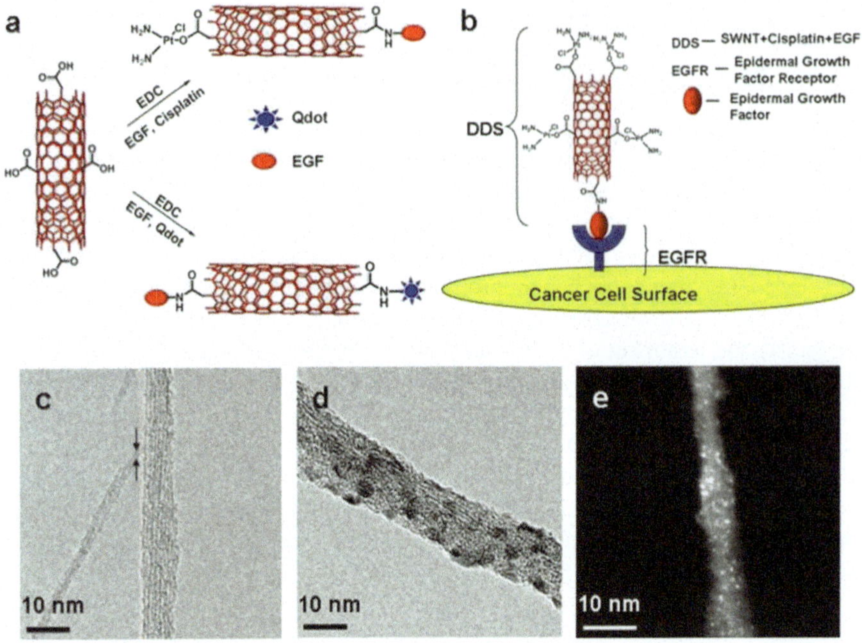


**Fig. 8.9** The delivery system based on nanotubes. (**a**) Description of chemical reactions in which EGF, cisplatin, and Qdots were linked to carboxylated SWCNTs (in red) using EDC as the coupling agent. (**b**) The scheme showed SWCNT bundles bioconjugated with EGF and cisplatin and targeted a single HNSCC cell via cell surface receptor EGFR. (**c**) TEM image of oxidized SWCNT bundles; arrow was shown as a single SWCNT. (**d**) SWCNT-Qdot-EGF bioconjugate bundle. (**e**) STEM micrographs of SWCNT bundle; the bright spots were presented as cisplatin (scale bar = 10 nm) [46]

nanosystem, which could efficiently kill MDA-MB231 tumor cells (>70%) via combining photodynamic and photothermal effects to produce sufficient heat. More importantly, the cGdots were equally effective in the model of rat bearing with MDA-MB231 xenografted tumor [47]. Zhang et al. developed a multifunctional tumor-targeting drug transportation system HA-MWCNTs/Tf@ART which achieved significantly enhanced anti-cancer effects both in cultured MCF-7 cells in vitro and in a tumor-bearing mouse model in vivo. The improvement of anti-tumor efficacy benefits most from the enhancement of intracellular accumulation of ART in cancer cells and the simultaneous transmission of Tf and ART [6]. Mo et al. constructed a SWCNT-CHI-HA-DOX nanosystem and carried out a cell viability experiment, revealing that the nanosystem was highly toxic to HeLa cells and could kill them efficiently. However, it was less toxic to fibroblasts as a model of normal cells and tissues. They also performed a combination of histological examinations and the blood's characteristics analysis and found no severe damage to the important organs of male Sprague-Dawley mice [43] (Figs. 8.9, 8.10, 8.11, 8.12, and 8.13).

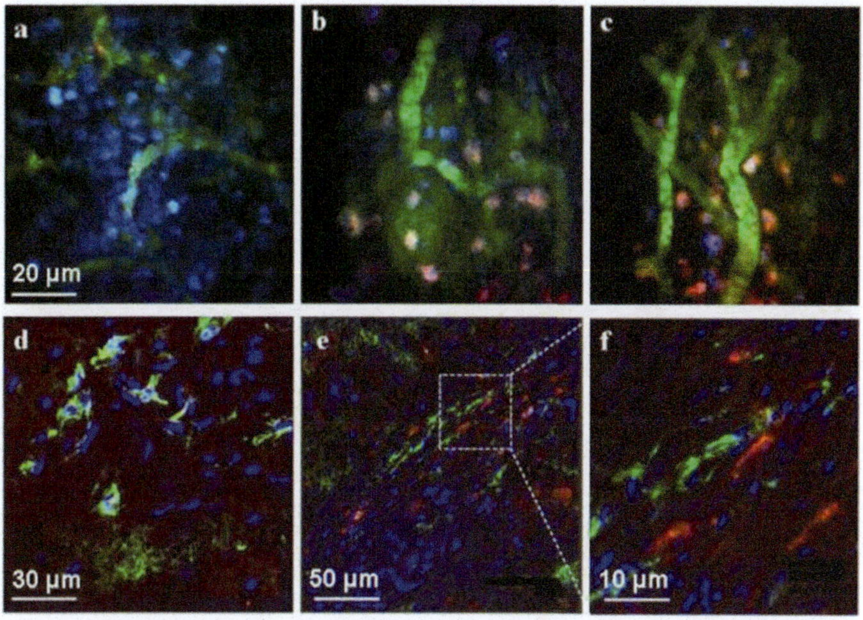


**Fig. 8.10** Nanotube bioconjugates were detected in tumors in vivo. (**a–c**) were representative images from time-delay videos captured through three-color, intravital two-photon microscopy. The HN12 xenografts-born rats were anesthetized and administrated with SQ or SQE (red) bioconjugates. Cell nuclei were stained with Höechst (blue) and blood vessels with 500 kDa FITC-dextran (green): For SQ alone with no EGF (**a**), very little or no red fluorescence characterizing the Qdot signal was detected in the tumor mass 45 min after injection. Figure (**b** and **c**) revealed two different views after treatment of SQE, achieving red fluorescence 45 min post injection in the tumor micro-environment. The red SQE bioconjugate was located close to the nuclei, suggesting it was internalized in the tumor cells in the xenograft (scale bar in **a–c** was 20 μm). Figure (**d–f**) represented confocal microscopy images of fixed xenograft cryosections. In the SQ treated tumor sections (**d**), only cell nuclei (blue) stained by Höechst and dextran (green) labeled by FITC in blood vessels were visible (scale bar 30 μm). (**e**) In SQE-treated mice, characteristic red fluorescence was widely distributed in the tumor microenvironment (scale bar 50 μm). (**f**) Magnified dotted area of (**e**), revealing internalization of SEQ bioconjugates in the cells of the tumor mass (scale bar 10 μm) [46]

### *8.4.3 Acute Hyperglycemia*

Excessive sugar intake can lead to acute hyperglycemia, which has also been proposed to be a better predictor of future cardiovascular disease (CVD) events than that of fasting blood sugar in healthy people and diabetic patients. In fact, this acute hyperglycemic stress is considered to be one of the main causes of vascular dysfunction, which is one of the major precursors of CVD [48]. The treatment of acute hyperglycemia is another application field of carbon nanoparticles. Fabian et al. developed a novel carbon nanoparticle in which hydrophilic carbon clusters were bound to PEG (PEG-HCCs). These particles were high-capacity superoxide

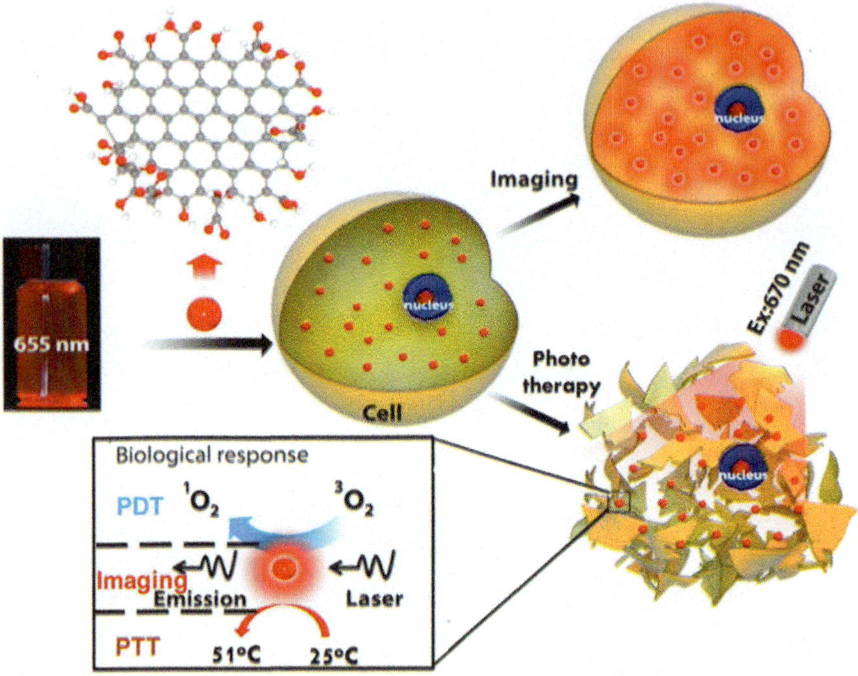


**Fig. 8.11** Typical chemical structure mode of near-infrared cGdots for optical imaging and treatment [47]

dismutase mimics which were effective for hydroxyl radical, recovering the balance between nitric oxide and superoxide in the blood vessels. Under hyperglycemic conditions, the effects of PEG-HCCs used during reperfusion after transient middle cerebral artery occlusion (tMCAO) via suture in the rat were analyzed. PEG-HCCs had a protective effect on hydrogen peroxide in culture models. In vivo improvement was also observed after PEG-HCCs with 90-min ischemia with reduction of infarct size (42%), hemisphere swelling (46%), hemorrhage score (53%), and advancement in Bederson score (70%) ($p = 0.068–0.001$) [49].

## 8.5 Biodistribution of Carbon Nanomaterials

### *8.5.1 Brain Distribution*

Neurological diseases involving the brain are often dangerous and fatal. Because of the existence of blood-brain barrier, it is difficult for drugs to reach brain lesions effectively. Importantly, carbon-based nanomaterials are considered to be effective carriers of drugs into the brain. The study for distribution of carbon-based nanomaterials in the brain is very important to confirm whether the material crosses

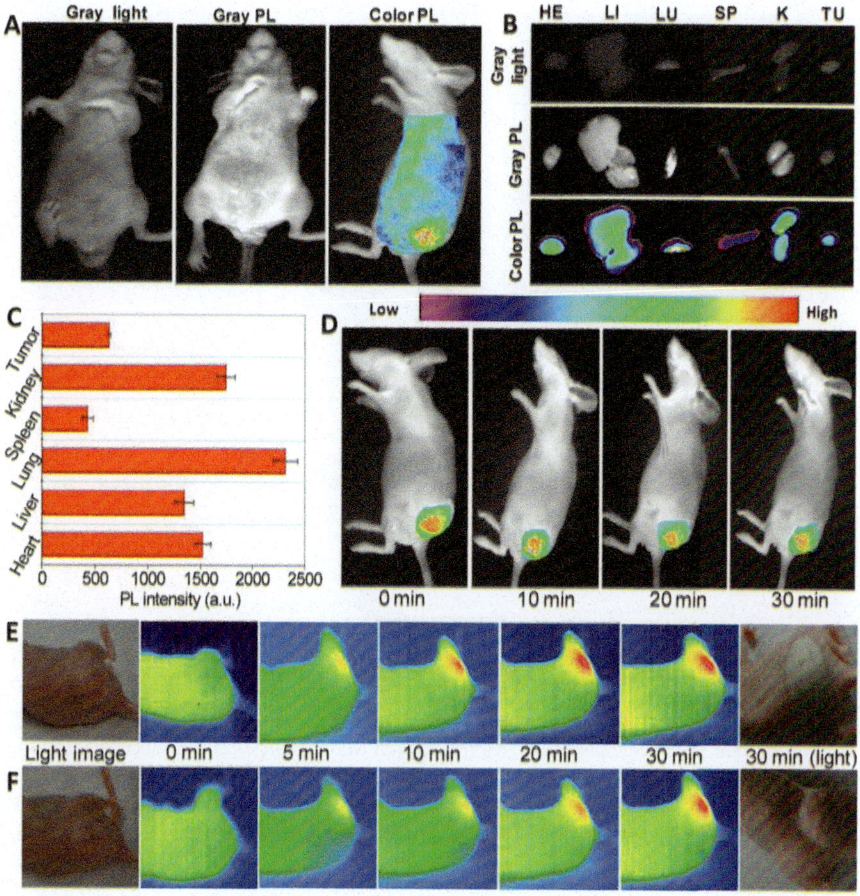


**Fig. 8.12** The optical imaging of rats having MDA-MB231 cells injected with cGdots in vitro and in vivo. (**a**) Noninvasive images of MDA-MB231 breast cancer cell-bearing mice treated with cGdots. (**b**) Ex vivo presentations of each harvested organ from MDA-MB231 cells-bearing mice with cGdots. HE, LI, LU, SP, K, and TU represented heart, liver, lung, spleen, kidney, and MDA-MB231 cells, respectively. (**c**) Fluorescence intensity of each harvested organ obtained from PL analyzer (n = 4). (**d**) In vivo photoquenching curve of cGdots after 30 min laser irradiation via intratumoral injection of cGdots. (**e**, **f**) The effect of light dosage on skin was analyzed under 0.6 and 0.3 W/cm$^2$ [47]

the blood-brain barrier or not. Wang et al. applied various qualitative and quantitative methods to analyze brain distribution of f-MWCNT after intravenous injection. The results showed the f-MWCNT existed in parenchyma of mice brain as well as in brain endothelium [33]. Trusel et al. injected fluorescently labeled CNOs (fluo-CNOs) stereotaxically into hippocampus of mice to analyze the possible uses of graphitic CNOs for diagnostic and therapeutic interventions on CNS diseases. The results showed that fluo-CNOs could diffuse from the injection zone to the

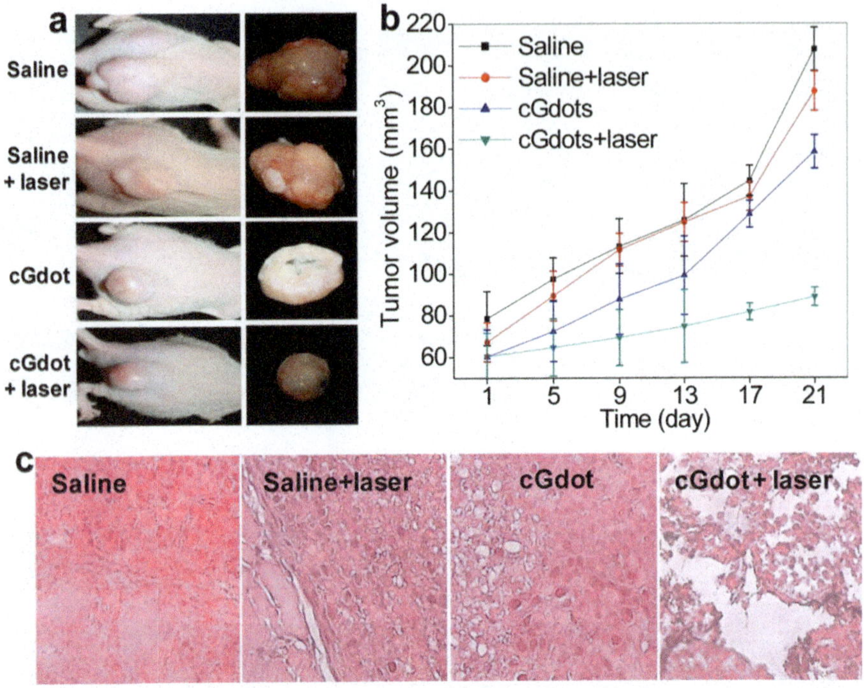


**Fig. 8.13** In vivo photothermal therapy of rats bearing MDA-MB231 breast cancer cell after cGdots injection and laser exposure. (**a**) Morphology of MDA-MB231 cells in subcutaneous tissue at 21 days after injection of cGdots and laser exposure. (**b**) Plot of tumor size volume of rats. Data are indicated as mean ± SEM (n = 4). (**c**) Histological analysis of tumor tissue at 21 days after injection of cGdots and laser exposure. Magnification: × 400 [47]

surrounding tissue, and then penetrate the neuronal membrane. Finally, they were widely settled inside the neurons cytoplasm, including both endoplasmic vesicles and free-floating in the cytoplasm [50].

### 8.5.2 Body Distribution

The purpose of developing drug carriers is to transport them to certain cell types or ideal areas of the body. The study of histological distribution of nanomaterials is helpful to analyze the targeting of nanomaterials and it is useful to identify whether the the drug has been delivered into the desired sites in vivo. The remove of material could also be estimated by the study of histological distribution. Moore et al. developed a biocompatible block-co-polymer-coated CNT, which was composed

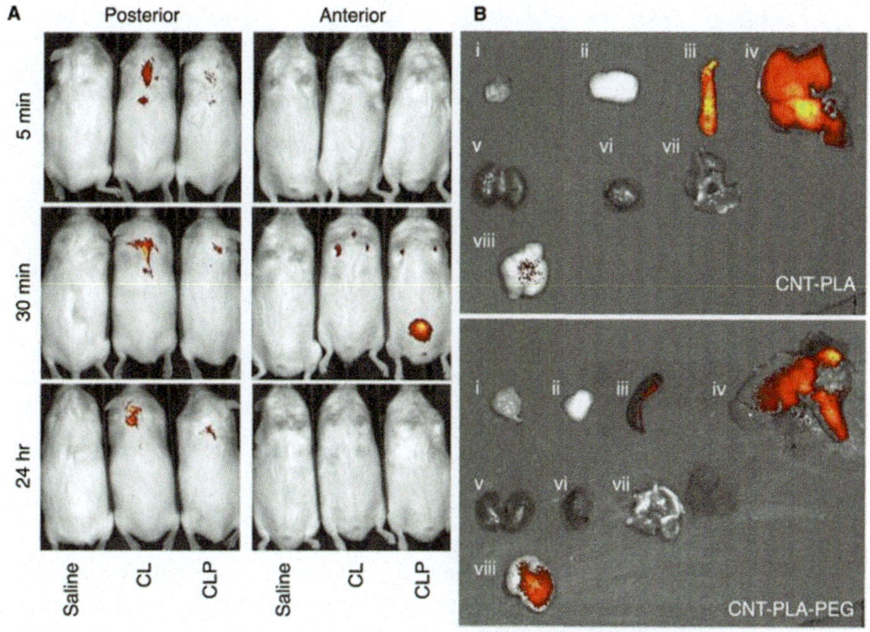


**Fig. 8.14** (**a**) Representational images of live Balb/c rat injected intravenously with saline, fluorescently loaded CNT-PLA (CL), or fluorescently loaded CNT-PLA-PEG (CLP) revealed clearance of coated CNT via the bladder (n = 3). (**b**) Fluorescent imaging of transplanted organs after 24 h suggested accumulation of fluorescent signals in the spleen, liver, and brain. Organs demonstrated were the (*i*) muscle, (*ii*) fat, (*iii*) spleen, (*iv*) liver, (*v*) kidneys, (*vi*) heart, (*vii*) lung, and (*viii*) brain, respectively [36]

of PLA-PEG to sustain drug release of PTX and prevent aggregation. The biodistribution and histology studies in vivo revealed CLP accumulated more in the brain and less in the spleen than the CNT-PLA (CL) formulation, resulting in a lower degree of aggregation in tissues. Biodistribution studies in Balb/c mice also showed fluorescent-loaded CLP was cleared via the kidneys and bladder, as well as the reticular endothelial system, such as liver and spleen. The histological results revealed the CNT aggregated in lungs, liver and spleen. Meanwhile CL also aggregated in the liver, lung and brain. However, although CLP particles were found in the brain and lungs, the images showed no CLP aggregation [36]. Bhirde et al. analyzed the important organs and tumors of rats injected with the SWCNT bioconjugates to observe their short-term biodistribution. The results revealed that the uptake of SWCNT bioconjugates within the tumor section increased significantly when the targeting ligand EGF was incorporated into the bioconjugates. In addition, a smaller amount of SWCNT bioconjugates were found in the spleen, lung, liver, kidney, and heart [46] (Figs. 8.14 and 8.15).

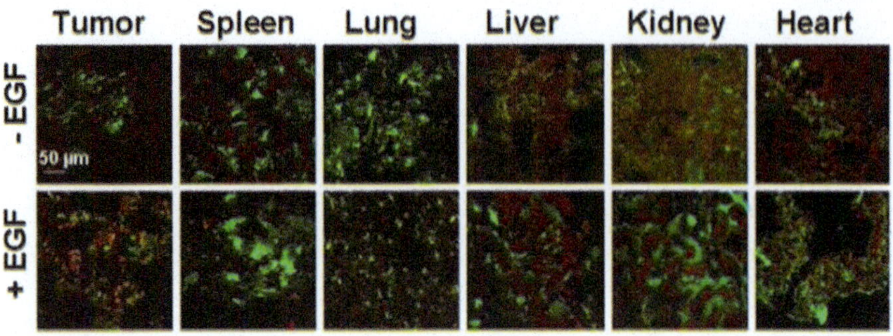


**Fig. 8.15** Distribution analysis of nanotube bioconjugates in vivo. It was only when EGF was on nanotubes that the uptake of biological conjugates increased in tumor tissue, showing a red color. Whether EGF exists or not, the spleen, liver, kidney and heart display some red fluorescence characteristic of the SWCNT-Qdots. Scale bar is 50 μm [46]

## 8.6 Summary

With the progress of carbon-based techniques, more useful and promising biomedical applications of carbon-based nanomaterials or even novel carbon nanostructures are expected to burst onto the scene, toward the development of more human-friendly and environmentally friendly products and technologies in the future.

## References

1. Lim D-J, Sim M, Oh L, Lim K, Park H. Carbon-based drug delivery carriers for cancer therapy. Arch Pharm Res. 2014;37(1):43–52.
2. Tripisciano C, Kraemer K, Taylor A, Borowiak-Palen E, Borowiak-Palena E. Single-wall carbon nanotubes based anticancer drug delivery system. Chem Phys Lett. 2009;478 (4-6):200–5.
3. Naderi N, Madani SY, Mosahebi A, Seifalian AM. Octa-ammonium POSS-conjugated single-walled carbon nanotubes as vehicles for targeted delivery of paclitaxel. Nanotechnol Rev. 2015;6:28297.
4. Pan Q, Lv Y, Williams GR, Tao L, Yang H, Li H, Zhu L. Lactobionic acid and carboxymethyl chitosan functionalized graphene oxide nanocomposites as targeted anticancer drug delivery systems. Carbohydr Polym. 2016;151:812–20.
5. Zhou T, Zhou X, Xing D. Controlled release of doxorubicin from graphene oxide based charge-reversal nanocarrier. Biomaterials. 2014;35(13):4185–94.
6. Zhang H, Ji Y, Chen Q, Jiao X, Hou L, Zhu X, Zhang Z. Enhancement of cytotoxicity of artemisinin toward cancer cells by transferrin-mediated carbon nanotubes nanoparticles. J Drug Target. 2015;23(6):552–67.
7. Zhang L, Xia J, Zhao Q, Liu L, Zhang Z. Functional graphene oxide as a nanocarrier for controlled loading and targeted delivery of mixed anticancer drugs. Small. 2010;6(4):537–44.
8. Qin W, Yang K, Tang H, Tan L, Xie Q, Ma M, Zhang Y, Yao S. Improved GFP gene transfection mediated by polyamidoamine dendrimer-functionalized multi-walled carbon nanotubes with high biocompatibility. Colloids Surf B: Biointerfaces. 2011;84(1):206–13.

9. Karmakar A, Bratton SM, Dervishi E, Ghosh A, Mahmood M, Yang X, Saeed LM, Mustafa T, Casciano D, Radominska-Pandya A, Biris AS. Ethylenediamine functionalized-single-walled nanotube (f-SWNT)-assisted in vitro delivery of the oncogene suppressor p53 gene to breast cancer MCF-7 cells. Int J Nanomedicine. 2011;6:1045–55.
10. Hao Y, Xu P, He C, Yang X, Huang M, Xing J, Chen J. Impact of carbodiimide crosslinker used for magnetic carbon nanotube mediated GFP plasmid delivery. Nanotechnology. 2011;22(28):285103.
11. Inoue Y, Fujimoto H, Ogino T, Iwata H. Site-specific gene transfer with high efficiency onto a carbon nanotube-loaded electrode. J R Soc Interface. 2008;5(25):909–18.
12. Wang T, Upponi JR, Torchilin VP. Design of multifunctional non-viral gene vectors to overcome physiological barriers: dilemmas and strategies. Int J Pharm. 2012;427(1):3–20.
13. Al-Jamal KT, Gherardini L, Bardi G, Nunes A, Guo C, Bussy C, Antonia Herrero M, Bianco A, Prato M, Kostarelos K, Pizzorusso T. Functional motor recovery from brain ischemic insult by carbon nanotube-mediated siRNA silencing. Proc Natl Acad Sci U S A. 2011;108(27):10952–7.
14. Dong H, Ding L, Yan F, Ji H, Ju H. The use of polyethylenimine-grafted graphene nanoribbon for cellular delivery of locked nucleic acid modified molecular beacon for recognition of microRNA. Biomaterials. 2011;32(15):3875–82.
15. Feng L, Yang X, Shi X, Tan X, Peng R, Wang J, Liu Z. Polyethylene glycol and polyethylenimine dual-functionalized nano-graphene oxide for photothermally enhanced gene delivery. Small. 2013;9(11):1989–97.
16. Liu X, Zhang Y, Ma D, Tang H, Tan L, Xie Q, Yao S. Biocompatible multi-walled carbon nanotube-chitosan-folic acid nanoparticle hybrids as GFP gene delivery materials. Colloids Surf B: Biointerfaces. 2013;111:224–31.
17. Nunes A, Amsharov N, Guo C, Van den Bossche J, Santhosh P, Karachalios TK, Nitodas SF, Burghard M, Kostarelos K, Al-Jamal KT. Hybrid polymer-grafted multiwalled carbon nanotubes for in vitro gene delivery. Small. 2010;6(20):2281–91.
18. Corr SJ, Raoof M, Cisneros BT, Orbaek AW, Cheney MA, Law JJ, Lara NC, Barron AR, Wilson LJ, Curley SA. Radiofrequency electric-field heating behaviors of highly enriched semiconducting and metallic single-walled carbon nanotubes. Nano Res. 2015;8:2859–70.
19. Santos T, Fang X, Chen M-T, Wang W, Ferreira R, Jhaveri N, Gundersen M, Zhou C, Pagnini P, Hofman FM, Chen TC. Sequential administration of carbon nanotubes and near-infrared radiation for the treatment of gliomas. Front Oncol. 2014;4:180.
20. Zhang B, Wang H, Shen S, She X, Shi W, Chen J, Zhang Q, Hu Y, Pang Z, Jiang X. Fibrin-targeting peptide CREKA-conjugated multi-walled carbon nanotubes for self-amplified photothermal therapy of tumor. Biomaterials. 2016;79:46–55.
21. Zhou F, Wu S, Yuan Y, Chen WR, Xing D. Mitochondria-targeting photoacoustic therapy using single-walled carbon nanotubes. Small. 2012;8(10):1543–50.
22. Taratula O, Patel M, Schumann C, Naleway MA, Pang AJ, He H, Taratula O. Phthalocyanine-loaded graphene nanoplatform for imaging-guided combinatorial phototherapy. Int J Nanomedicine. 2015;10:2347–62.
23. Battogtokh G, Ko YT. Graphene oxide-incorporated pH-responsive folate-albumin-photosensitizer nanocomplex as image-guided dual therapeutics. J Control Release. 2016;234:10–20.
24. Al Faraj A, Shaik AS, Al Sayed B, Halwani R, Al Jammaz I. Specific targeting and noninvasive imaging of breast cancer stem cells using single-walled carbon nanotubes as novel multimodality nanoprobes. Nanomedicine. 2016;11(1):31–46.
25. Welsher K, Liu Z, Sherlock SP, Robinson JT, Chen Z, Daranciang D, Dai H. A route to brightly fluorescent carbon nanotubes for near-infrared imaging in mice. Nat Nanotechnol. 2009;4(11):773–80.
26. Hong G, Lee JC, Robinson JT, Raaz U, Xie L, Huang NF, Cooke JP, Dai H. Multifunctional in vivo vascular imaging using near-infrared II fluorescence. Nat Med. 2012;18(12):1841–6.
27. Yi H, Ghosh D, Ham M-h, Qi J, Barone PW, Strano MS, Belcher AM. M13 phage-functionalized single-walled carbon nanotubes as nanoprobes for second near-infrared window fluorescence imaging of targeted tumors. Nano Lett. 2012;12(3):1176–83.

28. Ghosh D, Bagley AF, Na YJ, Birrer MJ, Bhatia SN, Belcher AM. Deep, noninvasive imaging and surgical guidance of submillimeter tumors using targeted M13-stabilized single-walled carbon nanotubes. Proc Natl Acad Sci U S A. 2014;111(38):13948–53.
29. Xie L, Wang G, Zhou H, Zhang F, Guo Z, Liu C, Zhang X, Zhu L. Functional long circulating single walled carbon nanotubes for fluorescent/photoacoustic imaging-guided enhanced phototherapy. Biomaterials. 2016;103:219–28.
30. Kafa H, Wang JT-W, Rubio N, Klippstein R, Costa PM, Hassan HAFM, Sosabowski JK, Bansal SS, Preston JE, Joan Abbott N, Al-Jamala KT. Translocation of LRP1 targeted carbon nanotubes of different diameters across the blood-brain barrier in vitro and in vivo. J Control Release. 2016;225:217–29.
31. Zhao D, Alizadeh D, Zhang L, Liu W, Farrukh O, Manuel E, Diamond DJ, Badie B. Carbon nanotubes enhance CpG uptake and potentiate antiglioma immunity. Clin Cancer Res. 2011;17(4):771–82.
32. Xue X, Yang J-Y, He Y, Wang L-R, Liu P, Yu L-S, Bi G-H, Zhu M-M, Liu Y-Y, Xiang R-W, Yang X-T, Fan X-Y, Wang X-M, Qi J, Zhang H-J, Wei T, Cui W, Ge G-L, Xi Z-X, Wu C-F, Liang X-J. Aggregated single-walled carbon nanotubes attenuate the behavioural and neurochemical effects of methamphetamine in mice. Nat Nanotechnol. 2016;11(7):613–20.
33. Wang JT-W, Rubio N, Kafa H, Venturelli E, Fabbro C, Ménard-Moyon C, Da Ros T, Sosabowski JK, Lawson AD, Robinson MK, Prato M, Bianco A, Festy F, Preston JE, Kostarelos K, Al-Jamala KT. Kinetics of functionalised carbon nanotube distribution in mouse brain after systemic injection: spatial to ultra-structural analyses. J Control Release. 2016;224:22–32.
34. Yang K, Wan J, Zhang S, Zhang Y, Lee S-T, Liu Z. In vivo pharmacokinetics, long-term biodistribution, and toxicology of PEGylated graphene in mice. ACS Nano. 2011;5(1):516–22.
35. Koromilas ND, Lainioti GC, Gialeli C, Barbouri D, Kouravelou KB, Karamanos NK, Voyiatzis GA, Kallitsis JK. Preparation and toxicological assessment of functionalized carbon nanotube-polymer hybrids. PLoS One. 2014;9(9):e107029.
36. Moore TL, Pitzer JE, Podila R, Wang X, Lewis RL, Grimes SW, Wilson JR, Skjervold E, Brown JM, Rao A, Alexis F. Multifunctional polymer-coated carbon nanotubes for safe drug delivery. Part Part Syst Charact. 2013;30(4):365–73.
37. Lacerda L, Ali-Boucetta H, Herrero MA, Pastorin G, Bianco A, Prato M, Kostarelos K. Tissue histology and physiology following intravenous administration of different types of functionalized multiwalled carbon nanotubes. Nanomedicine. 2008;3(2):149–61.
38. Kostarelos K. Carbon nanotubes: Fibrillar pharmacology. Nat Mater. 2010;9(10):793–5.
39. Ruggiero A, Villa CH, Bander E, Rey DA, Bergkvist M, Batt CA, Manova-Todorova K, Deen WM, Scheinberg DA, McDevitta MR. Paradoxical glomerular filtration of carbon nanotubes. Proc Natl Acad Sci. 2010;107(27):12369–74.
40. Singh R, Pantarotto D, Lacerda L, Pastorin G, Klumpp C, Prato M, Bianco A, Kostarelos K. Tissue biodistribution and blood clearance rates of intravenously administered carbon nanotube radiotracers. Proc Natl Acad Sci. 2006;103(9):3357–62.
41. Lacerda L, Herrero MA, Venner K, Bianco A, Prato M, Kostarelos K. Carbon-nanotube shape and individualization critical for renal excretion. Small. 2008;4(8):1130–2.
42. Zhang X, Meng L, Lu Q, Fei Z, Dyson PJ. Targeted delivery and controlled release of doxorubicin to cancer cells using modified single wall carbon nanotubes. Biomaterials. 2009;30(30):6041–7.
43. Mo Y, Wang H, Liu J, Lan Y, Guo R, Zhang Y, Xue W, Zhang Y. Controlled release and targeted delivery to cancer cells of doxorubicin from polysaccharide-functionalised single-walled carbon nanotubes. J Mater Chem B. 2015;3(9):1846–55.
44. Meihua Tan J, Saifullah B, Umar Kura A, Fakurazi S, Hussein MZ. Incorporation of levodopa into biopolymer coatings based on carboxylated carbon nanotubes for pH-dependent sustained release drug delivery. Nanomaterials (Basel). 2018;8(6):pii: E389.
45. Yang Z, Zhang Y, Yang Y, Sun L, Han D, Li H, Wang C. Pharmacological and toxicological target organelles and safe use of single-walled carbon nanotubes as drug carriers in treating Alzheimer disease. Nanomedicine. 2010;6(3):427–41.

46. Bhirde AA, Patel V, Gavard J, Zhang G, Sousa AA, Masedunskas A, Leapman RD, Weigert R, Gutkind JS, Rusling JF. Targeted killing of cancer cells in vivo and in vitro with EGF-directed carbon nanotube-based drug delivery. ACS Nano. 2009;3(2):307–16.
47. Md N, Khatun Z, Reeck GR, Lee DY, Lee Y-k. Photoluminescent graphene nanoparticles for cancer phototherapy and imaging. ACS Appl Mater Interfaces. 2014;6(15):12413–21.
48. Loader J, Montero D, Lorenzen C, Watts R, Méziat C, Reboul C, Stewart S, Walther G. Acute hyperglycemia impairs vascular function in healthy and cardiometabolic diseased subjects. Arterioscler Thromb Vasc Biol. 2015;35(9):2060–72.
49. Fabian RH, Derry PJ, Rea HC, Dalmeida WV, Nilewski LG, Sikkema WKA, Mandava P, Tsai A-L, Mendoza K, Berka V, Tour JM, Kent TA. Efficacy of novel carbon nanoparticle antioxidant therapy in a severe model of reversible middle cerebral artery stroke in acutely hyperglycemic rats. Front Neurol. 2018;9:199.
50. Trusel M, Baldrighi M, Marotta R, Gatto F, Pesce M, Frasconi M, Catelani T, Papaleo F, Pompa PP, Tonini R, Giordani S. Internalization of carbon nano-onions by hippocampal cells preserves neuronal circuit function and recognition memory. ACS Appl Mater Interfaces. 2018;10(20):16952–63.

# Chapter 9
# Polymeric Nanomedicine

**Yu Zhao, Chunxiong Zheng, and Yang Liu**

**Abstract** Advances in the engineering of polymeric nanomaterials and their applications in nanomedicine are enabling new strategies that have great potential to help improve our understanding and treatment of brain diseases. Based on distinctive polymeric materials, nanomedicine has been developed to an impressive stage with the ability to perform targeted delivery with temporal and spatial control. In this chapter, the various polymeric nanoparticles by which therapeutics can be delivered into the brain are introduced, and some key challenges facing translation of the researches to bedside are highlighted.

**Keywords** Polymeric nanomaterials · Nanomedicine · Brain diseases · Targeted delivery · Blood-brain barrier

## 9.1 Introduction: Polymeric Nanomedicine and Polymeric Nanoparticles

Developing novel and effective strategies for treating brain diseases is one of the most challenging and expensive market positions for many pharmaceutical companies. For example, during the process of developing new drugs, the costs for reaching phase I clinical trials may go up to US$100 million and around US$1 billion before reaching the consumer [1]. Unfortunately, although a great number of free drugs (e.g., chemotherapeutic agents, peptides, proteins, and genes) have been reported in an effort to combat brain diseases, a large percentage of those drugs have proved ineffective in vivo. The poor stability in biological fluids, rapid enzymatic degradation, inadequate release, and unfavorable pharmacokinetic properties might be the main barriers for failing to achieve clinical efficacy [2].

Y. Zhao · C. Zheng · Y. Liu (✉)
State Key Laboratory of Medicinal Chemical Biology, Key Laboratory of Functional Polymer Materials of Ministry of Education, College of Chemistry, Nankai University, Tianjin, China
e-mail: yliu@nankai.edu.cn



X. Xue (ed.), *Nanomedicine in Brain Diseases*,
https://doi.org/10.1007/978-981-13-8731-9_9

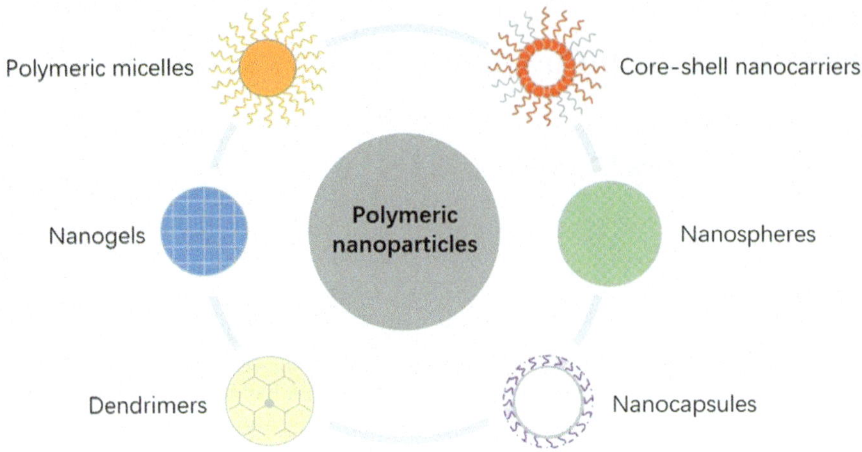


**Fig. 9.1** Various polymeric nanoparticles

In order to overcome these barriers, nanomaterials are increasingly being developed to protect and deliver the drugs, both small molecules and biomacromolecules, which are ineffective on their own. Despite a large variety of nanomaterials developed so far, it is noteworthy that only amphiphilic molecule-formed liposomes and polymeric nanoparticles (size range, from 10 to 1000 nm) have been extensively exploited for brain drug delivery, due to the rapid development in polymer chemistry and nanotechnology [3, 4]. Synthetic polymeric nanoparticles may be prepared using polymers such as poly(ε-caprolactone) (PCL), polyethylene (PEG), polyesters, poly(ethylenimine) (PEI), poly(lactic-co-glycolic acid) (PLGA), poly(lactic acid) (PLA), poly(alkylcyanoacrylates) (PACAs), and poly(amidoamine) dendrimers (PAMAMs). Natural polymeric nanoparticles are produced from natural polymers, such as polysaccharides (chitosan and alginate), amino acids (e.g., poly(lysine), poly(aspartic acid) (PASA)), and proteins (gelatin and albumin) [5]. There are so many different types of polymeric nanoparticles, each of which boasts unique pharmacodynamic characteristics (Fig. 9.1). For example, natural nanoparticles have the advantage of providing biological signals to interact with specific receptors/transporters expressed by endothelial cells, such as hyaluronic acid and CD (cluster of differentiation) 44. The various polymeric nanoparticles and their unique properties will be introduced with emphasis on the particles for brain drug delivery in the following chapters (Sects. 9.2 and 9.3, respectively).

The drugs usually associate with polymeric nanoparticles by being adsorbed, dissolved, encapsulated, and bound covalently [6, 7]. Taking advantages of these polymeric nanomaterials, the nanomedicine can achieve [8] (1) improved solubility of poorly water-soluble (hydrophobic) drugs, (2) prolonged half-life of drugs by reducing immunogenicity, (3) transcytosis across tight epithelial and endothelial barriers (blood-brain barrier), (4) intracellular delivery, (5) tissue targeting, (6) controllable release of the drugs loaded in nanoparticles at a sustained rate or in a way that responds to specific environment, (7) co-delivery of two or more therapeutic drugs for combination therapy, and (8) high visualization and precision with imaging modalities.

Blood-brain barrier (BBB) is a highly regulated brain endothelial structure, which could provide a shelter to the brain. The BBB could regulate the brain homeostasis and impede passive influx of most molecules or microorganisms, which are harmful for brain function and development. However, the pharmaceutical transport from blood to brain is also hampered by this almost impermeable and highly selective barrier [9]. The polymeric nanoparticles with the ability to carry a range of drugs and their convenient surface modification make themselves an attractive alternative for transporting drug across the BBB. Most polymeric nanoparticles that exhibit some degree of efficacy for drug delivery across the BBB in vivo adhere to a general set of guidelines that are listed below [10]. The ideal polymeric nanoparticle will be the one which can fulfil all these characteristics:

- Biodegradable, biocompatible, and nontoxic
- Particle size less than 100 nm (except if transport via monocytes or macrophages, in which case the vesicle size can be up to ≈1 μm)
- Stable in blood (limited aggregation and dissociation)
- Prolonged blood circulation time and non-immunogenic (avoid being taken up by immune cells)
- Well-maintained parent drug stability (chemical degradation and protein denaturation)
- BBB-targeted moiety (receptor- or adsorptive-mediated mechanism or uptake by monocytes or macrophages)
- Tunable drug release profiles
- A scalable and cost-effective process

## 9.2 Various Polymeric Nanoparticles

Homopolymers, copolymers (e.g., amphiphilic block polymers), and natural polymers have been used for the fabrication of a variety of polymeric nanoparticle systems for drug delivery to brain, including micelles, core-shell nanocarriers, nanospheres, nanocapsules, dendrimers, and nanogels [11, 12]. Some of them that have been applied for brain drug delivery will be introduced below.

### *9.2.1 Polymeric Micelles*

Polymeric micelles are basically spherical aggregates of amphiphilic surfactant molecules dispersing in aqueous solution with their hydrophilic “head” groups on the surface and their hydrophobic “tails” collected inside. Polymeric micelles, also

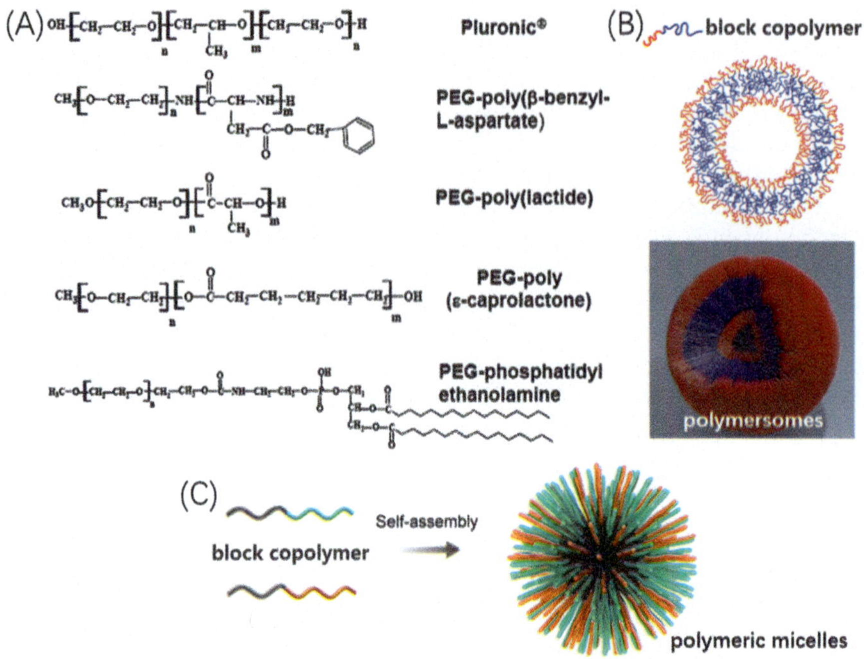


**Fig. 9.2** (**a**) Selected examples of PEG containing block polymers which can be self-assembled to form polymeric micelles for drug delivery. (**b**) A representative structure of polymersomers. (**c**) A representative structure of polymeric micelles

known as polymersomes, have considerable stability, high loading efficiency, and sustained drug release profiles. One of the most important properties of micelles is their ability to increase the solubility and bioavailability of poorly soluble pharmaceuticals (the poor water-soluble drugs can be entrapped into the hydrophobic cores of micelles). Polymeric micelles are self-assembled particles composed of block copolymer amphiphiles [13], such as poly(ethylene glycol)-block-poly(lactic acid) (PEG-b-PLA), poly(ethylene glycol)-block-poly(ε-caprolactone) (PEG-b-PCL), poly(2-methacryloyloxyethyl phosphorylcholine)-block-poly(ε-caprolactone) (PMPC-b-PCL), and so on [14] (Fig. 9.2).

As a promising replacement of traditional pharmaceutical excipients, the amphiphilic block copolymers exhibit many flexible and elaborate properties to bring much convenience to design for various drug therapeutics [15]. Although block copolymers have the same basic amphiphilic property as liposomes, the former consist of distinct polymer chains covalently linked in a series of two or more segments [3]. This is because polymer molecular weights can be orders of magnitude greater than those of lipids, leading to higher stability of the self-assembled structures. In addition, the polymers also offer the possibility of designing the ideal drug delivery systems benefiting from their variety of block component and block length.

The most common hydrophobic core-forming polymers are polyamines or poly (amine ester)s, poly(D,L-lactide), poly(caprolactone), poly(propylene oxide), poly (propylene glycol), and poly(amino acid)s. In addition, cholesterol included in membrane phospholipids could stabilize the membrane and reduce the membrane permeability toward encapsulated materials. The cholesterol shows a strong ability to assist self-assembly of cholesterol-containing polymers in aqueous solution and a high affinity to cholesterol receptors on the cell surface. Therefore, cholesterol can also be employed as the hydrophobic core of the micelles [16].

For example, Liu et al. developed a design of polymeric micelles (smaller than 200 nm) that consisted of cholesterol-conjugated PEG and modified with cell-penetrating peptides (TAT, YGRKKRRQRRR). Polymeric micelles could cross the BBB both in human astrocyte cell model and in rat model [11] and showed a sustained drug release over the course of 5–6 h [17]. The results confirmed that an effective brain penetration of polymeric micelles was achievable, subsequently enabling the drugs to be delivered into astrocytes and neurons.

In another study, small polymeric micelles (14 nm) anchored with angiopep-2 (Ang) peptide (PE-b-PEG-Ang) were found capable of transporting amphotericin B (AmB), a hydrophobic antifungal agent that demonstrates poor brain penetration, across the BBB in rat and mouse models for fungi intracerebral infection therapy (brain uptake of functionalized micelles was 1.6-fold higher than unmodified micelles and threefold higher than free AmB) [18]. The increased brain uptake was attributed to receptor-mediated transcytosis, since Ang was known to mediate transcytosis across the BBB via low-density lipoprotein receptor-related protein (LRP-1) binding [19]. In the biodistribution study, compared to free drugs, AmB-loaded micelles demonstrated much lower affinities for the liver and spleen, resulting in a more prolonged plasma circulation. Moreover, encapsulation of AmB into polymeric micelles leads to a significant reduction in cytotoxicity and hemolysis. This was because the aggregated form of AmB showed obvious cytotoxicity to mammalian and fungal cells, while the monomeric AmB was nontoxic [20]. Micellar drug loading might lead to the selective release of monomeric AmB and improved toxicity.

### *9.2.2 Core-Shell Nanocarriers*

Most studied brain-targeted systems are liposomes and nanoparticles. These delivery systems with a core-shell structure can protect loaded drugs from biological or chemical degradation in the blood, control their release in a suitable manner, modify the targeting ligands on the surface, and allow for steric PEGylation. For example, liposomes consist of natural lipids with structural similarity to cell membrane. Therefore, liposomes are considered to be biologically compatible with negligible toxicity (Fig. 9.3). However, without the use of surface modification by a hydrophilic polymer including PEG or PMPC, the half-lives of liposomes in the blood are very short due to a series of factors, such as the tendency of the liposome to exchange

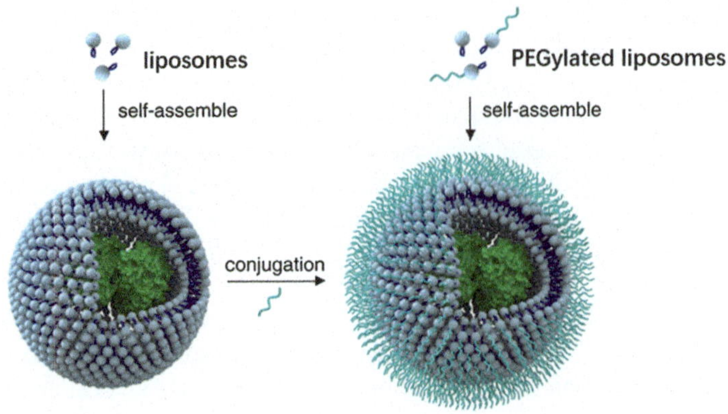


**Fig. 9.3** The representative structures of liposomes and PEGylated liposomes (core-shell nanocarriers)

lipid with cell membranes and their uptake by monocytes. After intravenous injection of unmodified liposomes or nanoparticles, they adsorb plasma components (opsonins) rapidly and result in a clearance from the blood by the macrophages which are mainly located in the liver (Kupffer cells) and spleen. Some research suggests that the clearance of nanoparticles after intravenous (IV) application can occur already within 5 min after the injection [21]. However, the grafting of PEG to the surface of liposomes to form a core-shell structure significantly changes their fate and prolongs their residence time in vivo [22–24]. The reason for the prolonged plasma circulation is suggested to be due to the highly flexible and hydrated PEG chain attached to the liposomes surface, which is assumed to have an effective opsonin-resistant property via its steric repulsion effect.

The PEGylation of core-shell structure has also been applied to other drug delivery systems, including macromolecular conjugates and dendrimers. In addition, PEG-containing surfactants, such as poly(oxy-ethylene)-poly(oxy-propylene) block copolymers (Pluronic®F108 and Tetronic®) could also be used to prolong the blood circulation time of particles [25].

In addition to offering the steric stabilization, PEG on the surface of the nanoparticles also allows for the preparation of particles by conjugating bioactive ligands on the surface for brain-targeted delivery. The obvious reasons for the design of ligand modified (core-shell structure) long-circulating drug carriers were listed as follows:

- Ligands (e.g., antibodies, proteins, peptides, and carbohydrates) modified to the surface of particles may increase the elimination from the blood by the liver and spleen, while the presence of PEG may compensate for this imperfection.
- Long plasma circulation of the targeted ligand-conjugating nanoparticles may allow for their efficient accumulation in target tissue with a diminished blood flow and a low concentration of the surface antigens or markers.

Covalent linkage of targeting ligands on the surface of nanoparticles is typically made by a coupling reaction via amine or thiol groups. To achieve more significant brain targeting, the ligands must be attached to the particles via the PEG spacer arm, so that the targeting ligands are extended and exposed to exclude steric hindrance for their binding affinities to the target receptors [26].

### *9.2.3 Nanospheres*

Polymeric nanospheres are composed of a dense polymer matrix that enables the dispersion, adsorption, or binding of drugs. These nanoparticles are currently being investigated for their potential utility for brain drug delivery applications. Chitosan, poly(lactic-co-glycolic acid (PLGA)/poly(lactic acid) PLA and poly (butylcyanoacrylate) (PBCA) are among the most common polymers used in biomedical applications, and these are discussed below.

#### 9.2.3.1 Chitosan

Chitosan is a natural linear polysaccharide obtained by alkaline deacetylation of chitin, which is composed of randomly distributed β-(1→4)-linked D-glucosamine and N-acetyl-D-glucosamine (Fig. 9.4). Chitosan is a kind of highly biocompatible and biodegradable polymer that can function as a stand-alone nanoparticle [27]. As a cationic polysaccharide, the molecular structure of chitosan contains free amino

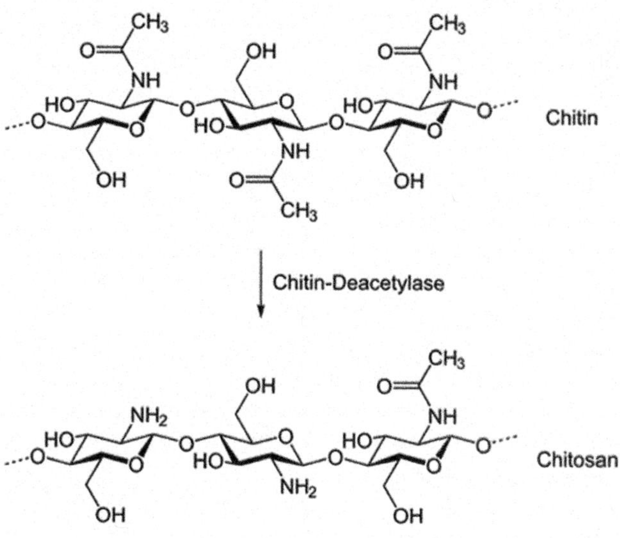


**Fig. 9.4** The preparation and chemical structure of chitosan

groups that render it insoluble in neutral or basic pH conditions. However, the free amino groups undergo protonation, making chitosan selectively soluble in water under acidic conditions [28]. One potential application of this property includes a drug delivery system that is able to maintain its integrity in neutral or basic environments but swell and degrade in acidic environments, thus releasing the drugs to the target tissues (e.g., the significant pH changes in glioblastoma) [29].

Collectively, benefits associated with chitosan nanoparticles include their controllable drug release, the linear polyamine structure featuring free amine groups capable of cross-linking, their biocompatibility with living tissues (degrade slowly into harmless, absorbable products), and their mucoadhesive nature, which increases residual time at absorption sites [28]. There has been a mass of attempts to incorporate these benefits abovementioned into drug delivery systems, including brain drug delivery [30].

Chitosan nanoparticles loaded with neuroactive drugs have demonstrated efficacy in crossing the BBB and acting therapeutically at target sites after intravenous administration, especially when surface-modified with protective and targeting moieties [31]. Yemişci et al. reported a design of PEG-chitosan nanoparticles loading with caspase-3 inhibitor (caspase-3 is a protease enzyme that plays an important role in mediating neuronal cell death) [31]. The anti-TfR monoclonal antibodies were conjugated to the ends of the PEG chains to promote the effect on crossing the BBB. The results in animal model showed that the PEG-chitosan nanoparticles conjugated with anti-TfR monoclonal antibodies were able to penetrate the brain to a much greater extent, compared to those unconjugated anti-TfR. As a result, caspase-3 activity was significantly inhibited, and the neuroprotective effects were evident in mice treated with mAbs-modified PEG-chitosan nanoparticles containing a high dose of caspase-3 inhibitor.

Interestingly, due to their cationic surface charge, bare chitosan nanoparticles can also induce transport across the BBB via adsorptive-mediated transcytosis (AMT) [32]. This phenomenon might benefit from the electrostatic interaction between the positively charged chitosan nanoparticles and the negatively charged membranes. For example, Wang et al. reported N-trimethyl chitosan nanoparticles (TMCNPs; zeta potential, $\approx$30 mV) and studied their ability to deliver anti-neuroexcitation peptides (ANEP), a potential therapeutic peptide for epilepsy, across the BBB [32]. In mouse studies, the cationic surface charge rendered the TMCNPs more capable of retention in the brain after intravenous administration than free drug (the maximum brain concentration of ANEP transported by TMCNPs was 1.3 μg/mL of brain tissue and shows twofold higher than free drug).

In addition, intranasal delivery may offer an alternate, safer, quicker, and more effective pathway by which chitosan nanoparticles can be delivered to the brain. In studies of mice models, chitosan nanoparticles have been found to accumulate in the brain to a greater extent and within a shorter timeframe via intranasal delivery than via intravenous injection. These results for intranasal administration of nanoparticles were achieved while maintaining lower levels of nontarget drug dispersion (e.g., in the liver and lungs) than observed via intravenous injection of nanoparticles.

### 9.2.3.2 PLGA Nanoparticles

Poly(lactic-co-glycolic acid) (PLGA) is one of the most common polymers applied in the formulation of polymeric nanospheres. PLGA is a copolymer with high biocompatibility, biodegradability, and controllable release of the drugs both in vitro and in vivo [27]. The release properties of PLGA nanoparticles can be modulated via changing the molar ratio between lactic acid and glycolic acid as well as the polymer molecular mass [33]. For the treatment of intracranial disease, PLGA nanoparticles showed superior release kinetics than liposomes and polymeric micelles.

Drug encapsulation efficiency in PLGA nanoparticles depends on many factors, such as solid-state drug solubility, drug-polymer interaction, molecular weight of polymers, and surface functional groups. PLGA is a hydrophobic copolymer; compared to hydrophilic drugs, the hydrophobic ones are easier to be encapsuled. The average entrapment amount could range from 5% to 10% (w/w) [34], and the amount can vary substantially depending upon the property of the drugs and the preparation procedure (in some cases, the drug content can constitute up to 50%).

PLGA nanoparticles can also be conjugated with protective and brain-targeting ligands to enhance BBB penetration. Without surface modification, a large number of PLGA nanoparticles are typically cleared from circulation by systemic organs. For example, around 40% of nanoparticles accumulated in the liver as a result of the reticuloendothelial system (RES), while another 25% accumulated in the kidney. Fortunately, even with the high rates of PLGA nanoparticle accumulation in the liver or kidney, the nanoparticles tend to exhibit negligible or no toxic effects.

Current challenges associated with PLGA nanoparticles mainly include irreversible adsorption of proteins to the polymer matrix (known as the "protein corona") [35]. However, after functionalization, PLGA nanoparticles have been used to deliver various drugs into the brain successfully and safely [36]. In the established reports, PLGA nanoparticles were formulated with various stabilizers, such as PEG, polyvinyl alcohol (PVA), and human serum albumin (HSA), and coated with surfactants, including poloxamer 188 and surfactants-polysorbate 80 (PS 80) [37]. For example, the modified PLGA nanoparticles loading doxorubicin have been reported for their potential to exert antitumor effects in glioblastoma. The loperamide-loaded nanoparticles have been investigated for their potential to induce central analgesic effects in mice (significant drug passage across the BBB compared to unmodified nanoparticles that were incapable of achieving transit).

Interestingly, it was found that the efficacy of the surfactants could vary depending upon the type of nanoparticles used. For instance, for poly(butyl cyanoacrylate) (PBCA) nanoparticles, both poloxamer 188 and PS 80 induced similar, distinctive pharmacological effects when delivering either loperamide [38] or doxorubicin [39] across BBB. This finding suggests that the core properties of the polymer materials are able to influence the therapeutic efficacy of surface functionalization strategies. In addition, also striking was the fact that upon comparing pharmacological profiles for PBCA and PLGA formulations, PBCA was suggested to be a faster degraded material since the effects of loperamide administration were sustained for a much shorter period than when PLGA nanoparticles were used [37].

#### 9.2.3.3 PBCA Nanoparticles

PS 80-coated poly(butylcyanoacrylate) (PBCA) nanoparticles have been successfully used to transport drugs, including rivastigmine, tacrine, and enkephalin, for crossing the BBB after intravenous administration [40, 41]. Furthermore, PBCA nanoparticles showed negligible cytotoxic or inflammatory effect at therapeutic concentrations and incubation times owing to their biodegradable and biocompatible PBCA polymer. It was reported that PBCA could collaborate with plasma apolipoprotein E (ApoE) to facilitate BBB crossing [42], without inducing nonspecific BBB disruption.

### 9.2.4 *Nanocapsules*

Polymeric nanocapsules consist of a core-shell arrangement; the shell is typically polymeric and encapsulates an inner aqueous or oily core. For example, an interface-adherence polyer shell initiated by in situ polymerization was proposed and developed in blossoming for protein and RNA therapeutic (Fig. 9.5) [43]. This novel approach first appeared in 2006, in which a single enzyme was encapsulated in the nanogel to enhance protein biocatalytic activity and stability [44]. According to this approach, a number of protein and nucleic acid nanocapsules were developed for cancer or metabolic disease treatments.

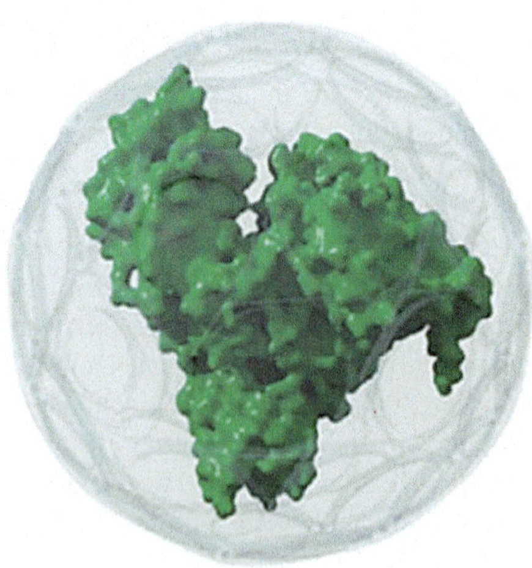

**Fig. 9.5** A representative structure of protein nanocapsules

Advantages of nanocapsules are emphasized: (1) the polymer shell stabilizes the protein and nucleic acid against proteolysis and nuclease; (2) the properties of the nanocapsule such as the surface charge, hydrophibility, and degradability can be regulated via the choice of different functional monomers or cross-linkers; and (3) the size of a single nanocapsule with a diameter of 20 nm is favorable for prolonged plasma circulation [45].

For example, in order to maximize the protein efficiency, the degradable nanocapsules have been constructed to release the wrapped proteins. Zhao et al. encapsulated recombinant maltose-binding protein-fused apoptin (MBP-APO) via in situ polymerization using a disulfide bond containing degradable-cross-linker so that polymer shell could be degraded when the nanocapsules were exposed to the reducing environment in tumor cells. Similar to Zhao's work, a series of degradable protein nanocapsules taking advantage of environmental responsive manner have been developed for efficient intracellular delivery or tunable controlled release of protein. SiRNA can also be wrapped into the nanocapsule. Liu et al. developed a miRNA nanocapsule to enhance miRNA (AS-miR-21) stability and effective intracellular delivery, suppressing the angiogenesis and tumor retrogression in mouse models [46].

In addition, to minimize the immunogenicity of therapeutic biomacromolecules and prolong the plasma circulation, zwitterionic 2-methacryloyloxyethyl phosphorylcholine (MPC) was employed as the monomer during in situ polymerization owing to their evasion of mononuclear phagocyte system (MPS). The protein and nucleic acid coated with the zwitterionic polymer shell could shield serum protein adsorption under blood flow, thus prolonging the half-life in vivo. These nanocapsules have enormous potential for the brain disease treatment, although the implications for brain drug delivery still have yet to be fully explored.

### 9.2.5 Dendrimers

Dendrimers emerged from the "cascade molecules" initially reported by Vögtle et al. at the end of 1970. Dendrimers are three-dimensional, repetitively branched polymers that adopt a spheroidal and symmetrical morphology in aqueous solution (Fig. 9.6) [27]. They comprise three domains: (1) a multivalent surface, containing several potentially reactive functional groups on their periphery; (2) radially concentric interior shells, resembling tree-like branching from the core (called "generations" (G)); and (3) the core itself, to which dendrons attach via focal points [47]. The most researched dendrimers as nanocarriers for brain disease treatments are the poly(amidoamine) (PAMAM) dendrimers, other vectors including poly(ether)-copoly(ester) (PEPE), poly(ether imine) (PETIM), poly(L-lysine)-based (PLL), poly(propylene imine) (PPI), gallic acid-triethylene glycol (GATG), carbosilane, and phosphorus dendrimers have also been developed.

These dendrimers present appealing properties to mediate the delivery of different drugs to the brain due to their unique structural and physicochemical properties: (1) a

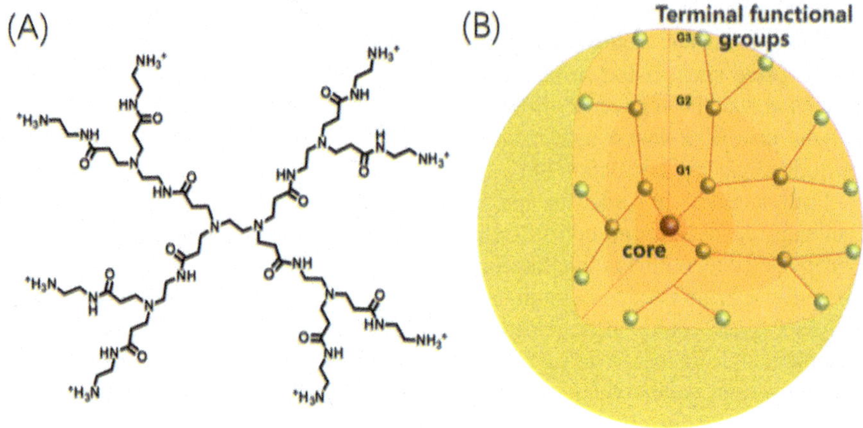


**Fig. 9.6** (**a**) Chemical structure of the generation 1 cationic PAMAM dendrime. (**b**) Schematic representation of a dendrimer structure with three generations (G)

globular, well-defined, and high-branched structure; (2) monodispersity, low viscosity, adaptable solubility, and high loading capacity; and (3) controllable nanosize which is important when defining the capacity to overcome the biological barriers including BBB [27]. Moreover, the abundance of peripheral groups allows the controllable and specific tethering of various bioactive ligands (e.g., small molecules, peptides, and proteins), imitating the multivalency existing in several biological systems. In reality, this characteristic multivalency is the greater superiority of dendrimers. Importantly, the molecular structure is tightly packed at the periphery but loosely packed in the core, leaving spaces that can facilitate drug entrapment [48].

The drugs can associate with dendrimers in three main ways: (1) can be covalently attached to the peripheries, forming dendrimer prodrugs; (2) can interact with the outer functional groups via ionic interactions; and (3) can be encapsulated within dendrimers via formation of dendrimer-drug supramolecular assemblies [49]. For example, functionalization of PAMAM dendrimers has proven to be an effective approach to enhance the penetration of cells in the neurovascular unit (NVU), both via intravenous and neurosurgical administration [50]. Ke et al. reported that a PAMAM-PEG-angiopep/DNA nanoparticle consisted of angiopep (a ligand-targeting LRP-1)-modified PEG-PAMAM and DNA. After intravenous injection, the angiopep-modified nanoparticles were found to be successful in crossing the BBB via clathrin-/caveloae-mediated endocytosis and partial macropinocytosis [51]. The interaction between angiopep and LRP-1 was suspected to be the primary mechanism of BBB crossing. Brain uptake was able to be enhanced by increasing the ratio of angiopep incorporated in the NP formulation. In addition, dendrimers like PAMAM have also been successfully modified with other ligands for transcytosis across the BBB, such as lactoferrin [52] and transferrin [53]; all of these were found to demonstrate higher BBB crossing capability than their unmodified dendrimers.

### 9.2.6 Nanogels

Nanogels are cross-linked polymeric nanoparticles with three-dimensional (3D) configuration (diameters ranging from 1 to 1000 nm). The potential application of polymeric nanogels, based on both natural and synthetic polymers, has emerged as one of the most significant trends in nanomedicine (e.g., drug delivery) [54]. Natural polymers, such as chitosan and alginate, have been studied widely for preparation of nanogels. In addition, nanogels based on poly(ethylene oxide) (PEO), poly (ethylenimine) (PEI), poly(vinyl alcohol) (PVA), poly(vinyl pyrrolidone) (PVP), and poly-N-isopropylacrylamide (PNIPAAm) have also been reported with different features for drug delivery. Nanogels usually possess high water content, high-specific surface area, biocompatibility, and desirable mechanical properties. They offer unique advantages for nanomedicine, especially the polymer-based drug delivery system: (1) a tunable size, (2) a large surface area for multivalent or functional bioconjugation, (3) an interior network for the incorporation of biomolecules, (4) stability for prolonged circulation in the bloodstream, (5) biodegradability for sustained release of drugs for a desired period of time, and (6) facile removal of the empty vehicles. In addition, nanogels also can be considered as "hydrogels" if they are composed of water-soluble/water-swellable polymer chains [55]. Their affinity to absorb water is attributed to the presence of hydrophilic groups such as -OH, -CONH-, -$CONH_2$-, and -$SO_3H$ in polymers.

Nanogels can be first synthesized in the absence of the drug (chemically and physically induced cross-linking, e.g., covalent bonds, hydrogen binding, van der Waals interactions, or physical entanglements), swollen in water, and then loaded with the drug. Drug loading occurs spontaneously and results in decrease of the solvent volume, which is followed by the gel collapse and formation of dense nanoscale particles. However, the release mechanisms of the loaded drugs from nanogels are complex resulting from three main vectors, i.e., drug diffusion, nanogel matrix swelling, and chemical reactivity of the drug/matrix [56].

Nanogels, also called hydrogels, offer the prospect of drug transport across the intact BBB. Some reports have shown that nanogels with suitable surface charge exhibited better internalization property on cell membrane, compared to the neutral ones. For example, Gil et al. reported cationic polysaccharide-based nanogels containing β-cyclodextrin and poly(β-amino ester) for transporting doxorubicin and insulin across the BBB (enhanced permeability of insulin across the BBB model by 20% in vitro).

Surface modification of nanogels can also be used to accelerate encapsulated drugs across the BBB. In a report by Azadia et al., they prepared methotrexate (MTX)-loaded nanogels via an ionic gelation process using chitosan and sodium tripolyphosphate (TPP). The surface of the MTX-loaded nanogels were coated with PS 80 to improve brain drug delivery. The cumulative release profiles in vitro indicated a non-Fickian diffusion kinetic, apparently governed by and the swelling/disintegration of the polymeric networks in nanogels. After intravenous administration, remarkably higher brain concentrations of MTX were achieved with the nanogel formulations in comparison to the free drug [57].

Being hydrophilic carriers, nanogels are favorable in uploading aqueously soluble drugs as well as proteins or nuclear acids. Brain-specific anion transporter 1 (BSAT1) and connexin 43 (Cx43) are the promising targeted antigens for drug delivery to the brain (e.g., gliomas). For example, they have been conjugated on the surface of the cisplatin-loaded nanogels prepared by poly(ethylene glycol) (PEG)-block-poly(methyl methylacrylate) (PMAA) and $CaCl_2$ using 1-(3-dimethylaminopropyl)-3-ethylcarbodiimide hydrochloride (EDC) as cross-linkers, for the treatment of intracranial gliomas. The survival of rats treated with nanogels conjugated with targeted mAbs against extracellular loops of BSAT1 and Cx43 were 26.6 and 27 days longer than control, respectively, which indicated the effectiveness of the ligands in increasing BBB-penetrating efficiency of the nanogels. In addition, modification of transferrin (Tf) or insulin with nanogels of cross-linked PEG and PEI for the delivery of oligonucleotide in the brain also showed similar results [58].

## 9.3 The Application of Polymeric Nanomedicine in Brain Disease

The concept of polymeric nanoparticle therapeutics can be traced back to the 1950s, and a variety of drug delivery systems have been reported and developed in the following decades. In this section, we introduce a series of representative examples to highlight the advanced applications with polymeric nanoparticles in therapy for brain diseases, including brain cancers, Alzheimer's Disease (AD), Parkinson's Disease (PD), Cerebral Amyloid Angiopathy (CAA), Stroke and Multiple Sclerosis, etc.

### *9.3.1 Brain Cancers*

Cancer/tumor is one of the major causes of death worldwide. Although cancer therapies are developing, the treatment remains one of the most challenging problems. Brain cancers belong to the most dangerous CNS diseases and most aggressive human cancers with an average survival of 30 weeks for IV grade tumors [59]. For example, patients with intracranial glioblastoma, the most aggressive and prevalent glioma variant, have an average survival of only 15 months [59]. Although radiation therapy has been more successful in eliminating childhood glioblastoma, with 5-year survival rates as high as 70–80% [60], the long-term side effects of these conventional treatment modalities can be severe. Among the approaches used for the treatment of gliomas, chemotherapy just plays a role of an adjuvant, the potential being limited by its inability to overcome the BBB in vivo. Thus, the employment of polymeric nanoparticles to achieve brain delivery for noninvasive systemic chemotherapy of gliomas appears to be an ideal option [7].

It is well known that the amount of drug delivered to the brain is equally proportional to the BBB permeability and the area under the curve of plasma concentration versus time (AUC) [61]. Accordingly, it has been shown that PEGylation or surfactant coating of the nanoparticles can improve their uptake in the brain effectively. For example, surfactant coatings were effective for brain delivery of different types of polymeric nanoparticles such as PBCA (core-shell structure). The effectiveness of brain delivery by the PS 80-coated PBCA nanoparticles was most clearly demonstrated by the remarkable efficacy of the nanoparticle-bound doxorubicin against intracranial 101/8 glioblastoma in rats (Fig. 9.7). It should be noted that the tumor model employed, 101/8 rats glioblastoma, is morphologically similar to human glioblastomas, which are characterized by a transplantation success of >95% and reproducible growth pattern [62].

Active targeting is a noninvasive method to transport drugs to targeted tissue using site-specific ligands. Endogenous and chimeric ligands binding to the

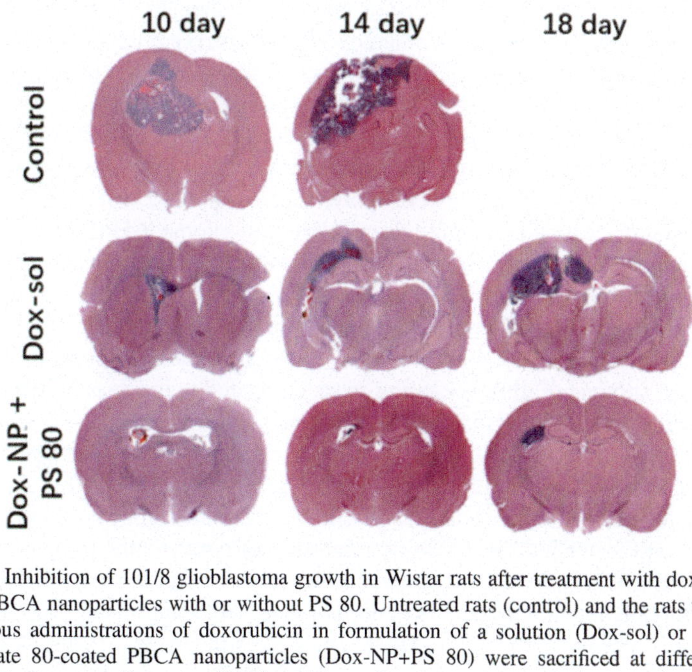


**Fig. 9.7** Inhibition of 101/8 glioblastoma growth in Wistar rats after treatment with doxorubicin-loaded PBCA nanoparticles with or without PS 80. Untreated rats (control) and the rats treated by intravenous administrations of doxorubicin in formulation of a solution (Dox-sol) or loaded in polysorbate 80-coated PBCA nanoparticles (Dox-NP+PS 80) were sacrificed at different time points. The hematoxylin-eosin (H&E) staining of representative brains is shown. Note the decreased tumor size, especially in the Dox-NP + PS 80 group

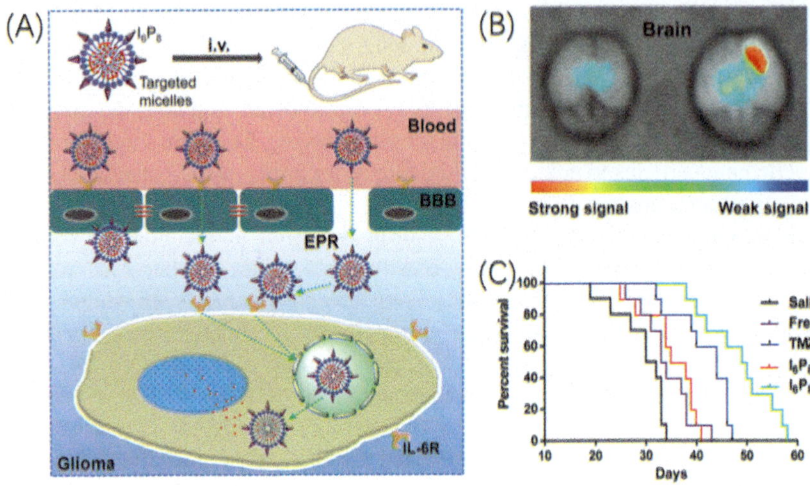


**Fig. 9.8** (**a**) Transportation of targeted micelles after intravenously injected to glioma-bearing mice. (**b**) Fluorescent images of brains of glioma-bearing nude mice treated with ICG-loaded $I_6P_8$scr-modified micelle (left) and $I_6P_8$-modified micelle (right). The brains were excised 4 h after treatments. The fluorescence signal of near-infrared dye ICG was recorded. (**c**) The survival profile of glioma-bearing mice with different treatments

receptors of BBB have been directly or indirectly conjugated to nanoparticles and shown promising potential in oncotherapy via recognizing brain capillary endothelial cells and cerebral tumor cells [63]. Shi et al. reported that a new interleukin-6 (IL-6) receptor-mediated polymeric micelle system, where a short peptide $I_6P_8$ (LSLITRL was reported to possess similar targeting function for selectively binding with IL-6R to mediate nanoparticles into glioma cells [64]) was conjugated to biodegradable poly(ethylene glycol)-poly(lactic-co-glycolic acid) (PEG-PLGA), was prepared for cascade-targeting drug delivery to glioma [65] (Fig. 9.8). In vitro results indicated that the $I_6P_8$-modified doxorubicin-loaded micelle ($I_6P_8$-D-M) could transport across the BBB significantly and subsequently target the glioma, which was superior to the scrambled peptide-modified formulations. In vivo results showed that this $I_6P_8$-D-M could introduce the highest level of apoptosis in glioma and longest survival of glioma-bearing mice, compared to the ones in other groups. All of the results suggested that the $I_6P_8$-D-M polymeric micelle is an ideal cascade-targeting therapeutic strategy, which could overcome the BBB for glioma therapy.

The chemotherapy of glioblastoma (e.g., temozolomide (TMZ)) is severely hindered by the immunosuppressive tumor microenvironment, especially the tumor growth factor β (TGF-β), an immunosuppressive cytokine [66]. In the study reported by Qiao et al., they employed RNA interference (RNAi)-based immunomodulation to remodel the tumor immune microenvironment and sensitized the effect of chemotherapy [67]. Therefore, a nano-theranostic system (Angiopep LipoPCB(Temozolomide + BAP/siTGF-β), ALBTA) with ROS response is established for glioblastoma treatment (Fig. 9.9). This kind of polymeric

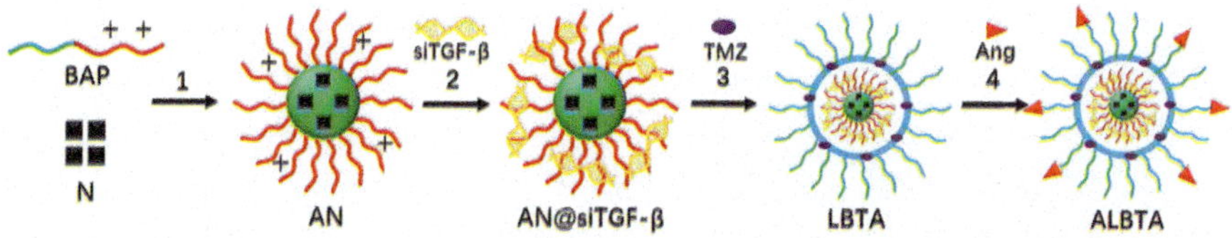


**Fig. 9.9** The preparation of the nanoparticles. (1) BAP polymers self-assemble into nanoparticles (AN) via encapsulating SPIONs in the hydrophobic core. (2) The positively charged AN load negatively charged siTGF-β by electrostatic interactions. (3) The zwitterionic lipid-based envelopes were coated to the siTGF-β-loaded nanoparticles. (4) The angiopep-2 was conjugated onto the nanoparticles by the coupled reaction between the sulfhydryl groups of angiopep-2 and maleimide groups of DSPE-PCB-Mal

nanoparticles exhibit strong small interfering RNA (siRNA) condensation, good serum stability and magnetic property. They were able to cross the BBB efficiently and target glioblastoma via receptor-mediated transcytosis (LRP-mediated). The zwitterionic lipid (distearoyl phosphoethanol-amine-polycarboxybetaine lipid) in ALBTA promoted endosomal/lysosomal escape and thus enhanced the cytotoxicity of TMZ and improved gene silencing efficiency of siTGF-β. ALBTA significantly improve the immunosuppressive microenvironment and improve the survival rates of glioma-bearing mice. Furthermore, ALBTA could also be traced by magnetic resonance imaging (MRI) accurately in brain tumors via encasing the superparamagnetic iron nanocubes (SPIONs) in the polymeric nanoparticles. This study indicated that this polymeric nanomedical platform can serve as a flexible and powerful synergistic system for treatment of brain tumors as well as other brain diseases in the central nervous system (CNS).

In addition, aberrant mesenchymal-epithelial transition factor (MET) activation occurs in approximately 30% of glioma patients and correlates with poor prognosis, elevated invasion, and increased drug resistance [68]. In another report, Wu et al. developed a novel nano-inhibitor by conjugating MET-targeting cMET-binding peptide (cMBP) on the G4 dendrimer [69] (Fig. 9.10). The binding affinity of the nano-inhibitor to MET increased three orders of magnitude to $1.32 \times 10^{-10}$ M than free peptide ($K_D = 3.96 \times 10^{-7}$ M). This nano-inhibitor reduced the proliferation and invasion of human glioblastoma U87 MG cells efficiently in vitro via blocking the MET signal pathway with remarkably attenuated levels of phosphorylated MET (pMET) as well as the downstream signaling proteins, including pAKT and pERK1/2. In vivo magnetic resonance imaging (MRI) indicated a significant delay in tumor growth after intravenous administration of the nano-inhibitor. Furthermore, the medium survival was extended by 59%, which was similar to the therapeutic effects of PF-04217903, a MET inhibitor in clinical trials. The pMET and its downstream proteins pAKT and pERK1/2, which were involved in the survival and invasion of cancer cells, significantly decreased in the nano-inhibitor-treated group, compared with those in the control group. This may be the first report of therapeutic inhibition of glioma growth by blocking MET signaling with a novel nano-inhibitor made by polymeric nanomaterials. Compared to antibodies and chemical inhibitors in clinical

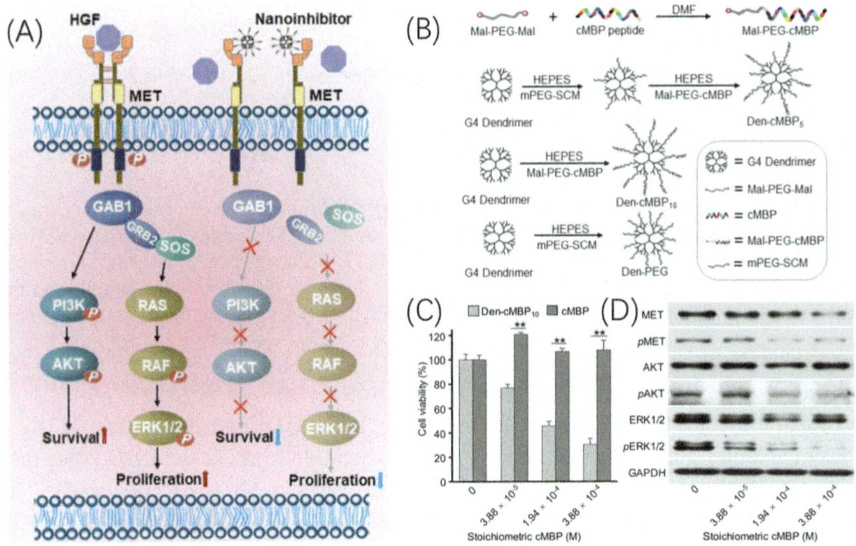


**Fig. 9.10** (**a**) Illustration of signaling pathways of HGF-MET axis in the presence or absence of nano-inhibitor. (**b**) Synthetic scheme of nano-inhibitor Den-cMBP and control nanoparticles Den-PEG. (**c**) Viabilities of U87 MG cells at 24 h post treatment of cMBP or Den-cMBP$_{10}$. $**P < 0.01$. (**d**) Western blot shows decreased phosphorylation levels of MET, AKT, and ERK1/2 in U87 MG cells as a function of Den-cMBP$_{10}$ concentration

trials, the nano-inhibitor prepared by G4 dendrimer blocked MET signaling and provided a new approach for the treatment of glioma with the advantages of high efficiency, affordability, and most importantly, potentially reduced drug resistance.

### *9.3.2 Alzheimer's Disease*

Alzheimer's disease (AD) is a chronic irreversible neurodegenerative disorder with long preclinical and prodromal phases (20 years) and an average clinical duration of 8~10 years [70]. AD can increase the risk including progressive mental, behavioral, and functional decline and loss of the ability to learn [71]. AD has an estimated prevalence of 10~30% in the population >65 years of age with an incidence of 1~3%. It is forecasted that at the current rate, 1 in 85 persons worldwide will be living with AD by 2050. The two hallmarks of AD are (1) the buildup of cortical and cerebrovascular deposits or aggregates of amyloid-β (Aβ) peptide (CSF $A\beta_{1-42}$ levels are more than 10 times that of plasma levels) and (2) the accumulation of unnatural levels of hyper-phosphorylated, microtubule-associated tau protein. Most AD patients (>95%) have the sporadic form that is characterized by a late onset (>80 years of age) and is the consequence of the failure to eliminate the Aβ peptide or plaque in cerebrospinal fluid (CSF). A small proportion of patients (<1%) have

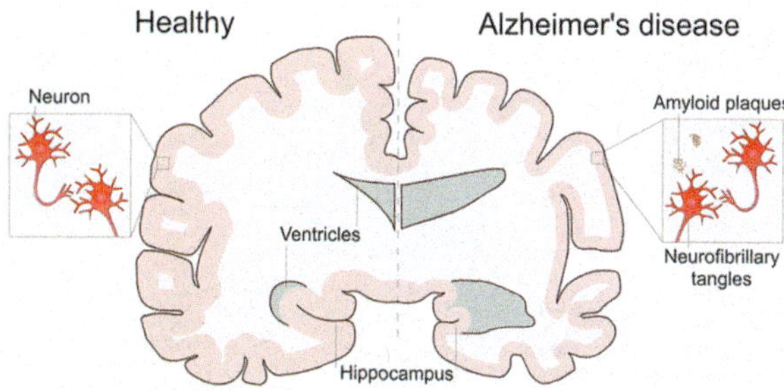


**Fig. 9.11** The main hallmarks of AD are the presence of Aβ plaques and neurofibrillary tangles in neurons, which culminates in severe neurodegeneration, shrinkage of the cerebral cortices and hippocampus, and enlargement of the ventricles

inherited mutations in genes that affect processing of Aβ and develop the disease at a much younger age (mean age of ~45 years). Apart from them, various deranged mechanisms such as inflammation, chronic oxidative stress, neurofibrillary tangles accumulation, mitochondrial dysfunction, hormone imbalance, calcium mishandling, and genetic components also play important roles in the disease process [71]. Several approved drugs ameliorate some of the symptoms of Alzheimer's disease; however, no current interventions can modify the underlying disease mechanisms. As a result, the development of new interventions that can delay, halt, or reverse disease onset and progression is an area of research receiving intense interest (Fig. 9.11).

Molecular chaperones [72], the endogenous proteins in the human body, play a non-substitutable role in regulating undesired protein misfolding and maintaining the intricate homeostasis of protein metabolism in vivo, which act as an ideal candidate for AD treatment. For example, Huang et al. developed a novel artificial chaperone consisting of mixed-shell polymeric micelles (MSPMs) via hydrophobic interaction to reduce Aβ-induced neurotoxicity (Fig. 9.12) [73]. The MSPMs were prepared by the self-assembly of two amphiphilic diblock copolymers, including poly(ε-caprolactone)-block-poly(N-isopropylacrylamide) (PCL-b-PNIPAM) and poly(ε-caprolactone)-block-poly(ethylene oxide) (PCL-b-PEG) in aqueous solution. The hydrophobic domains on the surface of micelles act as cavities which could interact with hydrophobic Aβ monomers, while the hydrophilic PEG chains provide a protective layer to prevent the MSPMs from aggregating after the adsorption of Aβ. The results proved that the MSPMs with appropriate weight ratio of PNIPAM and PEG (PCL-b-PEG/PCL-b-PNIPAM = 3/7) could serve as an excellent suppressor of AD and show enhanced therapeutic effects in PC12 cells. Therefore, these results could provide new insights into the development of artificial chaperone (MSPMs) as an ideal candidate for AD treatment.

Previous studies had suggested that the KLVFF ($A\beta^{16-20}$) peptide could effectively stabilize the soluble Aβ conformation and destabilize the altered conformer due to the strong affinity with Aβ, eventually inhibit Aβ aggregation or dissolve Aβ fibrils [74]. Combining KLVFF peptide and mixed-shell polymeric micelles, Qu et al. developed a highly efficient system to achieve the synergy both disaggregating Aβ fibrils and reducing Aβ-mediated neurotoxicity (Fig. 9.13) [75]. First, the KLVFF sequence were modified onto the classical MSPMs, which consisted of

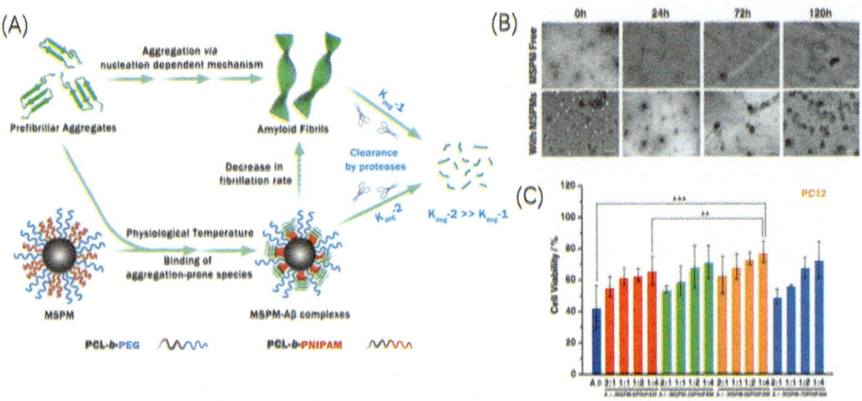


**Fig. 9.12** Maintenance of Aβ peptide homeostasis by mixed-shell polymeric micelles: (**a**) the schematic diagram of the formation of amyloid fibrils and MSPMs-Aβ complexes; (**b**) TEM images of Aβ incubated with and without MSPMs at different time points, respectively (The PCL-b-PEG/ PCL-b-PNIPAM ratio in the MSPMs was 3/7. The concentration of Aβ was 20 μM. The concentration of MSPMs was 0.4 mg/mL. 10 mM PBS, pH = 7.4. Scale bar, 200 nm); and (**c**) PC12 cells viability measured by MTT assay. Significance levels are expressed by asterisks: $**P < 0.01$ and $***P < 0.001$

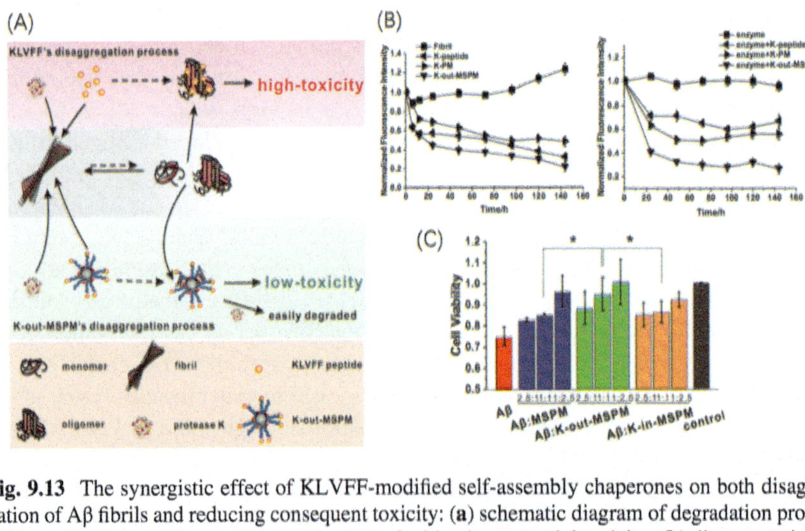


**Fig. 9.13** The synergistic effect of KLVFF-modified self-assembly chaperones on both disaggregation of Aβ fibrils and reducing consequent toxicity: (**a**) schematic diagram of degradation process of K-out-MSPM and the possible mechanism of mitigating potential toxicity; (**b**) disaggregation of Aβ fibrils by K-peptides, K-PMs, and K-out-MSPMs with or without proteases measured via the ThT fluorescence assay; and (**c**) mitigation of Aβ toxicity by different concentrations of MSPMs separately. Significance levels are expressed as asterisks: $*P < 0.05$

poly(ε-caprolactone)-block-poly(β-amino ester) (PCL-b-PAE) and (PCL-b-PEG), to make sure that the peptide could interact with hydrophobic Aβ monomers in aqueous solution. When the KLVFF disaggregated the Aβ fibrils into fragments, the

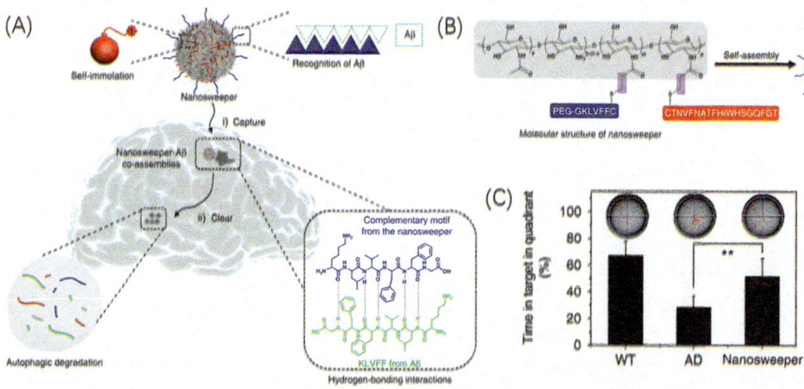


**Fig. 9.14** (**a**) The schematic of the nanosweeper mechanism of action. The nanosweeper captures Aβ or Aβ aggregates via hydrogen-bonded co-assembly and internalizes by neuronal cells carrying Aβ. Then, the nanosweeper induces the autophagic response, resulting in the degradation of Aβ. (**b**) Molecular structures of the nanosweeper, K and B were the abbreviation of HS-CFFVLKG-PEG (capture component) and HS-CTNVFNATFHIWHSGQFGT (clearance component), respectively. (**c**) Morris water maze (MWM) experiment. Statistical significance is indicated as $*p < 0.05$ and $**p < 0.01$

hydrophobic domains on the surface of the MSPMs could bind them promptly, resulting in reduction of consequent toxicity. Besides disaggregating the fresh fibrils, KLVFF-modified MSPMs could also assist protease (e.g., protease K) in degrading tangled Aβ fibrils to maintain the healthy proteostasis. Thus, KLVFF-modified MSPM was able to effectively solve the synchronism problem between degrading fibrils and mitigating the neurotoxicity caused by newly formed Aβ fragments during the disaggregation process. The unique MSPMs may provide new thoughts on the treatment of AD in a safer way.

The imbalance between Aβ production and clearance resulting in cerebral Aβ accumulation is one of the most important factors in the formation of AD. In order to preserve the maintenance of Aβ homeostasis, achieving a method to clear up Aβ plaques out of the brain becomes an emerging task. Herein, Luo et al. described a self-destructive nanosweeper based on multifunctional peptide-polymers that is capable of capturing and carrying Aβ into cells, eventually clearing extracellular Aβ via cellular autophagy for the effective treatment of AD [76]. The nanosweeper was composed of a cationic chitosan (CS) core which was decorated with PEGylated-GKLVFF and Beclin-1 (an autophagy–inducing peptide, TGFQGSHWIHFTANFVNT) (Fig. 9.14). KLVFF could recognize and co-assemble with Aβ monomers via hydrogen-bonding interactions. Beclin-1 could induce autophagy to degrade and eliminate Aβ thoroughly. The PEG increases the stability of the nanosweepers in water, providing the appropriate biocompatibility. As a result, the nanosweeper decreases the cytotoxicity of Aβ and rescues memory deficits of AD transgenic mice. We believe that this resourceful and synergistic approach has valuable potential as an AD treatment strategy.

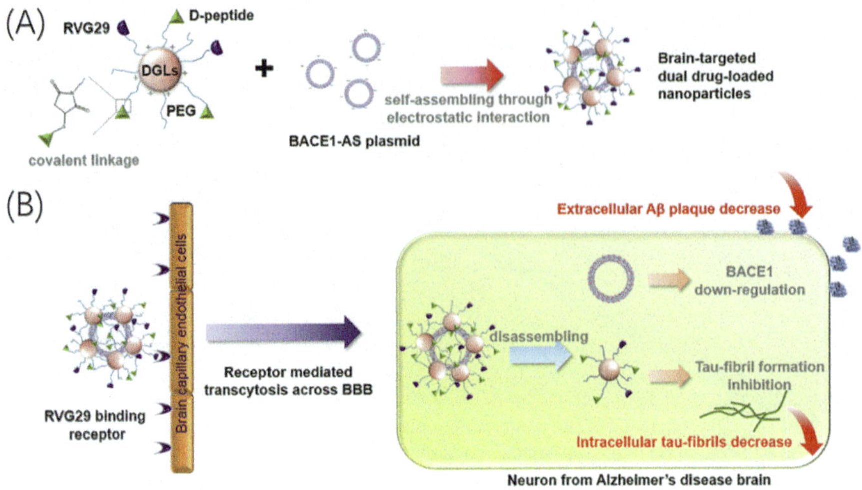


**Fig. 9.15** Schematic of therapeutic gene and peptide co-delivery system. (**a**) The therapeutic gene (pBACE1-AS) and peptide loading using brain-targeted DGL vector. Briefly, the therapeutic peptide was covalently conjugated to the DGLs, and therapeutic plasmid was encapsulated within DGLs via electrostatic interactions. (**b**) After intravenous injection, the DGLs-PEG-RVG29-D-peptide/DNA NPs were able to cross the BBB through RVG29-binding receptor-mediated transcytosis. Then, the NPs were uptaken by neurons in the AD brain through endocytosis. After endosomal escape, the released therapeutic plasmid downregulated BACE1 expression, leading to a reduction of extracellular Aβ plaques. The D-peptides inhibited tau-fibril formation, resulting in a decrease of tau-fibrils in the cytoplasm.

Aβ is generated from the amyloid precursor protein (APP) via sequential proteolytic cleavages by β-site APP cleaving enzyme 1 (BACE1) [77]. Some studies show that BACE1 activity had been upregulated in sporadic AD cases, particularly in neuronal cells around the Aβ plaques. Thus, efficient and specific delivery of therapeutic gene targeting to silence the BACE1 mRNA in brain parenchyma cells is promising to decrease the Aβ burden. On the other hand, the inhibition of tau-related fibril formation in AD brain provides another effective strategy to prevent or delay the progress of AD [78]. Liu et al. developed a multifunctional nanocarrier for AD treatment by achieving therapeutic gene and peptide co-delivery to the brain based on PEGylated dendrigraft poly(L-lysines) (DGLs) via systemic administration (Fig. 9.15) [79]. The dendritic amine-rich structure of DGLs provided plenty reaction sites and positive charge for drug loading. Successful co-delivery of drugs overcoming the BBB through brain-targeted ligand modification (e.g., rabies virus glycoprotein, RVG29) was demonstrated both in vitro and in vivo. Downregulation of the key enzyme BACE1 for Aβ formation was achieved via delivering the noncoding RNA plasmid. Furthermore, simultaneous delivery of therapeutic peptide into brain results in a significant reduction of tau-related neurofibrillary tangles (NFT). As expected, the memory loss rescue in AD mice model was also observed.

In another report, Liu et al. explored a B6 peptide (CGHKAKGPRK, a peptide showing high affinity to transferrin receptor)-conjugated PEG-PLA nanoparticles

(NP) to enhance the delivery of neuroprotective peptides across the BBB for AD treatment [80]. B6-modified NP (B6-NP) exhibited significantly higher accumulation in brain capillary endothelial cells via lipid raft-mediated and clathrin-mediated endocytosis. In vivo experiments showed that the fluorescently labeled B6-NP exhibited much higher brain accumulation than the NP without B6 modification, indicating a successful BBB crossing. Administration of B6-NP encapsulated neuroprotective peptide (NAPVSIPQ, NAP) to AD mouse models showed excellent amelioration in learning impairments, cholinergic disruption, and loss of hippocampal neurons even at a lower dose.

### *9.3.3 Cerebral Amyloid Angiopathy (CAA)*

Cerebral amyloid angiopathy (CAA) is characterized by the deposition of amyloid beta (Aβ) proteins within the walls of small- to medium-sized blood vessels of the brain and leptomeninges with subsequent aggressive vascular inflammation leading to recurrent hemorrhagic strokes. About 80% of AD patients manifest some degree of CAA [81]. Furthermore, CAA currently affects one third of the aging population over 60 years of age. Furthermore, in addition to causing cerebrovascular inflammation, CAA triggers vascular dysfunction that may accelerate the AD progression [82]. Regrettably, there is neither a treatment nor a definitive pre-mortem diagnosis available for CAA.

Agyare et al. developed a theranostic nanovehicle (TNV) which was capable of (1) targeting cerebrovascular amyloid; (2) treating cerebrovascular inflammation resulting from CAA; and (3) providing magnetic resonance imaging (MRI) contrast for the early detection of CAA. The TNVs comprised a polymeric core made from Magnevist® (gadopentetate dimeglumine, MRI contrast agent)-modified chitosan. The core of TNV was also loaded with cyclophosphamide (CYC), an immunosuppressant that could reduce the cerebrovascular inflammation. Meanwhile, putrescine- modified F(ab′)2 fragment of anti-amyloid antibody, IgG4.1 (pF(ab′)24.1), was conjugated to the surface of the core so as to target cerebrovascular amyloid. The leakage of Magnevist® from the TNVs was a modest 0.2% over 4 days, and the CYC release from the TNVs followed Higuchi's model that describes sustained drug release from polymeric matrices. The studies conducted both in polarized human microvascular endothelial cell monolayers (hCMEC/D3) and in mouse models have demonstrated the ability of TNVs to target cerebrovascular amyloid. Moreover, the TNVs could provide the contrast for imaging cerebrovascular amyloid by MRI and single photon emission computed tomography (SPECT). The TNVs were also shown to reduce pro-inflammatory cytokine secretion by the Aβ-challenged blood-brain barrier (BBB) endothelium, which were more effective than the free cyclophosphamide [83].

### 9.3.4 Parkinson's Disease (PD)

Parkinson's disease (PD) is a chronic, progressive neurodegenerative disorder associated with the selective loss of dopaminergic (DA) neurons in the substantia nigra (SN), which presents with classic motor impairments including tremors, rigidity, slow movement (bradykinesia), poor balance, and difficulty in walking (Parkinsonian gait), and it affects about 1~1.5% of the population over 60 years of age [84]. Similar to other CNS diseases, dysfunction of the BBB has been observed in PD. The inflammatory mediator-induced BBB disruption can remodel the junctional proteins and lead to short "opening" of the endothelial cells [85]. Furthermore, P-gp is significantly down-regulated to prevent the outflow of many toxic substrates from the CNS [86]. As a consequence, dopaminergic neurons (dopamine-producing nerve cells) usually located in the substantia nigra are disrupted [87, 88]. By the time PD patients experience the onset of clinical symptoms, they have typically lost more than 80% of dopaminergic neurons. At present, there is no cure for PD, and current management is limited to supportive care that partially alleviates disease signs and symptoms and treatment that does not slow or halt disease progression. Pharmacological treatment has been focused on restoring dopaminergic neurotransmission, because dopamine is an essential CNS neurotransmitter and peripheral chemical messenger.

For example, the dopamine precursor L-3,4-dihydroxyphenylalanine (L-DOPA) remains the gold standard for the treatment of PD. Other non-dopaminergic treatment strategies, including drugs targeting adenosine, adrenergic, and glutamate and serotonin receptors, as well as anti-inflammatories, iron chelators, glucagon-like peptide (GLP)-1 agonists, calcium channel blockers, neurotrophic factors, and gene therapies, have also been developed recently [89]. However, the main limitation remains the development of a carrier suitable for a variety of drugs. In addition, systems involving sustained release of specific drugs also show beneficial pharmacological effects [90]. Therefore, the development of effective and efficient specific drug delivery carriers poses a challenge, and the polymeric nanoparticles are an attractive alternative.

Pahuja et al. have developed dopamine-loaded PLGA nanoparticles (DA NPs) (size: 120 nm; zeta potential: −3 mV) that improved animal behavior, without showing any signs of cardiac-related alterations or sudden changes in the brain. DA NPs could cross the BBB mainly in the SN and striatum (PD-altered regions) of 6-hydroxydopamine (6-OHDA) mice. Furthermore, a slow and controlled release of DA from NPs as well as a reduced plasma clearance were observed within 6 h after injection. DA NPs could increase the DA level and its metabolites homovanillic acid (HVA) and dihydroxyphenylacetic acid (DOPAC) to levels similar to those of control mice. This study demonstrates that DA NPs can prevent toxicity associated with large amount of dopamine and can provide a novel strategy to combat PD [91].

The PBCA nanoparticles coated with PS 80 proved to be an effective delivery vector system of nerve growth factor (NGF) [92]. The success of the NGF transport to the brain was determined by measuring the protein concentration in the brain of

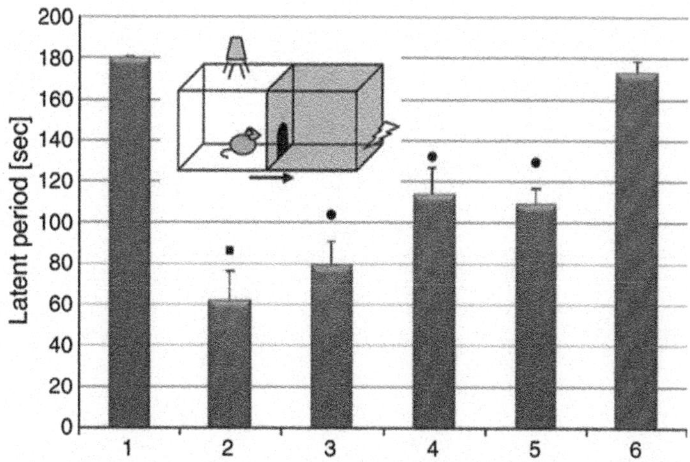


**Fig. 9.16** Using the following formulations, the average incubation period in the passive avoidance reflex test (PAR test): *1*, 0.9% NaCl solution intravenously (passive control); *2*, scopolamine (1.5 mg/kg, s.c.) (control of amnesia); *3*, scopolamine (1.5 mg/kg, s.c.) and NGF solution (5 μg/mouse, i.v.); *4*, scopolamine (1.5 mg/kg, s.c.) and NGF in 1% PS 80 solution (5 μg/mouse, i.v.); *5*, scopolamine (1.5 mg/kg, s.c.) and NGF adsorbed on PBCA nanoparticles (5 μg/mouse, i.v.); *6*, scopolamine (1.5 mg/kg, s.c.) and NGF adsorbed on PBCA nanoparticles coated with PS 80 (5 μg mouse, i.v.). Significance levels: ▪ statistically reliable difference from groups 1 and 6 ($p \leq 0.01$); ● statistically reliable difference from groups 1 and 6 ($p \leq 0.05$)

mice, reversing a scopolamine-induced amnesia and in many Parkinson's disease models. The passive avoidance reflex (PAR) test showed that administration of NGF intravenously adsorbed on PS 80-coated PBCA nanoparticles was successful in reversing scopolamine-induced amnesia and improving the recognition and memory in mice model of acute amnesia (Fig. 9.16). In addition, this preparation showed a significant reduction in the underlying symptoms of Parkinsonism (oligokinesia, rigidity, or tremor). The results indicated that PS 80-coated PBCA nanoparticles are an effective carrier system for NGF transport to the CNS after intravenous administration and may improve the treatment of age-related neurodegenerative diseases [93].

In addition, excessive iron deposition in the brain often leads to the damage and necrosis of dopaminergic neurons in the substantia nigra associated with oxidative stress, which has been reported to be one of the major risk factors in PD. The iron chelation therapy using deferoxamine (DFO) may suppress this nigrostriatal degeneration and prevent PD. However, DFO had shown a very short half-life and hardly penetrated the BBB in vivo. Therefore, the development of DFO formulations for safe and efficient intracerebral drug delivery is very meaningful. You et al. developed a polymer nanoparticle conjugated with brain-targeting peptide rabies virus glycoprotein (RVG) 29, which could deliver DFO into the brain (Fig. 9.17). The nanoparticle system may penetrate the BBB via specific receptor-mediated endocytosis triggered by RVG29 peptide. Administration of these polymeric nanoparticles

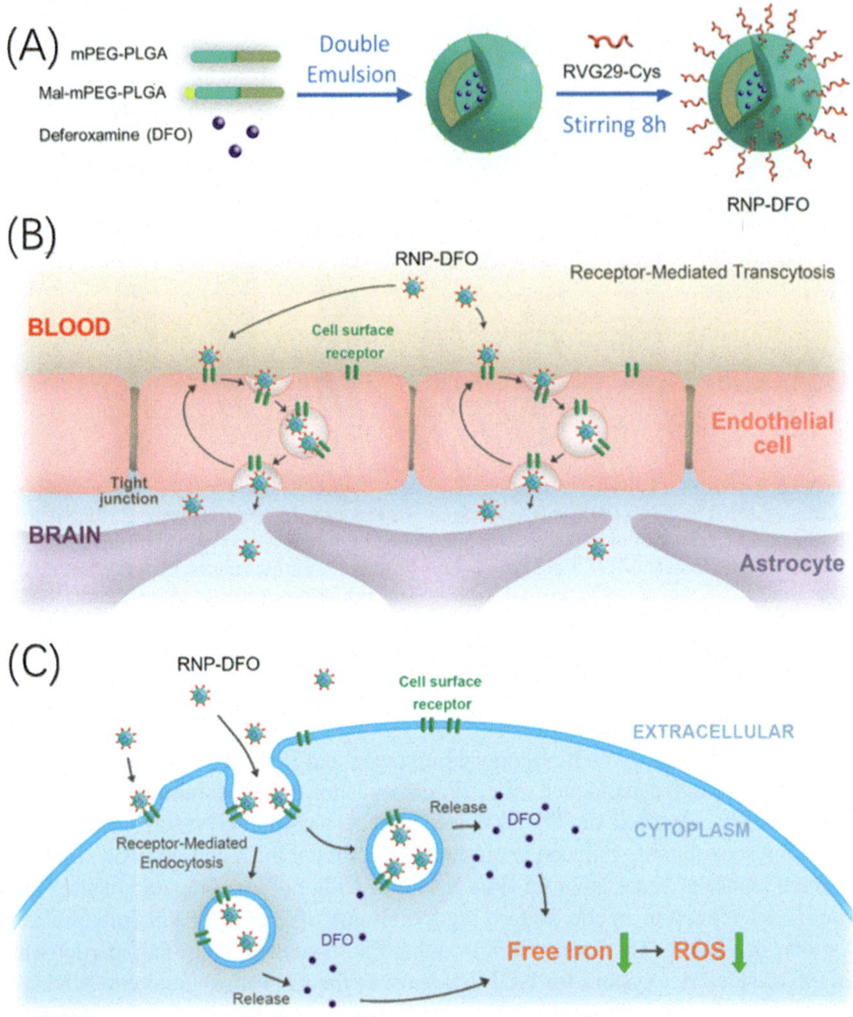


**Fig. 9.17** (**a**) Schematic of the synthesis of RNP-DFO nanoparticle. Polymer nanoparticles were produced by a double emulsion method, and RVG29-Cys was attached to the surface of nanoparticle by a maleimide-thiol coupling reaction. (**b**) Schematic of the proposed mechanism of RNP-DFO to penetrate the BBB. RVG29 peptide binds to cell surface receptors on brain endothelial cells that form the BBB, which are further mediated by nanoparticles across the cell layer of transcytosis. (**c**) Proposed role of RNP-DFO in iron-loaded neuronal cells. Internalization of RNP-DFO allows the removal of excess cellular iron via the released DFO and reduces iron-related oxidative stress

could effectively reduce iron content and oxidative stress levels in the substantia nigra and striatum of PD mice and significantly reduce the damage of dopaminergic neuron and reverse their neurobehavioral defects, causing negligible side effects to the brain and other organs [94].

RNA interference (RNAi)-based strategies that mediate target gene-specific knockdown by administration of small interfering RNAs (siRNAs) could be used to treat neurodegenerative diseases that are currently incurable, such as PD. However, the inefficient delivery of siRNA to neurons hinders application of RNAi in vivo. Helmschrodt et al. developed a 4~12 kDa-branched poly (ethylenimine) (PEI) F25-LMW which has excellent transfection efficiency and can deliver siRNA in vivo. The results suggested that siRNA complexed with this PEI was widely distributed in the lumbar spinal cord in the CNS after a single intracerebroventricular infusion. siRNA against α-synuclein (SNCA), a presynaptic protein aggregating in PD, was complexed with PEI and injected into the lateral ventricle in mice overexpressing human wild-type SNCA. Five days after a single injection of 0.75 mg, PEI/siRNA, SNCA mRNA expression in the striatum was reduced by 65%, while SNCA protein was reduced by ~50%, without any toxicity and adverse effects. In addition, ependymocytes and brain parenchyma was fully intact, with no immune cell invasion, astrogliosis, or microglial activation. These results supported the efficacy and safety of PEI nanoparticle-mediated delivery of siRNA to the brain for therapeutic intervention [95].

### 9.3.5 Stroke

Stroke was ranked among the leading causes of years of life lost in most regions around the world, accounting for more than 11% of total deaths globally [96]. During the onset of stroke, the brain loses blood supply as a result of bleeding vessel (hemorrhagic stroke) or occlusion of blood vessels due to thrombosis (ischemic stroke) (Fig. 9.18) [97]. In both cases there is a deprivation of oxygen and nutrients, resulting in neuronal cell death and neuroinflammation that can culminate in the loss of neurological functions and ultimately death in patients. Cerebral ischemia, accounting for 80~85% of stroke cases [98], is primarily explored in this section.

The occurrence of cerebral ischemia actually occurs after cerebral vessel occlusion by a clot (thrombus or embolus). Vascular obstruction leads to a decrease in downstream cerebral blood flow (CBF), and the resulting lack of blood flow rapidly triggers an ischemic cascade involving neuronal cell death and neuronal inflammation [99]. Neuroinflammation is thought to be closely related to this neuronal injury, as well as the acute and delayed downstream effects of cerebral infarction, and a key hallmark of the pathophysiology of this condition is BBB destruction.

There are usually two stages of BBB disorder after cerebral ischemia: (1) an initial and acute destruction of the BBB 3–5 h after the ischemic event and (2) a widespread increase in BBB permeability at 48 h, likely as a result of cerebral ischemia reperfusion, accompanied by further expansion of the cerebral infarct size [100]. Specifically, it contributes to the reopening of BBB which is attributed to endothelium activation, ROS production, leukocyte recruitment, cytokine production, and edema formation [101].

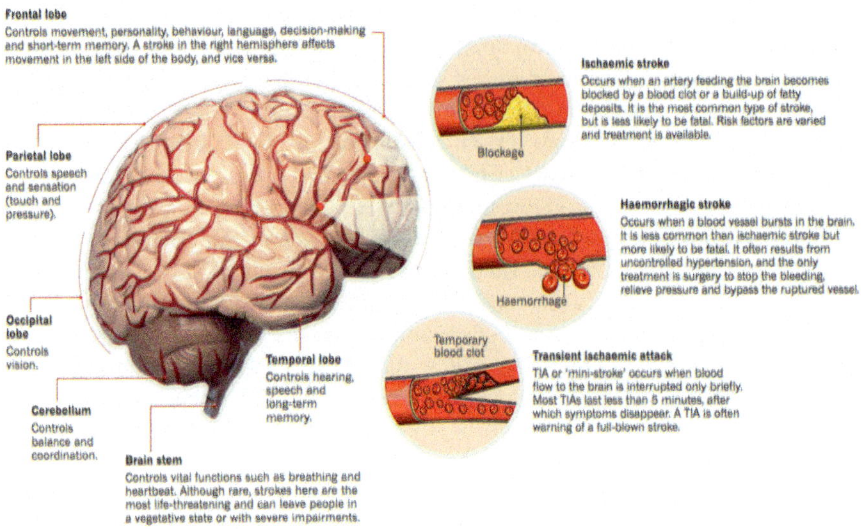


**Fig. 9.18** Schematic overview of the pathways that can lead to stroke and the regions of the brain that can be affected as a result [97]

Drug delivery in cases of stroke should take into consideration the impaired tight junctions of endothelial cells (reopening of BBB) and the initial and late opening of the BBB. The leaky BBB or some receptors high-expressed on the luminal side of endothelial cells may increase the rate of NPs to cross the BBB. Therefore, the BBB itself can be a promising target to improve drug delivery to the ischemic brain.

Polymeric nanoparticles have been used to deliver neuroprotective drugs that in their free forms just pass the BBB in very small amounts and are rapidly cleared by the reticuloendothelial system [102]. For example, Karatas et al. reported a chitosan nanoparticle that modified with the transferrin receptor antibodies and loaded with the specific caspase-3 inhibitors (ZDEVD-FMK), showing promising results. This nanoparticle was capable of bypassing the BBB (peak levels after 75 min), and significantly reduced infarct volume (40~45%) and neurological deficits induced by ischemia in a mice model of stroke via reducing caspase-3 activity [103]. Moreover, these NPs were able to further decrease the infarct volume and to improve the motor function deficit scores of middle cerebral artery occlusion (MCAO) mice when loaded with both Z-DEVD-FMK and basic fibroblast growth factor (bFGF) providing a 3 h therapeutic window [104].

As everyone knows, in the brain after the cerebral ischemia reperfusion, huge toxic reactive oxygen species (ROS) in the ischemic site are upregulated, resulting in neuronal damage [105]. However, from the point of view of drug delivery design, the increase of the ROS level in the ischemic site may be used as an intelligent sensitive trigger to control drug release, which could be applied to the development

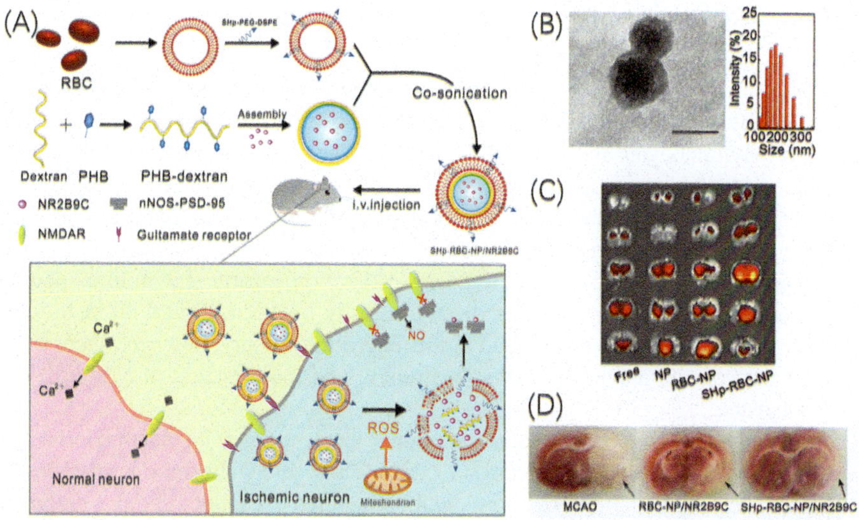


**Fig. 9.19** (**a**) Schematic of the SHp-RBC-NP/NR2B9C. After intravenous administration, the SHp-RBC-NP/NR2B9C could prolong the life cycle with the RBC mimetic properties and then target the ischemic brain through stroke homing peptide-mediated transcytosis. When internalized as ischemic neurons, the NR2B9C is released from the PHB-dextran polymer nanoparticles due to the high levels of intracellular ROS, and then PMD-95 selectively destroys NMDAR to prevent the excessive production of nitric oxide (toxic signaling agent). (**b**) TEM images and hydrodynamic size distribution of RBC-coated nanoparticles. (**c**) Ex vivo fluorescent image of rhodamine-labeled free NR2B9C, NP, RBC-NP, and SHp-RBC-NP in the ischemic brain sections at 2 h, respectively. (**d**) Representative tissue slices showing that RBC-NP/NR2B9C and SHp-RBC-NP/NR2B9C groups can significantly reduce the infarct volume; the arrows indicate the infarct region as observed

of drug delivery systems. In view of this, Lv et al. developed a bioengineered ROS-response nanoparticle for stroke-specific delivery of a neuroprotective agent, NR2B9C, for the treatment of ischemic brain injury (Fig. 9.19). The nanoparticle consisted of a dextran polymer core which was conjugated with ROS-responsive boronic ester and a red blood cell (RBC) membrane shell modified with a stroke homing peptide (SHp). Thus, these targeted "core-shell" nanoparticles (SHp-RBC-NP) could control the release of NR2B9C triggered by a high intracellular ROS in ischemic neurons after reaching to the ischemic brain tissues. The potential of the SHp-RBC-NP for the treatment of ischemic stroke was systematically evaluated both in vitro and in rat MCAO models. In vitro experiments showed that SHp-RBC-NP had strong protective effects on cytotoxicity induced by glutamate in PC-12 cells. Furthermore, in vivo pharmacokinetic (PK) and pharmacodynamic (PD) experiments proved that the bioengineered nanoparticles can significantly prolong the systemic circulation of NR2B9C, enhance the active targeting of ischemic region of MCAO rats, and reduce ischemic brain injury [106].

### 9.3.6 *Multiple Sclerosis*

Multiple sclerosis (MS) is a chronic neuroinflammatory disease characterized most strikingly by the demyelination of neurons and the CNS invasion of immune cells, especially lymphocytes and macrophages [107]. The causes of the MS have not yet been completely determined. However, it is thought to involve a relationship between environmental factors (such as low levels of vitamin D), smoking and genetic factors (including dangerous gene variants, such as the well-known HLADR2 allele) involved [108]. By one estimation, nearly 2.3 million people worldwide will be living with this disease, mostly between 20 and 40 years old. Although the disease was recognized in the nineteenth century, however, there were limited data on the diagnosis, pathophysiology, and management of MS until the past 50 years.

Nanomedicine is a strategy that integrates therapeutics with nanomaterials to develop new treatments with enhanced efficacy and safety. In these systems, it is effective for PEGylation of core-shell structure. Schmidt et al. prepared prednisolone-loaded PEG-liposomes with a diameter of 90~100 nm for the treatment of MS [109]. In the experimental autoimmune encephalomyelitis (MS animal model) mice injected intravenously, the liposomes were observed to accumulate within 2 h at high levels in the central nervous system. Treatment with drug-loaded liposomes restored BBB integrity and reduced inflammation and macrophage infiltration. This treatment was considered superior to the administration of free glucocorticosteroids, a traditional treatment for MS.

## References

1. Saraiva C, Praca C, Ferreira R, Santos T, Ferreira L, Bernardino L. Nanoparticle-mediated brain drug delivery: overcoming blood-brain barrier to treat neurodegenerative diseases. J Control Release. 2016;235:34–47.
2. Olivier JC. Drug transport to brain with targeted nanoparticles. NeuroRx. 2005;2(1):108–19.
3. Chen Y, Liu L. Modern methods for delivery of drugs across the blood-brain barrier. Adv Drug Deliv Rev. 2012;64(7):640–65.
4. Garcia-Garcia E, Andrieux K, Gil S, Couvreur P. Colloidal carriers and blood-brain barrier (BBB) translocation: a way to deliver drugs to the brain? Int J Pharm. 2005;298(2):274–92.
5. Aktas Y, Yemisci M, Andrieux K, Gursoy RN, Alonso MJ, Fernandez-Megia E, Novoa-Carballal R, Quinoa E, Riguera R, Sargon MF, et al. Development and brain delivery of chitosan-PEG nanoparticles functionalized with the monoclonal antibody OX26. Bioconjug Chem. 2005;16(6):1503–11.
6. Barbu E, Molnar E, Tsibouklis J, Gorecki DC. The potential for nanoparticle-based drug delivery to the brain: overcoming the blood-brain barrier. Expert Opin Drug Deliv. 2009;6(6):553–65.
7. Wohlfart S, Gelperina S, Kreuter J. Transport of drugs across the blood-brain barrier by nanoparticles. J Control Release. 2012;161(2):264–73.
8. Chun-Xiong Zheng YZ, Liu Y. Recent advances in self-assembled nano-therapeutics. Chinese J. Polym. Sci. 2018;36:322–46.

9. Zhao Z, Nelson AR, Betsholtz C, Zlokovic BV. Establishment and dysfunction of the blood-brain barrier. Cell. 2015;163(5):1064–78.
10. Koo YE, Reddy GR, Bhojani M, Schneider R, Philbert MA, Rehemtulla A, Ross BD, Kopelman R. Brain cancer diagnosis and therapy with nanoplatforms. Adv Drug Deliv Rev. 2006;58(14):1556–77.
11. Furtado D, Bjornmalm M, Ayton S, Bush AI, Kempe K, Caruso F. Overcoming the blood-brain barrier: the role of nanomaterials in treating neurological diseases. Adv Mater. 2018;30(46):e1801362.
12. Letchford K, Burt H. A review of the formation and classification of amphiphilic block copolymer nanoparticulate structures: micelles, nanospheres, nanocapsules and polymersomes. Eur J Pharm Biopharm. 2007;65(3):259–69.
13. Discher DE, Ahmed F. Polymersomes. Annu Rev Biomed Eng. 2006;8:323–41.
14. Ou H, Cheng T, Zhang Y, Liu J, Ding Y, Zhen J, Shen W, Xu Y, Yang W, Niu P, et al. Surface-adaptive zwitterionic nanoparticles for prolonged blood circulation time and enhanced cellular uptake in tumor cells. Acta Biomater. 2018;65:339–48.
15. Chen Y, Liang G. Enzymatic self-assembly of nanostructures for theranostics. Theranostics. 2012;2(2):139–47.
16. Liu XM, Pramoda KP, Yang YY, Chow SY, He C. Cholesteryl-grafted functional amphiphilic poly(N-isopropylacrylamide-co-N-hydroxylmethylacrylamide): synthesis, temperature-sensitivity, self-assembly and encapsulation of a hydrophobic agent. Biomaterials. 2004;25 (13):2619–28.
17. Liu L, Guo K, Lu J, Venkatraman SS, Luo D, Ng KC, Ling EA, Moochhala S, Yang YY. Biologically active core/shell nanoparticles self-assembled from cholesterol-terminated PEG-TAT for drug delivery across the blood-brain barrier. Biomaterials. 2008;29 (10):1509–17.
18. Shao K, Huang R, Li J, Han L, Ye L, Lou J, Jiang C. Angiopep-2 modified PE-PEG based polymeric micelles for amphotericin B delivery targeted to the brain. J Control Release. 2010;147(1):118–26.
19. Demeule M, Regina A, Che C, Poirier J, Nguyen T, Gabathuler R, Castaigne JP, Beliveau R. Identification and design of peptides as a new drug delivery system for the brain. J Pharmacol Exp Ther. 2008;324(3):1064–72.
20. Bolard J, Legrand P, Heitz F, Cybulska B. One-sided action of amphotericin B on cholesterol-containing membranes is determined by its self-association in the medium. Biochemistry. 1991;30(23):5707–15.
21. Pardridge WM. Recent developments in peptide drug delivery to the brain. Pharmacol Toxicol. 1992;71(1):3–10.
22. Hatakeyama H, Akita H, Ishida E, Hashimoto K, Kobayashi H, Aoki T, Yasuda J, Obata K, Kikuchi H, Ishida T, et al. Tumor targeting of doxorubicin by anti-MT1-MMP antibody-modified PEG liposomes. Int J Pharm. 2007;342(1-2):194–200.
23. Li X, Ding L, Xu Y, Wang Y, Ping Q. Targeted delivery of doxorubicin using stealth liposomes modified with transferrin. Int J Pharm. 2009;373(1-2):116–23.
24. Liu Y, Li J, Lu Y. Enzyme therapeutics for systemic detoxification[J]. Advanced drug delivery reviews. 2015;90:24–39.
25. Coombes AG, Scholes PD, Davies MC, Illum L, Davis SS. Resorbable polymeric microspheres for drug delivery–production and simultaneous surface modification using PEO-PPO surfactants. Biomaterials. 1994;15(9):673–80.
26. Chen Y, Dalwadi G, Benson HA. Drug delivery across the blood-brain barrier. Curr Drug Deliv. 2004;1(4):361–76.
27. Jain KK. Nanobiotechnology-based strategies for crossing the blood-brain barrier. Nanomedicine (Lond). 2012;7(8):1225–33.
28. Agnihotri SA, Mallikarjuna NN, Aminabhavi TM. Recent advances on chitosan-based micro- and nanoparticles in drug delivery. J Control Release. 2004;100(1):5–28.
29. Gerweck LE, Seetharaman K. Cellular pH gradient in tumor versus normal tissue: potential exploitation for the treatment of cancer. Cancer Res. 1996;56(6):1194–8.

30. Mura S, Nicolas J, Couvreur P. Stimuli-responsive nanocarriers for drug delivery. Nat Mater. 2013;12(11):991–1003.
31. Yemisci M, Gursoy-Ozdemir Y, Caban S, Bodur E, Capan Y, Dalkara T. Transport of a caspase inhibitor across the blood-brain barrier by chitosan nanoparticles. Methods Enzymol. 2012;508:253–69.
32. Wang S, Jiang T, Ma M, Hu Y, Zhang J. Preparation and evaluation of anti-neuroexcitation peptide (ANEP) loaded N-trimethyl chitosan chloride nanoparticles for brain-targeting. Int J Pharm. 2010;386(1-2):249–55.
33. Yi X, Manickam DS, Brynskikh A, Kabanov AV. Agile delivery of protein therapeutics to CNS. J Control Release. 2014;190:637–63.
34. Govender T, Stolnik S, Garnett MC, Illum L, Davis SS. PLGA nanoparticles prepared by nanoprecipitation: drug loading and release studies of a water soluble drug. J Control Release. 1999;57(2):171–85.
35. Cheng J, Teply BA, Sherifi I, Sung J, Luther G, Gu FX, Levy-Nissenbaum E, Radovic-Moreno AF, Langer R, Farokhzad OC. Formulation of functionalized PLGA-PEG nanoparticles for in vivo targeted drug delivery. Biomaterials. 2007;28(5):869–76.
36. Zhang X, Chen G, Wen L, Yang F, Shao AL, Li X, Long W, Mu L. Novel multiple agents loaded PLGA nanoparticles for brain delivery via inner ear administration: in vitro and in vivo evaluation. Eur J Pharm Sci. 2013;48(4-5):595–603.
37. Gelperina S, Maksimenko O, Khalansky A, Vanchugova L, Shipulo E, Abbasova K, Berdiev R, Wohlfart S, Chepurnova N, Kreuter J. Drug delivery to the brain using surfactant-coated poly(lactide-co-glycolide) nanoparticles: influence of the formulation parameters. Eur J Pharm Biopharm. 2010;74(2):157–63.
38. Hekmatara T, Gelperina S, Vogel V, Yang SR, Kreuter J. Encapsulation of water-insoluble drugs in poly(butyl cyanoacrylate) nanoparticles. J Nanosci Nanotechnol. 2009;9(8):5091–8.
39. Petri B, Bootz A, Khalansky A, Hekmatara T, Muller R, Uhl R, Kreuter J, Gelperina S. Chemotherapy of brain tumour using doxorubicin bound to surfactant-coated poly(butyl cyanoacrylate) nanoparticles: revisiting the role of surfactants. J Control Release. 2007;117(1):51–8.
40. Schroeder U, Sommerfeld P, Ulrich S, Sabel BA. Nanoparticle technology for delivery of drugs across the blood-brain barrier. J Pharm Sci. 1998;87(11):1305–7.
41. Wilson B, Samanta MK, Santhi K, Kumar KP, Paramakrishnan N, Suresh B. Poly (n-butylcyanoacrylate) nanoparticles coated with polysorbate 80 for the targeted delivery of rivastigmine into the brain to treat Alzheimer's disease. Brain Res. 2008;1200:159–68.
42. Kolter M, Ott M, Hauer C, Reimold I, Fricker G. Nanotoxicity of poly(n-butylcyano-acrylate) nanoparticles at the blood-brain barrier, in human whole blood and in vivo. J Control Release. 2015;197:165–79.
43. Gu Z, Biswas A, Zhao M, Tang Y. Tailoring nanocarriers for intracellular protein delivery. Chem Soc Rev. 2011;40(7):3638–55.
44. Zhang L, Liu Y, Liu G, et al. Prolonging the plasma circulation of proteins by nano-encapsulation with phosphorylcholine-based polymer. Nano Research. 2016;9(8):2424–32.
45. Yan M, Du J, Gu Z, Liang M, Hu Y, Zhang W, Priceman S, Wu L, Zhou ZH, Liu Z, et al. A novel intracellular protein delivery platform based on single-protein nanocapsules. Nat Nanotechnol. 2010;5(1):48–53.
46. Liu C, Wen J, Meng Y, Zhang K, Zhu J, Ren Y, Qian X, Yuan X, Lu Y, Kang C. Efficient delivery of therapeutic miRNA nanocapsules for tumor suppression. Adv Mater. 2015;27(2):292–7.
47. Parveen S, Misra R, Sahoo SK. Nanoparticles: a boon to drug delivery, therapeutics, diagnostics and imaging. Nanomedicine. 2012;8(2):147–66.
48. Dhanikula RS, Hammady T, Hildgen P. On the mechanism and dynamics of uptake and permeation of polyether-copolyester dendrimers across an in vitro blood-brain barrier model. J Pharm Sci. 2009;98(10):3748–60.
49. Tekade RK, Dutta T, Gajbhiye V, Jain NK. Exploring dendrimer towards dual drug delivery: pH responsive simultaneous drug-release kinetics. J Microencapsul. 2009;26(4):287–96.

50. Albertazzi L, Gherardini L, Brondi M, Sulis Sato S, Bifone A, Pizzorusso T, Ratto GM, Bardi G. In vivo distribution and toxicity of PAMAM dendrimers in the central nervous system depend on their surface chemistry. Mol Pharm. 2013;10(1):249–60.
51. Ke W, Shao K, Huang R, Han L, Liu Y, Li J, Kuang Y, Ye L, Lou J, Jiang C. Gene delivery targeted to the brain using an Angiopep-conjugated polyethyleneglycol-modified polyamidoamine dendrimer. Biomaterials. 2009;30(36):6976–85.
52. Huang R, Ke W, Han L, Liu Y, Shao K, Ye L, Lou J, Jiang C, Pei Y. Brain-targeting mechanisms of lactoferrin-modified DNA-loaded nanoparticles. J Cereb Blood Flow Metab. 2009;29(12):1914–23.
53. Huang RQ, Qu YH, Ke WL, Zhu JH, Pei YY, Jiang C. Efficient gene delivery targeted to the brain using a transferrin-conjugated polyethyleneglycol-modified polyamidoamine dendrimer. Faseb j. 2007;21(4):1117–25.
54. Vinogradov SV. Nanogels in the race for drug delivery. Nanomedicine (Lond). 2010;5 (2):165–8.
55. Wagner V, Dullaart A, Bock AK, Zweck A. The emerging nanomedicine landscape. Nat Biotechnol. 2006;24(10):1211–7.
56. Hamidi M, Azadi A, Rafiei P. Hydrogel nanoparticles in drug delivery. Adv Drug Deliv Rev. 2008;60(15):1638–49.
57. Azadi A, Hamidi M, Rouini MR. Methotrexate-loaded chitosan nanogels as 'Trojan Horses' for drug delivery to brain: preparation and in vitro/in vivo characterization. Int J Biol Macromol. 2013;62:523–30.
58. Vinogradov SV, Batrakova EV, Kabanov AV. Nanogels for oligonucleotide delivery to the brain. Bioconjug Chem. 2004;15(1):50–60.
59. Meyers CA, Lamborn KR, Prados MD. In reference to lamborn et Al. (Neuro-oncology. 2008;10:162–170). Neuro Oncol. 2008;10(6):1171–2.
60. Gilbertson RJ. Medulloblastoma: signalling a change in treatment. Lancet Oncol. 2004;5 (4):209–18.
61. Pardridge WM. Non-invasive drug delivery to the human brain using endogenous blood-brain barrier transport systems. Pharm Sci Technolo Today. 1999;2(2):49–59.
62. Hekmatara T, Bernreuther C, Khalansky AS, Theisen A, Weissenberger J, Matschke J, Gelperina S, Kreuter J, Glatzel M. Efficient systemic therapy of rat glioblastoma by nanoparticle-bound doxorubicin is due to antiangiogenic effects. Clin Neuropathol. 2009;28 (3):153–64.
63. Beduneau A, Saulnier P, Benoit JP. Active targeting of brain tumors using nanocarriers. Biomaterials. 2007;28(33):4947–67.
64. Su JL, Lai KP, Chen CA, Yang CY, Chen PS, Chang CC, Chou CH, Hu CL, Kuo ML, Hsieh CY, et al. A novel peptide specifically binding to interleukin-6 receptor (gp80) inhibits angiogenesis and tumor growth. Cancer Res. 2005;65(11):4827–35.
65. Wei Shi XC, Shi J, Chen J, Yi W. Overcoming the blood–brain barrier for glioma-targeted therapy based on an interleukin-6 receptor-mediated micelle system. RSC Adv. 2017;7:27162–9.
66. Ksendzovsky A, Feinstein D, Zengou R, Sharp A, Polak P, Lichtor T, Glick RP. Investigation of immunosuppressive mechanisms in a mouse glioma model. J Neurooncol. 2009;93 (1):107–14.
67. Qiao C, Yang J, Shen Q, Liu R, Li Y, Shi Y, Chen J, Shen Y, Xiao Z, Weng J, et al. Traceable nanoparticles with dual targeting and ROS response for RNAi-based immunochemotherapy of intracranial Glioblastoma treatment. Adv Mater. 2018;30(18):e1705054.
68. Goyal L, Muzumdar MD, Zhu AX. Targeting the HGF/c-MET pathway in hepatocellular carcinoma. Clin Cancer Res. 2013;19(9):2310–8.
69. Wu Y, Fan Q, Zeng F, Zhu J, Chen J, Fan D, Li X, Duan W, Guo Q, Cao Z, et al. Peptide-functionalized nanoinhibitor restrains brain tumor growth by Abrogating Mesenchymal-Epithelial Transition factor (MET) signaling. Nano Lett. 2018;18:5488–98.
70. Hardy J, Selkoe DJ. The amyloid hypothesis of Alzheimer's disease: progress and problems on the road to therapeutics. Science. 2002;297(5580):353–6.

71. Anand R, Gill KD, Mahdi AA. Therapeutics of Alzheimer's disease: past, present and future. Neuropharmacology. 2014;76 Pt A:27–50.
72. Horwich AL. Molecular chaperones in cellular protein folding: the birth of a field. Cell. 2014;157(2):285–8.
73. Huang F, Wang J, Qu A, Shen L, Liu J, Liu J, Zhang Z, An Y, Shi L. Maintenance of amyloid beta peptide homeostasis by artificial chaperones based on mixed-shell polymeric micelles. Angew Chem Int Ed Engl. 2014;53(34):8985–90.
74. Vonghia L, Leggio L, Ferrulli A, Bertini M, Gasbarrini G, Addolorato G. Acute alcohol intoxication. Eur J Intern Med. 2008;19(8):561–7.
75. Liu LJ, Wang W, Zhong Z, Lin S, Lu L, Wang YT, Ma DL, Leung CH. Inhibition of TLR1/2 dimerization by enantiomers of metal complexes. Chem Commun (Camb). 2016;52 (83):12278–81.
76. Luo Q, Lin YX, Yang PP, Wang Y, Qi GB, Qiao ZY, Li BN, Zhang K, Zhang JP, Wang L, et al. A self-destructive nanosweeper that captures and clears amyloid beta-peptides. Nat Commun. 2018;9(1):1802.
77. O'Brien RJ, Wong PC. Amyloid precursor protein processing and Alzheimer's disease. Annu Rev Neurosci. 2011;34:185–204.
78. Brunden KR, Trojanowski JQ, Lee VM. Advances in tau-focused drug discovery for Alzheimer's disease and related tauopathies. Nat Rev Drug Discov. 2009;8(10):783–93.
79. Liu Y, An S, Li J, Kuang Y, He X, Guo Y, Ma H, Zhang Y, Ji B, Jiang C. Brain-targeted co-delivery of therapeutic gene and peptide by multifunctional nanoparticles in Alzheimer's disease mice. Biomaterials. 2016;80:33–45.
80. Liu Z, Gao X, Kang T, Jiang M, Miao D, Gu G, Hu Q, Song Q, Yao L, Tu Y, et al. B6 peptide-modified PEG-PLA nanoparticles for enhanced brain delivery of neuroprotective peptide. Bioconjug Chem. 2013;24(6):997–1007.
81. Vinters HV, Gilbert JJ. Cerebral amyloid angiopathy: incidence and complications in the aging brain. II. The distribution of amyloid vascular changes. Stroke. 1983;14(6):924–8.
82. Cordonnier C, van der Flier WM. Brain microbleeds and Alzheimer's disease: innocent observation or key player? Brain. 2011;134(Pt 2):335–44.
83. Agyare EK, Jaruszewski KM, Curran GL, Rosenberg JT, Grant SC, Lowe VJ, Ramakrishnan S, Paravastu AK, Poduslo JF, Kandimalla KK. Engineering theranostic nanovehicles capable of targeting cerebrovascular amyloid deposits. J Control Release. 2014;185:121–9.
84. de Lau LM, Breteler MM. Epidemiology of Parkinson's disease. Lancet Neurol. 2006;5 (6):525–35.
85. Desai BS, Monahan AJ, Carvey PM, Hendey B. Blood-brain barrier pathology in Alzheimer's and Parkinson's disease: implications for drug therapy. Cell Transplant. 2007;16(3):285–99.
86. Reale M, Iarlori C, Thomas A, Gambi D, Perfetti B, Di Nicola M, Onofrj M. Peripheral cytokines profile in Parkinson's disease. Brain Behav Immun. 2009;23(1):55–63.
87. Braak H, Del Tredici K, Rub U, de Vos RA, Jansen Steur EN, Braak E. Staging of brain pathology related to sporadic Parkinson's disease. Neurobiol Aging. 2003;24(2):197–211.
88. Youdim MB, Bakhle YS. Monoamine oxidase: isoforms and inhibitors in Parkinson's disease and depressive illness. Br J Pharmacol. 2006;147(Suppl 1):S287–96.
89. Kalia LV, Brotchie JM, Fox SH. Novel nondopaminergic targets for motor features of Parkinson's disease: review of recent trials. Mov Disord. 2013;28(2):131–44.
90. Stocchi F. Continuous dopaminergic stimulation and novel formulations of dopamine agonists. J Neurol. 2011;258(Suppl 2):S316–22.
91. Pahuja R, Seth K, Shukla A, Shukla RK, Bhatnagar P, Chauhan LK, Saxena PN, Arun J, Chaudhari BP, Patel DK, et al. Trans-blood brain barrier delivery of dopamine-loaded nanoparticles reverses functional deficits in parkinsonian rats. ACS Nano. 2015;9(5):4850–71.
92. Kurakhmaeva KB, Djindjikhashvili IA, Petrov VE, Balabanyan VU, Voronina TA, Trofimov SS, Kreuter J, Gelperina S, Begley D, Alyautdin RN. Brain targeting of nerve growth factor using poly(butyl cyanoacrylate) nanoparticles. J Drug Target. 2009;17(8):564–74.

93. Kurakhmaeva KB, Voronina TA, Kapica IG, Kreuter J, Nerobkova LN, Seredenin SB, Balabanian VY, Alyautdin RN. Antiparkinsonian effect of nerve growth factor adsorbed on polybutylcyanoacrylate nanoparticles coated with polysorbate-80. Bull Exp Biol Med. 2008;145(2):259–62.
94. You L, Wang J, Liu T, Zhang Y, Han X, Wang T, Guo S, Dong T, Xu J, Anderson GJ, et al. Targeted brain delivery of Rabies Virus Glycoprotein 29-modified deferoxamine-loaded nanoparticles reverses functional deficits in Parkinsonian Mice. ACS Nano. 2018;12 (5):4123–39.
95. Helmschrodt C, Hobel S, Schoniger S, Bauer A, Bonicelli J, Gringmuth M, Fietz SA, Aigner A, Richter A, Richter F. Polyethylenimine nanoparticle-mediated siRNA delivery to reduce alpha-Synuclein expression in a model of Parkinson's disease. Mol Ther Nucleic Acids. 2017;9:57–68.
96. Reddy KS. Global burden of disease study 2015 provides GPS for global health 2030. Lancet. 2016;388(10053):1448–9.
97. Corbyn Z. Statistics: a growing global burden. Nature. 2014;510(7506):S2–3.
98. Krol S, Macrez R, Docagne F, Defer G, Laurent S, Rahman M, Hajipour MJ, Kehoe PG, Mahmoudi M. Therapeutic benefits from nanoparticles: the potential significance of nanoscience in diseases with compromise to the blood brain barrier. Chem Rev. 2013;113 (3):1877–903.
99. Jin R, Yang G, Li G. Molecular insights and therapeutic targets for blood-brain barrier disruption in ischemic stroke: critical role of matrix metalloproteinases and tissue-type plasminogen activator. Neurobiol Dis. 2010;38(3):376–85.
100. Belayev L, Busto R, Zhao W, Ginsberg MD. Quantitative evaluation of blood-brain barrier permeability following middle cerebral artery occlusion in rats. Brain Res. 1996;739 (1-2):88–96.
101. Warach S, Latour LL. Evidence of reperfusion injury, exacerbated by thrombolytic therapy, in human focal brain ischemia using a novel imaging marker of early blood-brain barrier disruption. Stroke. 2004;35(11 Suppl 1):2659–61.
102. Santos T, Maia J, Agasse F, Xapelli S, Ferreira L, Bernardino L. Nanomedicine boosts neurogenesis: new strategies for brain repair. Integr Biol (Camb). 2012;4(9):973–81.
103. Karatas H, Aktas Y, Gursoy-Ozdemir Y, Bodur E, Yemisci M, Caban S, Vural A, Pinarbasli O, Capan Y, Fernandez-Megia E, et al. A nanomedicine transports a peptide caspase-3 inhibitor across the blood-brain barrier and provides neuroprotection. J Neurosci. 2009;29(44):13761–9.
104. Yemisci M, Caban S, Gursoy-Ozdemir Y, Lule S, Novoa-Carballal R, Riguera R, Fernandez-Megia E, Andrieux K, Couvreur P, Capan Y, et al. Systemically administered brain-targeted nanoparticles transport peptides across the blood-brain barrier and provide neuroprotection. J Cereb Blood Flow Metab. 2015;35(3):469–75.
105. Panagiotou S, Saha S. Therapeutic benefits of nanoparticles in stroke. Front Neurosci. 2015;9:182.
106. Lv W, Xu J, Wang X, Li X, Xu Q, Xin H. Bioengineered Boronic Ester modified Dextran polymer nanoparticles as reactive oxygen species responsive nanocarrier for ischemic stroke treatment. ACS Nano. 2018;21:5417–26.
107. Banks WA. From blood-brain barrier to blood-brain interface: new opportunities for CNS drug delivery. Nat Rev Drug Discov. 2016;15(4):275–92.
108. Brown C. Aetiology: neighbourhood watch. Nature. 2016;540(7631):S4–s6.
109. Schmidt J, Metselaar JM, Wauben MH, Toyka KV, Storm G, Gold R. Drug targeting by long-circulating liposomal glucocorticosteroids increases therapeutic efficacy in a model of multiple sclerosis. Brain 126 Pt. 2003;8:1895–904.

# Chapter 10
# Magnetic Nanomedicine

**M. Zubair Iqbal, Gohar Ijaz Dar, Israt Ali, and Aiguo Wu**

**Abstract** Nanotechnology emerged as a promising field of science with a diversity of applications in energy storage, biotechnology, medicine, sensing, and healthcare monitoring and in each aspect of nature. Owing to the significant characteristics of the smaller size, easy modification, and tunable physical and chemical properties, magnetic nanomaterials have gained potential fame in the nanomedicine field. In terms of treatment and diagnosis, magnetic nanoparticles (MNP) cannot be replaced with any other material. Surface functionalization and coating of ferromagnetic and superparamagnetic nanoparticles not only make them biocompatible but also effective for drug delivery and killing tumor cells. In this book chapter, we highlighted the emerging applications of magnetic nanoparticles, from synthesis to potential applications. Specifically, brief introduction of magnetic nanomaterials and their physical properties is discussed in detail. Further, the facile synthesis methods to prepare MNPs and recent developments in MNPs as magnetic hyperthermia agents, as a drug transporter, and use in magnetic resonance imaging as a contrast agents are also elaborated profoundly.

M. Z. Iqbal
Department of Materials Science and Engineering, Zhejiang Sci-Tech University, Hangzhou, People's Republic of China

Cixi Institute of Biomedical Engineering, CAS Key Laboratory of Magnetic Materials and Devices, & Key Laboratory of Additive Manufacturing Materials of Zhejiang Province, & Division of Functional Materials and Nanodevices, Ningbo Institute of Materials Technology and Engineering, Chinese Academy of Sciences, Ningbo, People's Republic of China

G. I. Dar · A. Wu (✉)
Cixi Institute of Biomedical Engineering, CAS Key Laboratory of Magnetic Materials and Devices, & Key Laboratory of Additive Manufacturing Materials of Zhejiang Province, & Division of Functional Materials and Nanodevices, Ningbo Institute of Materials Technology and Engineering, Chinese Academy of Sciences, Ningbo, People's Republic of China
e-mail: aiguo@nimte.ac.cn

I. Ali
Division of Polymer and Composite Materials, Ningbo Institute of Material Technology and Engineering, Chinese Academy of Sciences, Ningbo, People's Republic of China



X. Xue (ed.), *Nanomedicine in Brain Diseases*,
https://doi.org/10.1007/978-981-13-8731-9_10

**Keywords** Nanomaterials · Magnetic nanoparticles · Magnetic hyperthermia · Drug delivery · Magnetic resonance imaging

## 10.1 Nanomaterials

The concept of nanotechnology is coined by Richard P. Feynman (Nobel Laureate in Physics 1965) approximately 50 years ago. The nanotechnology word was first time used during the American Physical Society (APS) annual meeting which was held at California Institute of Technology (December 1959) [1]. However, researchers have not solidly identified the specific definition of nanomaterials. Generally, scientists are agreed that nanomaterials are partially described by their minute dimensions which can be measured in nanometers [2]. A nanoscale is recognized as one millionth of a millimeter, approximately 1/8000 the diameter of a human hair. Nano-sized phenomena also exist in nature such as the size of the bacterium, virus, antibody, and DNA molecules [3]. Nanoscale materials are the base of nanoscience and nanotechnology. Nanomaterial engineering is a broad and multidisciplinary research area which has been involved rapidly into human usage to comfort their living styles. Nanomaterials have already admirable commercial impacts in the form of nanomaterial-based industrial products and will definitely increase in the future [4].

Nanomaterials can be obtained from many products, such as carbon or different natural minerals. However, nanomaterials are restricted to have at least one dimension less than 100 nm. Special high-resolution microscopes are employed to examine the structural properties of nanomaterials rather than conventional lab microscopes. Michael Faraday presented one of the earlier scientific reports on gold nanoparticles in 1857 [5]. Later, metallic nanopowders were synthesized in the 1960s and 1970s and used for magnetic recording tapes. Nanotechnology took a prosperous consideration in the twentieth century, and scientists are either producing or utilizing nanomaterials or nanomaterial-based products [6].

Figure 10.1 shows the correlation between nano-sized materials and their comparison in living beings [7]. Nanomaterials can be in the form of single, aggregated, or fused forms with spherical, irregular, and tubular shapes. However, researchers have been paying attention to prepare many beautiful structures to change in their electrical, optical, mechanical, and magnetic properties for different purposes [8, 9]. The properties of the nano-sized materials are significantly different than the bulk materials (macro and micro) owing to their large fraction of surface atoms, spatial confinement, high surface energy, and reduced imperfections. Due to these attractive properties, nanomaterials are having a broad range of applications in the field of environmental protection, electronics, energy storage, agriculture, home appliances, food industry, and medicines [10, 11].

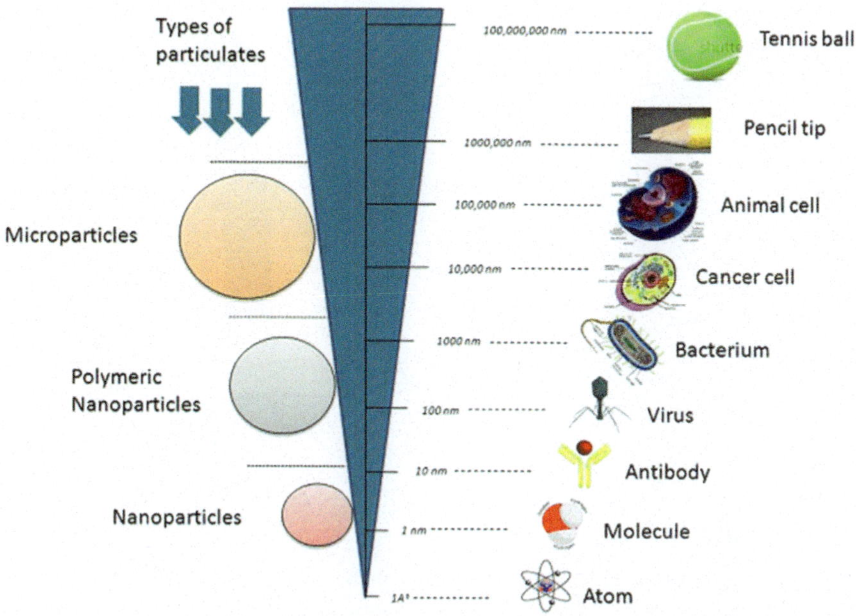


**Fig. 10.1** Scale representing the size between particles and main living units. (Reprinted with permission from reference [7])

## 10.2 Magnetic Nanoparticles

Nanoparticles of magnetic materials have been attaining significant attraction because of their promising applications in biomedicine, gyroscopic sensors, magnetic storage, catalysis, neural stimulation, and spintronics [12, 13]. In the past two decades, MNPs having one dimension in nanoscale are the imperative class of nanomaterials which has been growing rapidly, and their realm is expanding for the science and technology. The controllable size of the MNPs presents unique qualities in nanomedicine, and their sizes are comparable biological system such as a gene (2 nm wide and 10–100 nm long), a protein (5–50 nm), a virus (20–450 nm), and a cell (10–100 μm) [14, 15]. These kinds of special features are very hard to observe in bulk materials. Bulk magnetic materials show high saturation magnetism, but they cause intense toxicity which limits their use in clinical application. However, MNPs are considered to be safe and manipulate by external magnetic field. This process is very exciting in the sense of transferring energy with the help of external magnetic field to the MNPs. By this practice, MNPs can be heated and used as hyperthermia agents and chemotherapeutic agents, transporting the attached targeting molecules to the desired place. Therefore, these special physical properties

**Fig. 10.2** Schematic diagram of general spin canting process

of MNPs made them promising applicants in magnetic nanomedicine [16]. The volume of the magnetic core Vm, magnetic moment, and magnetization are important factors of MNPs for magnetic nanomedicine (MNM) and associated applications. The dimensions and shapes of the MNPs are strongly related with the magnetic moment and anisotropy constant. A strong applied magnetic field is favorable for drug/gene delivery-related applications, whereas higher magnetic moment is appropriate for molecular imaging and detection [17, 18].

The number of atoms increases onto the surface of nanoparticle as the size reduces in nanometer range. As a result, magnetic characteristics at the surface vary than the inner core, disturbance of the translational crystal symmetry of the nanoparticle, which leads to increase in the anisotropy constant and decrease in the saturation magnetizations of the MNPs. This effect is called spin-canting effect, and it only influenced on the core of the MNPs but have not been affected by organic coating [19, 20]. If the nanoparticle is divided into two parts such as the core and the surface layer which is considered as the shell, then spin-canting effect can be easily understood as shown in Fig. 10.2. In the presence of external magnetic field, aligned spins are observed in the core and canting spins in the shell layer.

The saturation magnetization of the spherical MNPs can be given [21].

$$M_s = M_{sb}\left(1 - {}^{2\delta}\!/\!_{D}\right)^3 \tag{10.1}$$

where saturation magnetization of the MNPs is denoted by $M_s$, $M_{sb}$ is the saturation magnetization of the bulk materials, $\delta$ is the thickness of the spin canting layer, and D is the diameter of the magnetic core.

The perceived anisotropy constants of MNPs are most of the time greater than their consistent bulk materials because of spin-canting effect.

$$K_{eff} = K_b + \left(\frac{6\Phi}{D}\right)K_s \tag{10.2}$$

where $K_b$ and $K_s$ are the bulk and surface anisotropy constants, respectively [22, 23].

The effective anisotropy including shape and surface anisotropy strongly depends on the shape, and shape anisotropy is insignificant as compared to surface anisotropy for the spherical-shaped MNPs. Researchers have examined the magnetic properties of magnetic nanomaterials using diameters of magnetic core by the abovementioned two models [20].

The magnetization of magnetic materials is magnetic moment per unit volume, and the relation is given in Eq. 10.3. Magnetization and magnetic moment determine that how a nucleus responds to the external magnetic field.

$$M = mB_o \tag{10.3}$$

where $B_o$ is the external applied magnetic field, m is the magnetic moment, and M is the magnetization.

Magnetic properties are strongly depending on the electronic structure of the material. A magnetic spin or domain (Weiss domain) refers to the dimensions of ferromagnetic material, where all magnetic dipoles have the same alignment with the exchange forces.

This concept of the materials spin or magnetic domain in the magnetic material is categorized into diamagnetic, ferromagnetic, paramagnetic, and superparamagnetic. Size-dependent magnetic behavior is also observed by domain structure of magnetic materials. The assessment of *m* for the substance is a multiplication of the magnetic moments of the atoms from the substance. Moreover, susceptibility ($\chi$), where $\chi$ = M/H, and permeability ($\mu$), where $\mu$ = B/H, are employed to measure the response of a material to magnetic field. For example, the iron atom shows high magnetic moment due to its unpaired electrons and presents different magnetic behaviors (Fig. 10.3).

Furthermore, large particles show similar behavior as bulk materials. The size-dependent magnetic properties can be observed in ferromagnetic materials. For example, at critical size Dc, the ferromagnetic material converts into a single domain. Generally, the critical size Dc is the range between 10 nm and 100 nm.

$$D_C \approx 18\frac{\sqrt{AK_{eff}}}{\mu_o M^2} \tag{10.4}$$

where $K_{eff}$ is the anisotropy constant, A the exchange constant, and $\mu_0$ the vacuum permeability. A single-domain particle is homogeneously magnetized and aligned parallel to the applied external magnetic field to give very high coercivity [25].

Many important parameters, such as magnetic saturation value, surface or domain wall energy, particle shape, the strength of the crystal anisotropy, and exchange forces, are influenced on the critical size of the single domain by the magnetic domain theory. Mostly materials possess magnetism in the presence of $B_o$. However,

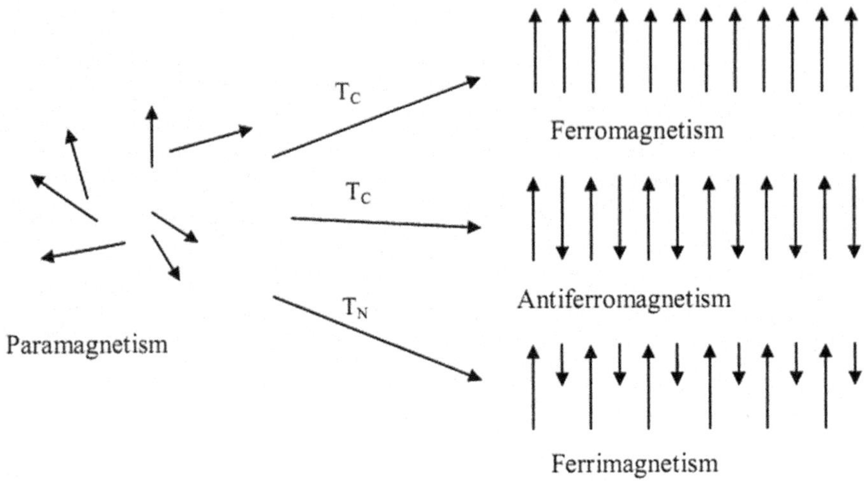


**Fig. 10.3** Arrangement of single atomic magnetic moment in different kinds of magnetic materials. (Reprinted with permission from reference [24])

some materials produced strong, and some show weak, magnetic behavior. Diamagnetism is very weak because electronic subshells of diamagnetic materials are filled, and magnetic moments are paired, canceling the overall effect. In case of ferromagnetic materials, magnetic effect is not zero because of magnetic moment of unpaired electrons under the applied field, and these materials have some remanence and coercivity termed as hysteresis loop. As the coercivity is strongly related to the size of the MNPs, such as by decreasing the particle size, the value of coercivity increases to a maximum and then reduces to zero. The relation between the nanoparticle size and magnetic domain structure is shown in Fig. 10.4.

In superparamagnetism, the coercivity value becomes zero because the dimensions of single-domain particle reduced or even less than critical diameter. Kittel presented the theoretical hypothesis in 1946 about superparamagnetism and the critical size of the particle having single magnetic domain. Superparamagnetism is a process in which MNPs indicate the same behavior as paramagnetism at temperatures below the Néel or the Curie temperature. This shows that superparamagnetism behavior is induced by thermal effects. Apart from other properties, superparamagnetic nanoparticles exhibit short relaxation time which is a very promising characteristic in hyperthermia applications. Their magnetic moments are rapidly reoriented in the presence of an alternating external magnetic field and return to a nonmagnetic state after removing the external magnet source; as a result MNPs release heat to surrounding atmosphere because of fast changes in the strength and frequency of magnetic field. This special feature of manipulation of MNPs only under magnetic force makes them promising when introduced to biological environments. Ferromagnetism can be observed in various crystalline materials other than Fe, Co, or Ni. However, iron is highly magnetic and naturally available mineral.

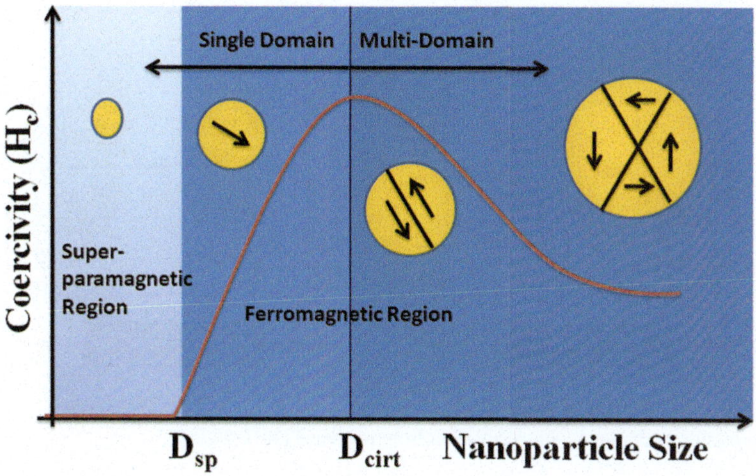


**Fig. 10.4** Graphical representation of the coercivity-size relationship with respect to MNPs

Therefore, it has commonly used to prepare superparamagnetic nanoparticles which widely employed as magnetic nanomedicine agents [26–29].

## 10.3 Synthesis Method of MNPs

Researchers have been paying attention to develop cost-effective materials synthesis techniques to produce fine nanoparticles with enhanced properties for industrial and technological applications. Presently, various chemical and physical methods have been employed for the preparation of nanomaterials with controlled dimensions [30–33].

Generally, two basic materials synthesis methods have been used frequently, top-down and bottom-up approaches. In the top-down process, a synthesis route starts from a large piece of particles and consequently reduces the size to create small particles (nanoparticles). Top-down process includes ball milling and bulk materials (or thin films). Alternatively, bottom-up technique uses self-assembly growth process on minute components of atomic or molecular dimensions to organize them into nanoparticles by using external force or natural physical principle [34, 35]. The schematic illustration of these two methodologies is shown in Fig. 10.5.

Both methods have benefits and limitations related to fast synthesis, reproducibility, cost, and easy and complicated handling. Frequently, monodisperse MNPs can be prepared by various chemical methods. In the last two decades, plentiful research has been carried out on the preparation of MNPs. MNPs have been prepared with various shapes, dimensions, and phases. These MNPs included magnetic alloys such as $CoPt_3$ and FePt [37, 38]; spinel-type ferromagnets such as $MgFe_2O_4$,

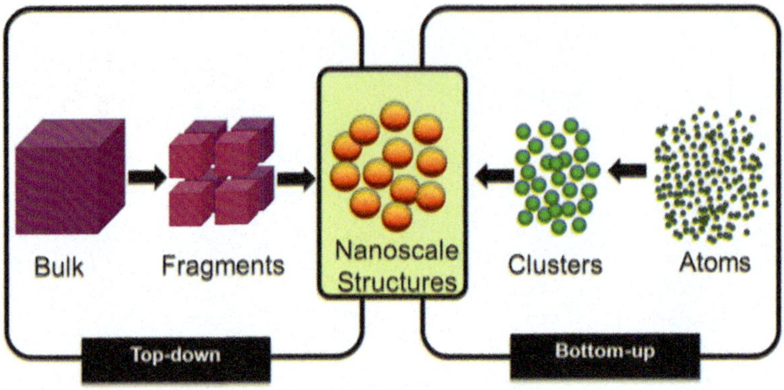


**Fig. 10.5** Schematic illustration of the formation of nanostructures via the top-down and bottom-up approaches. (Reprinted with permission from reference [36])

$MnFe_2O_4$, and $CoFe_2O_4$ [39, 40]; iron oxides such as $Fe_3O_4$ and $\gamma$-$Fe_2O_3$ [41]; and pure metals such as Fe and Co [42, 43]. The following section will describe some interesting and useful physical and chemical synthesis techniques for the fabrication of fine monodispersed MNPs.

## *10.3.1 Ball Milling Method*

The basic physical method in top-down approach is ball milling method which was developed by John Benjamin in 1970. In this technique, a sample is grounded to prepare the reduced size nanopowder [44]. This is a very promising technique among other physical methods to prepare the MNPs. Basically, there are three steps involved to fabricate the nanoparticles in ball milling method. Firstly, mechanical energy is delivered by balls to bulk materials and then collisions occurred between balls and bulk substance. The disruption density remains increasing with milling time. Due to this high energy transformation, physical and chemical properties of the materials change. Secondly, small particles are produced by rearrangement of dislocations. Finally, orientation of the grain became random, and small particles get rid from large materials. In case of MNPs, a magnet placed outside the tube which attracts the peeled of nanoparticles. Ball milling equipment is easy to handle but has long production cycle [45, 46]. However, there are some limitations associated with this method, such as contamination problems, polydisperse size distribution, and partially amorphous state. Also, the aggregation of particles is observed if the ball milling time increases, and likewise aggregation happens in small particles to reduce surface energy [47, 48].

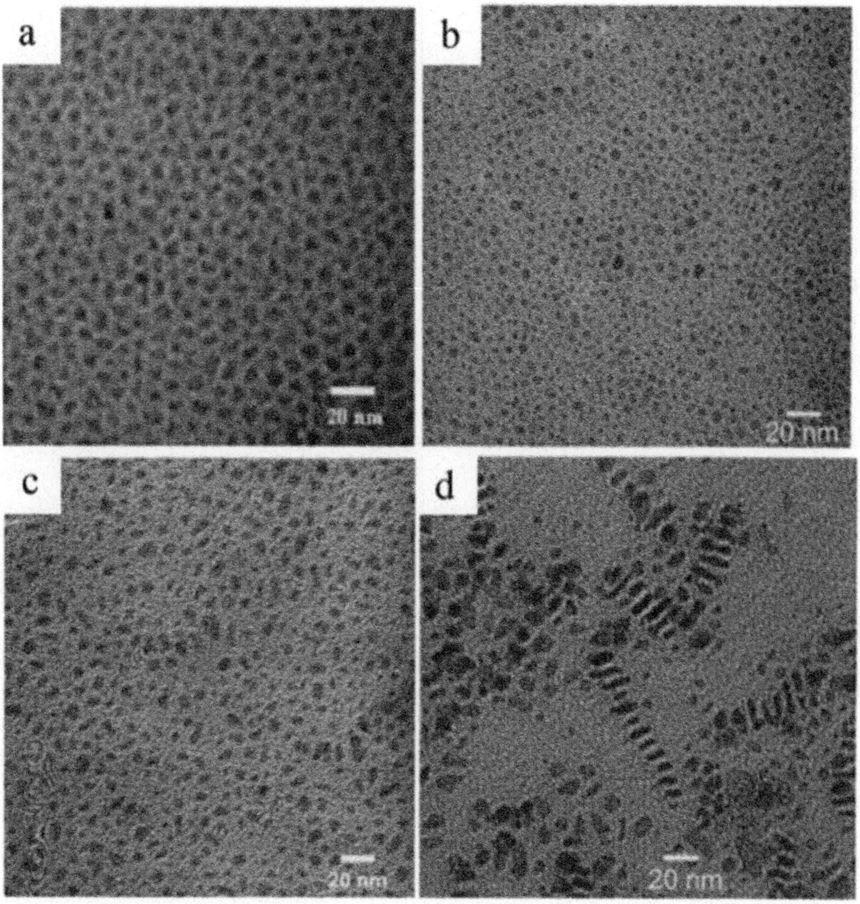


**Fig. 10.6** Transmission electron microscope images of the synthesized MNPs by ball milling method at different time. (Reprinted with permission from reference [49])

Researchers have been focusing on how to produce homogeneous size distribution of MNPs by using ball milling procedure. Some reports proposed the centrifugal separation method with different centrifuge speeds to collect different sizes of the NPs. In this way, magnetic nanoparticles and their composites are synthesized with narrower size distributions [50]. TEM pictures of Fe and $SmCo_5$ NPs synthesized by this method are shown in Fig. 10.6. 4–6 nm of Fe NPs and 3–13 nm of $SmCo_5$ NPs are obtained and shown in Fig. 10.6 (a, b). Also, Fig. 10.6 (c, d) shows the TEM images of $SnCo_5$ NPs at several milling times. Fe, Co, and FeCo NPs were fabricated by surfactant-assisted ball milling method. By controlling the parameters of ball milling equipment and applying size selection process, homogeneous sized 6 nm MNPs are obtained.

### 10.3.2 Chemical Methods

It is obvious that small NPs have high surface to volume ratio. Therefore, they agglomerate easily to reduce their surface energy. In case of superparamagnetic MNPs, aggregation effects are very high because of their colloidal nature. Fabrication of monodispersed MNPs is a complex process, and this is still a challenging task to synthesize MNPs with controlled size and reproducible at large scale without any difficult purification procedure. Various chemical routes can be utilized to prepare the magnetic nanoparticles for biomedical application such as sonochemical reactions [51], sol-gel method [52], microemulsion [53], flow injection synthesis [54], electrospray synthesis [55], electrospray synthesis [56], hydrothermal reactions [57], and thermal decomposition method [58]. While due to the colloidal complexion, the preparation of superparamagnetic nanoparticles is a little bit complicated process. Different methods have been utilized to synthesized magnetic nanoparticles with various phases and compositions, including pure metals like Co and Fe [59], iron oxides like $Fe_3O_4$ and $Fe_2O_3$, alloys such as FePt and CoPt [37, 60], as well as spinel-type ferromagnetic materials such as $MgFe_2O_4$ and $CoFe_2O_4$ [39, 61]. In the last few years, numerous efforts have been dedicated for the preparation of MNPs with control morphology, well stable and monodisperse through various chemical routes such as thermal decomposition, hydrothermal reactions, coprecipitation method, and laser pyrolysis techniques. Among them, the thermal decomposition method and hydrothermal reaction are extensively utilized to fabricate the MNPs with narrow size distribution and homogeneous composition. Here we attempt to sum up some discussion about the coprecipitation, thermal decomposition, and hydrothermal methods for the preparation of magnetic nanoparticles with an illustration of some examples.

### 10.3.3 Coprecipitation

The coprecipitation technique is very promising and efficient chemical method to prepare the MNPs especially superparamagnetic NPs. The biodegradable MNPs can be synthesized at large scale and highly aqueous dispersed by using this method. MNPs with different sizes, structures, and compositions can be obtained, but these properties depend on the initial magnetic salt precursor types, reducing agents, reaction time, and temperature. Once the optimized synthesis conditions are achieved, MNPs are reproducible by coprecipitation technique.

In case of iron oxide NPs, ferrous and ferric salts are mixed to prepare MNPs by an aging stoichiometric in aqueous medium. The chemical reaction involved in the synthesis of $Fe_3O_4$ NPs is given in Eq. 10.4.

$$M^{2+} + 2Fe^{3+} + 8OH^- \rightarrow MFe_2O_4 + 4H_2O \tag{10.5}$$

where M can be $Fe^{2+}$, $Mn^{2+}$, $Co^{2+}$, $Cu^{2+}$, $Mg^{2+}$, $Zn^{2+}$ and $Ni^{2+}$.

With a stoichiometric ratio of $Fe^{3+}$ and $M^{2+}$ is 2:1, the complete precipitation should be obtained at a pH levels between 8 and 14 in nonoxidizing atmosphere [62].

$Fe_3O_4$ is unstable at some circumstances and are easily oxidized to maghemite ($\gamma$-$Fe_2O_3$) in the presence of oxygen.

$$Fe_3O_4 + 2H^+ \rightarrow \gamma Fe_2O_3 + H_2O \tag{10.6}$$

Many other factors are also related to the oxidation of iron oxide apart from the air.

However, in the formation of MNPs, particularly $Fe_3O_4$ by coprecipitation method, there is still challengeable that how to obtain the narrow size distribution with controlled size and shape. Shen et al. prepared ultrasmall nanoparticles of iron oxide using coprecipitation and successfully applied in biomedical application [63]. Michael Barrow et al. also synthesized superparamagnetic iron oxide nanoparticles with different sizes by coprecipitation approach [64].

### 10.3.4 Thermal Decomposition

The considerable issue is to prepare the extremely fine particles with control size and distribution at nanoscale formulation. This problem is prominent when dealing with MNPs because of their continual growth with reduction of altogether surface energy as well as aggregation of particles. Furthermore, the *Ms* value is being significantly decreased than bulk materials at nanoscale systems [65–68]. While the preparation of MNPs with control size and homogeneous size distribution as well as with less *Ms* value is potentially important in the biomedical field, the thermal decomposition is a promising method to synthesize magnetic nanoparticles with the decomposition of organometallic compounds comprised of stabilizing surfactants and organic solvents without aggregation of particles [39, 69–71]. After the successful synthesis of various oxides and high conductive semiconductor nanocrystal through thermal decomposition [72, 73], corresponding method has been forward to the synthesis of MNPs with control and homogeneous morphology distribution. While the magnetic nanocrystals have been fabricated via thermal decomposition method by using organometallic compounds with high-boiling solvents including stabilizing surfactants [74], further, the metal acetylacetonates included in the organometallic precursors such as metal cupferronates [$M^xCup_x$] (M = metal ion; Cup = N-nitrosophenylamine, $C_6H_5N(NO)O$-) [75] [M(acac)n] (M = Mn, Fe, Ni, Cr, Co, n = 2 or 3, acac = acetylacetonate), oleic acid [76], hexadecylamine [77], and fatty acids [78] are frequently used as surfactants. Accordingly, the ratios of reagents, reaction time, and temperature, as well as aging period, are the determined parameters toward synthesizing of magnetic nanoparticles with tunable size having homogeneous morphological distribution. In the case of zerovalent metal in the precursor like in carbonyls, initially the formation of metal is being carried out through thermal decomposition, but oxide nanoparticles as well can be synthesized by a two-step procedure. Meanwhile, the precursors include cationic metal centers, the process of thermal decomposition directly produced the $Fe_3O_4$ NPs by decomposition of $[Fe_2(acac)_3]$ in the presence of various hydrophobic materials [79]. Peng et al. prepared manganese-based nanocrystals with controlled shapes and size using thermal decomposition method [78]. Maity and co-workers reported the synthesis of MNPs of iron oxide using acetylacetonate $[Fe(acac)_3]$ through thermal decomposition and studied the various effects on the structure of the $Fe_3O_4$ nanoparticles [80]. The particle size was well controlled with narrow size distribution, and *Ms* value is significantly enhanced with increasing the temperature and reaction time. Further, in this work saturation magnetization Ms. value of particles can be raised up through long reaction time or high reaction temperature, if the particle size can be conserved through either by usage of solvent-free thermal decomposition process or adhesion of oleic acid to the particle surface. The development performance and nucleation of nanoparticles formulations can be qualitatively illuminated by classical concepts. Ostwald ripening [81] and LaMer supersaturation [82] processes describe what is complied, specifically in the kinetically controlled synthetic procedures. Like other magnetic nanoparticles, superparamagnetic iron oxide (SPION) nanoparticles are synthesized by thermal decomposition methods with controlled sized distribution which are further suitable for numerous biomedical applications such as imaging [83], diagnostics [84], and targeting therapy [85]; SPIONs are considered to be nontoxic for intravenous investigations [86, 87]. Hufschmid et al. prepared the SPIONs with good dispersion by thermal decomposition approach and compared the three common iron-containing precursors such as iron(III) oleate ($Fe[C_{18}H_{33}O_2]_3$), iron pentacarbonyl ($Fe[CO]_5$), and iron oxyhydroxide (FEOOH) [88]. In this work, by using organic solvents, iron oxide particles were grown by thermal decomposition, and through generally excess surfactant, oleic acid is added to adjust size distribution and tailor the growth kinetics. By the addition of more oleic acid, the size of particles was altered from 10 to 25 nm and the quantity of oleic acid increased, resulting in increased sizes of particles. In another work, Iqbal et al. synthesized the ultrasmall silica-coated SPIONs by thermal decomposition of iron oleate complex with an average size of 4 nm with homogeneous size and monodispersion [89]. In this synthesis process, iron oleate was used as a precursor in the presence of organic solvents oleic acid and oleyl alcohol. The solution reaction color with respect to time and microscopic images of prepared SPIONs is shown in Fig. 10.7.

Characterization techniques revealed the good crystallinity and high dispersion of MNPs in organic and aqueous solution. Further, these nanoparticles are utilized as an excellent candidate for MRI imaging as $T_1$ contrast agent due to their smaller size and homogeneous distribution. Moreover, various pure composites of MNPs can be prepared by the thermal decomposition approach [90–93].

### 10.3.5 Hydrothermal Synthesis

There is a broad range of nanomaterials including magnetic nanoparticles that have been synthesized through the hydrothermal approach. To prepare ultrafine

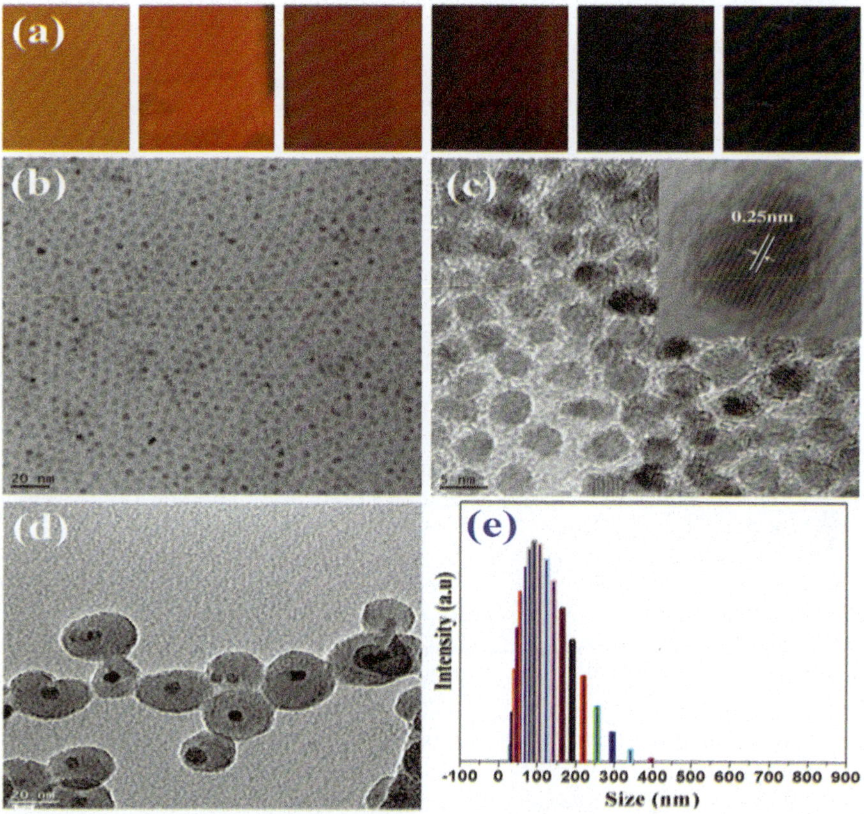


**Fig. 10.7** (**a**) Digital snapshots of solution reaction color with respect to time (**b-d**) TEM images of the ultrasmall nanoparticles and the $SiO_2$-coated $Fe_3O_4$ nanoparticles. (**e**) Particles size dispersion analyses of $Fe_3O_4@SiO_2$ in water. (Reprinted with permission from reference [89])

nanopowders and MNPs through hydrothermal reactions is widely reported in literature [93–95]. In the hydrothermal approach, the reactions take place in aqueous media in autoclaves or reactors where the temperature is more than 200 °C and pressure might be higher than 2000 psi. Li et al. described the typical synthesis route of hydrothermal reaction from a broad diversity of nanocrystals which has been synthesized through liquid-solid solution reaction [96]. The approach comprised of a water-ethanol solution, a metal linoleate (solid), and an ethanol-acid liquid phase under a hydrothermal atmosphere at various reaction temperatures. The fundamental point in this mechanism is phase transfer and separation strategy which happens at the interfaces of solid, liquid, and solution phases immediate throughout the synthesis. Generally, two mechanisms are defined for fabrication of ferrites through hydrothermal conditions, oxidation or nucleation and hydrolysis of hybrid metal hydroxides. There is quite a similarity that exists between these two mechanisms except the usage of ferrous salts in the second mechanism. In this approach, significantly effects are being occurred on resultant products by reaction conditions such as time and temperature of reaction and solvents [97]. Undeniably, it is

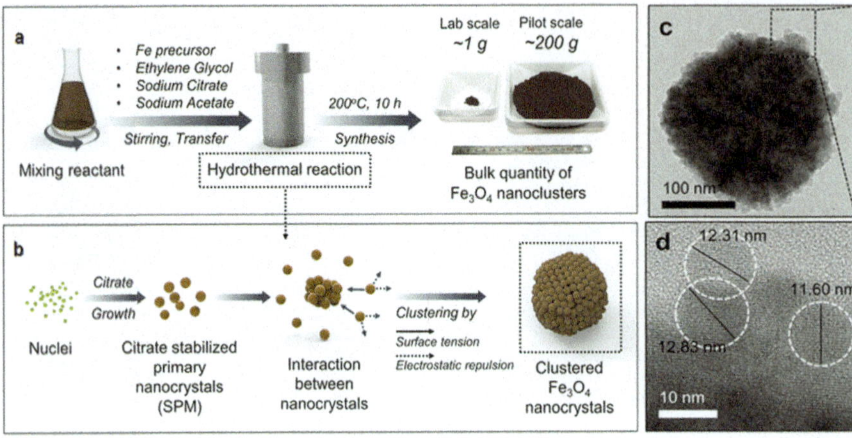


**Fig. 10.8** (**a–b**) Schematic presentation of the procedure for the gram-scale synthesis of MNCs. (**c** and **d**) A set of transmission electron microscopy (TEM) images of MNCs. (Reprinted with permission from reference [107])

observed that the precipitation of larger $Fe_3O_4$ nanoparticles resulted due to higher water content and the particle size increased due to prolonged reaction time. In the hydrothermal approach, the grain growth and the nucleation process rate mainly control the particle size in crystallization. By keeping other conditions unchanged, their process rate is determined by reaction temperature [97]. The solvo−/hydrothermal approach provides a cost-effective and versatile route for the fabrication of a large number of magnetic clusters with high crystallinity and controlled morphology [98, 99]. The hydrothermal approach is capable to prepare and reshape the nanoparticles via different processes to modify the magnetic behavior of resultant particles [100]. Recently, Kim et al. reported the gram scalable synthesis of superparamagnetic iron oxide nanoclusters $Fe_3O_4$(MNCs) of 11–13 nm composed of nanocrystals by the hydrothermal approach and modify the MNCs morphology by the diversity of molar ratios of surfactants and precursors [101]. The schematic presentation of the process of the preparation of $Fe_3O_4$ nanocluster is shown in Fig. 10.8. Usually, the ferrite substances might be categorized into three classes: garnet ferrites, hexagonal ferrites, and spinel ferrites [102–104]. In another work, Hedayati and co-workers reported the fabrication of $Fe_3O_4$ nanoparticles through hydrothermal synthesis and then investigated the magnetic properties [105]. Further, the outcome of the resultant morphology was investigated due to different surfactants such as cationic and anionic. Li et al. reported the synthesis of nanocomposites of α-$Fe_2O_3$/rGO through a simple hydrothermal approach with the observation of photocatalytic performance and negative MR effect [106].

In the preparation mechanism of MNPs through different approaches, the colloidal stability results from either electrostatic or steric repulsion, based on the polarity of the solvent and stabilizers such as amines or fatty acids used. Although the nanoparticles prepared through thermal decomposition are usually satirically stabilized by surfactants or fatty acids in an organic solvent [71].

## 10.4 Magnetic Nanoparticles as Hyperthermia Agent

Nanotechnology emerged as a prodigious field of science with a diversity of applications in biotechnology, medicine, sensing, and healthcare monitoring and in each and every aspect of nature [108–110]. Due to the smaller size, easy modification, and tunable physical and chemical properties, nano-sized materials have gained potential fame in the nanomedicine field [111–113]. In terms of treatment and diagnosis, magnetic nanoparticles (MNP) cannot be replaced with any other material [25, 114, 115]. Surface functionalization and coating of ferromagnetic and superparamagnetic nanoparticles not only make them biocompatible but also effective for killing tumor cells [116, 117]. By applying an alternative field, activation of MNP brings targeted therapeutic heating of tumor. Gilchrist et al. have explained firstly that by administration of $B_o$, the lymphatic metastases can be killed effectively due to the heating mechanism in 1951 [118]. Heating of MNPs by applying an external $B_o$ to induce killing of a tumor cell is a well-known term for magnetic hyperthermia [119]. According to Néel or Brown relaxation due to hysteresis losses, thermal energy is released into the environment [120, 121]. Basically, hysteresis loss is related to the nature and strength of applied external $B_o$ which further induced the increase in hyperthermia effect. Originally, various sizes of MNPs endure the applied external $B_o$; the alignment of MNPs against or with the direction of the applied magnetic field happens. The time delay phenomenon between the reverting back to normal state and magnetizable causes the generation of Brownian relaxation [122, 123]. The movement of these aligned states of nanoparticles and friction between them generates heat in the surrounding. This heat is used to kill or prolong the growth of tumor. So, to induce the magnetic hyperthermia, Brownian relaxation and generation of heat are the key parameters. In the past, there were different kinds of mechanisms that are involved to cure cancer like radiation therapy, chemotherapy, and surgery [124–126]. As all the therapeutic mechanisms have some drawbacks like cisplatin which is widely used to cure cancer but is harmful for the bone marrow, by combining different therapy mechanisms, therapeutic index can be improved [127]. In combination with other traditional cancer therapies, hyperthermia has shown promising effects without damaging other healthy tissues. Hyperthermia has been revealed promising not only as a singular therapy but also in conjunction with traditional cancer treatments such as chemotherapy and radiation. Since the quantity of heat produced can be observed during local hyperthermia, harm to nearby tissue can be evaded or minimized. Pain, discomfort, and burns are the most common effects that can be caused by hyperthermia [128, 129].

Fe, $Fe_3O_4$, Ni, strontium ferrite, and chromium oxide have been used extensively to produce MNPs [130]. The Curie temperature value along with other physical properties of such kind of materials containing a core of magnetic nanoparticles has shown great compatibility in animal studies. In terms of magnetic nanoparticles, iron oxide ($Fe_3O_4$) nanoparticles and their core-shell structures have gained fame in the nanomedicine field [131, 132]. There are different advantages of $Fe_3O_4$ nanoparticles over other magnetic nanoparticles; the most important one is biocompatibility and nontoxicity. Further, diversity of available synthetic procedure and ease in functionalization of iron oxide nanoparticles with organic and inorganic biocompatible chemicals make them perfect candidates for in vivo studies and biomedical applications [133]. Similarly, cobalt ferrite has also shown high magnetization along with constant heating effect [134]. The heating efficiency of cobalt ferrite predicted to be higher than iron oxide nanoparticles as they have not studied to large extent in comparison with iron oxide nanoparticles. Despite good heating efficiencies, cobalt and nickel nanomaterials have not been studied extensively for hyperthermia or some other biomedical applications because of difficulty in synthetic procedure [135, 136]. Along with it other types of magnetic nanoparticles in the form of nanocomposite have been utilized widely for magnetic hyperthermia therapy like $\gamma Fe_2O_3$, $NiFe_2O_4$, $MnFe_2O_4$, $CoFe_2O_4$, NiO, and $Co_3O_4$ [137]. Last but not the least, gold, silver, and copper nanomaterials with wide size range and multiplicity in structure have captured the attention of researchers and proved as new comers for nanoparticles' hyperthermia [128, 138].

### *10.4.1 Parameters Affecting the Heating Efficiency*

There are several factors influenced on the heating efficiency as discussed previous. The strength of external magnetic field plays a significant characteristic in magnetic hyperthermia, and the heating capacity of the MNPs also related to it. Generally, the external magnetic field intensity has direct relation with the power loss in magnetic hyperthermia as given in Eq. 10.7 [139]. The specific absorption rate normally increases with increasing the power of applied magnetic field [140]. Brezovich et al. observed that the product of an external applied magnetic field strength H and frequency f restricted to the value $H_f < 4.85 \times 108\ Am^{-1}\ s^{-1}$ which is not enough to provide excessive heating in patients [141].

$$P = 1/2\omega\mu 0\chi 0H2 * \omega\tau/1 + \omega 2\tau 2 \tag{10.7}$$

Nanoparticle diameter and anisotropy are also well concerned for the hyperthermia. The size of the particle is the major factor to determine that either Brownian-Néel relaxation is going to occur or hysteresis loss will cause the heat generation. In this regard, C. L. Dennis et al. studied extensively the particle size and hyperthermia effects generated by larger particle size [142]. As the particle size will be greater than 100 nm, it will not be suitable for biomedical application due to difficulty in excretion of a particle through the body, and it cannot disperse and penetrate well in the tumor. Hence, particle size should be smaller to about 15 nm which is a suitable size for the Brownian and Néel relaxation (for $Fe_3O_4$ nanoparticles). Heat generation below 15 nm size is due to Néel relaxation as explained in Fig. 10.9 [135].

The inferior plot demonstrates Brownian, Néel, and effective relaxation times for particles with anisotropy constants of $K = 40$, 25, and 10 $kJ/m^3$. To some extent, increasing particle concentration also affects the hyperthermia. High concentration

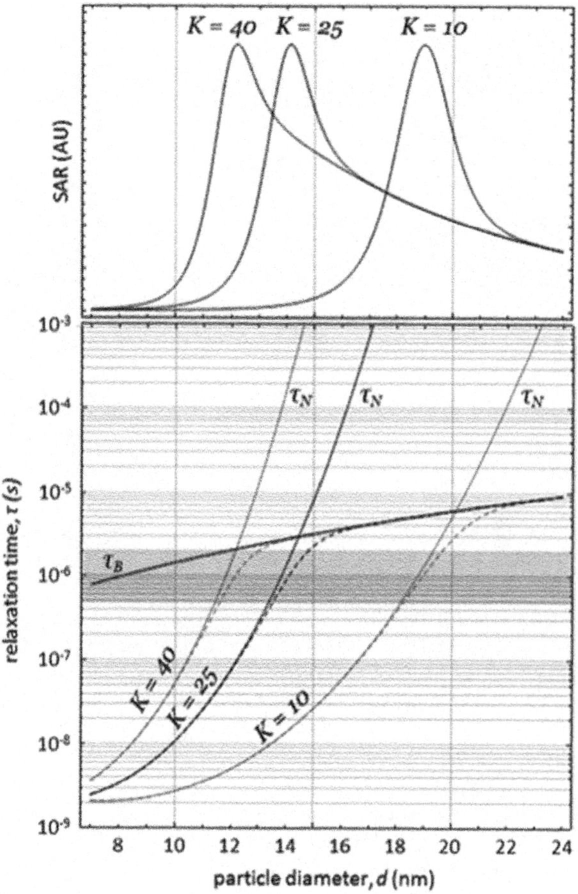


**Fig. 10.9** Effect of anisotropy on relaxation time and specific absorption rate. (Reprinted with permission from reference [135])

of particles causes agglomeration and aggregation which results in higher interaction between the nanoparticles and is rarely significant for hyperthermia, while lower concentration of particles causes homogeneous distribution of nanoparticles in the fluid and significantly influences the surface absorption rate [135].

Although diversity in the synthesis of MNPs for hyperthermia can be seen from 1951 onward, as mentioned earlier iron oxide NPs are used widely. In this regard, C. L. Dennis et al. synthesized ionized ferrite nanoparticles that are core-shell structures containing $Fe_3O_4$ and dextran by using high-pressure homogenization method conjugated with monoclonal antibodies and have shown good results by controlling tumor growth control in a mouse after applying H = 43.8 kA m$^{-1}$ at 150 kHz and 100% duty for 20 min [142].

In another aspect, Ching Jen Chen et al. showed that metal alloys can also be used for hyperthermia [143]. They have designed polyethylene glycol-coated magnetic Cu-Ni alloy material by polyol reduction method, and the Curie temperature measured for such an alloy system was about 43–46 °C. The material was very suitable

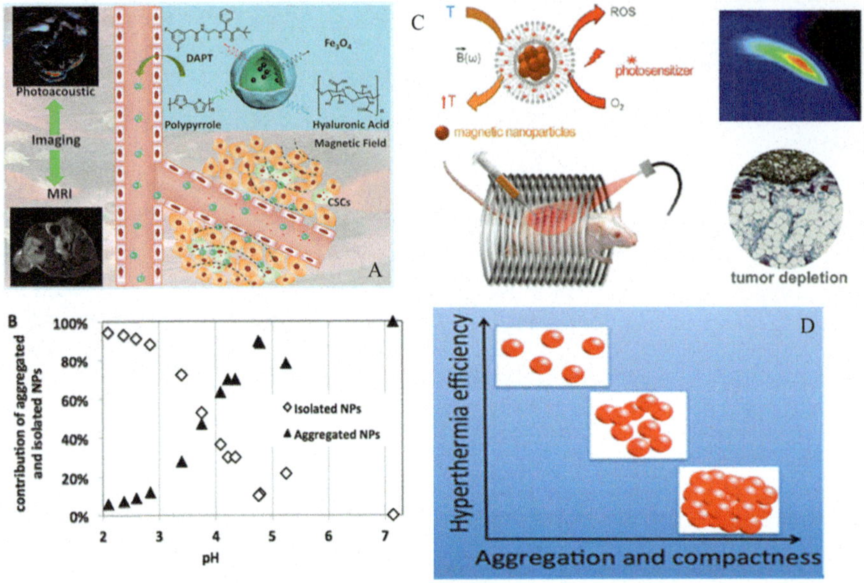


**Fig. 10.10** (**a**) Photoacoustic and magnetic resonance imaging of hyaluronic acid-containing magnetic nanoparticles, (**b**) effect of pH on the aggregation of PAA-coated iron oxide nanoparticles, (**c**) photodynamic therapy and hyperthermia studies of bilayer liposomal MNPs for drug delivery and hyperthermia therapy, and (**d**) decrease in hyperthermia efficiency due to increase in aggregation of nanoparticles. (Reprinted with permission from reference [155])

for the hyperthermia to cure cancer. To achieve dual therapeutic response from a single material has become an important need due to toxicity, less availability, and desire for new materials. Recently, to realize multi-functionalities from the iron oxide nanoparticles, Claire Wilhelm et al. designed iron oxide nanocubes for magnetic and photothermal therapy [128]. They have tested three types of cancer cells (SKOV3, A431, and PC3) and successfully achieved cell death due to high heating powers to about (specific loss powers SLP) 5000 W/g. The only dual action of nanocubes can cause cell death; only hyperthermia or photothermal effect alone just inhibits the growth of tumor cell (Fig. 10.10a). In vivo study has revealed that by using dual response, the heating capacity will be higher which will inhibit the growth of tumor to a high extent (Fig. 10.10c). For comparison between spherical and cubic nanoparticles' heating efficiency, H. Srikanth et al. have studied the FeO/$Fe_3O_4$ nanoparticles (20 nm) [144]. By changing the ratio of FeO/$Fe_3O_4$, they have changed the heating efficiency for hyperthermia. Their study revealed that although spherical nanoparticles have shown higher saturation magnetization, hyperthermia experiments have shown higher values for the cubic structures. Different target molecules or polymers have been used for the hyperthermia to get accurate targeting by avoiding the other tissues loss. In this regard, Y.Q. Wang et al. have synthesized folic acid-containing magnetic nanoparticles [145]. Folic acid served as an accurate targeting agent, that is, on combination with $Fe_3O_4$, can provide high saturation magnetization and fast magneto-temperature response that is necessary for hyperthermia therapy. Heat produced from such a material was recorded, and it reached up to 52.7 °F, i.e., enough to kill the tumor cell. By utilizing the magnetic response, they have studied the $T_2$ signal of KB cells by MRI and obtained promising results. Further after injecting the folic acid-containing magnetic nanoparticles in the mouse for KB cells, in just 4 h, the weakening of KB cells happens which was a good evidence of hyperthermia.

Recently, photothermal therapy by using some magnetic nanoparticles becomes promising therapeutic approach to cure different kinds of cancer [146]. Photothermal properties induced by black $TiO_2$ and other material have been used for cancer cure due to in situ heat generation [147]. Various efforts have been adopted to enhance the heating efficiency by designing new materials. Aiguo Wu et al. have synthesized poly (ethylene glycol)-modified iron oxide containing nanoflowers [148]. The in vivo and in vitro studies have shown promising results for photothermal therapy. As explained earlier the cube-shaped iron oxide nanoparticles have shown high heating efficiency, but the heating efficiency of synthesized iron oxide nanoflowers was proved to be better than iron oxide nanoparticles. Iron oxide nanoflowers have shown approximately 52 °C increase in temperature which was comparable to that of black $TiO_2$ core-shell nanostructures which revolutionized the biomedical field in both aspects of new nanomaterials synthesis and their advanced applications [149, 150]. To some extent, different kinds of core-shell nanomaterials have been synthesized yet for cancer, MRI, detection of various diseases, and other multifunctional tasks [151]. In this aspect to achieve application related to hyperthermia and killing cancer stem cell, Chunying Chen et al. synthesize hybrid magnetic nanomaterial for the chemo- and magnetohyperthermia therapy [152]. Hyaluronic acid was utilized to get an accumulation of this hybrid material containing Fe@PPA@HA. The use of PPr causes the photoacoustic effect, and $Fe_3O_4$ NPs used as MRI agent further PPr shell allowed a high amount of drug N-[N-(3,5-difluorophenacetyl-L-alanyl)]-S-phenylglycinet-butyles-ter (DAPT) loading (Fig. 10.10a). The accumulation of Fe@PPr@HA in the tumor cell and then easy drug release along with magnetic hyperthermia properties caused killing of cancer stem cells. The intraparticle interactions have been studied well by changing pH and ionic strength of the desired nanoparticles system. Poly(acrylic acid)-coated nanoparticles have shown good stability to a wide range of change of ionic strength induced by using different concentrations of ammonium chloride. For in vivo system, particle stability is a great factor to be considered. Similarly, pH variation also induced aggregation of bare and coated magnetic nanoparticles as can be seen in Fig. 10.10b. Interestingly vast variety of work has explained about multifunctionality of different materials in which utilization of properties of every material brought together in the form of one hybrid material. Claire Wilhelm et al. have designed a novel approach to utilize liposome property of hydrophilic and hydrophobic molecule transfer due to the lipid bilayer and magnetic property of magnetic ($Fe_3O_4$) nanoparticles [153]. The liposome has shown the photosensitizer action which on irradiation with NIR induced toxic reactive oxygen species that can induce photodynamic effect and application of magnetic field generated heating of magnetic nanoparticles to obtain magnetic hyperthermia. Both these effects have killed the tumor that has been studied experimentally by in vitro studies (Fig. 10.10c). Regardless of the work published on the magnetic nanoparticles for the biological system, the stability mechanism is still under consideration related to the biomedical application. Well-dispersed magnetic nanoparticles have gained fame in this field, but still heating efficiency is poorly understood [154]. For the aggregation system of magnetic nanoparticles, Jerôme Fresnais et al. have explained about poly(acrylic acid)-coated and poly(acrylic acid-co-maleic acid)-coated iron oxide nanoparticles [155]. Finally, they have concluded that aggregation of magnetic nanoparticles causes decrease in heating efficiency (Fig. 10.10d).

Magnetic hyperthermia for cancer cure is not only limited to iron oxide nanoparticles, but other parameters are also involved that cause controlled hyperthermia. Orestis Kalogirou et al. have synthesized carbon-based MNPs by solvothermal method with the size range from 150 to 250 nm [156]. The heating efficiency was controlled by synthesizing thinner layer of carbon which interestingly causes the increase in specific power loss and proved efficient for hyperthermia [157]. In an instant, ultrasmall iron oxide NPs may be excreted rapidly, and biodegradation may occur on direct exposure to the biological environment [158]. To cover this limitation, a coating of a suitable material is required. Ewa Borowiak-Palen et al. have designed a sol-gel approach to get magnetic silica nanotubes (Fig. 10.11a–e) [159]. These $Fe_2O_3$-containing silica nanotubes were not only successfully utilized for rhodamine B drug loading but also have been

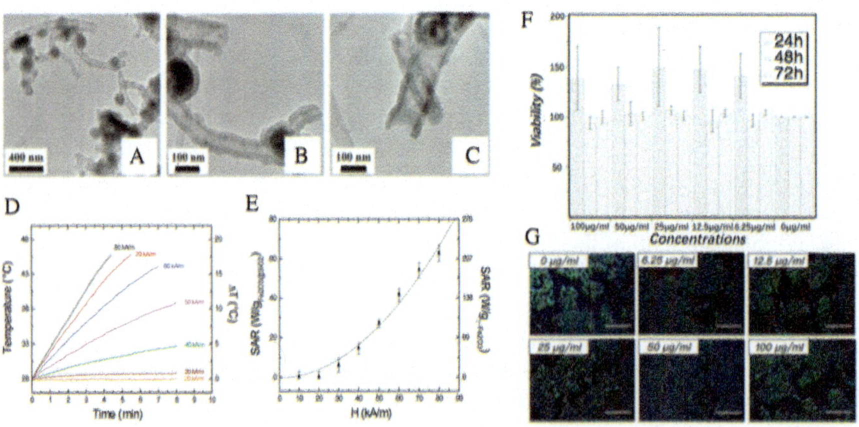


**Fig. 10.11** (**a–c**) Synthesized magnetic silica nanotubes, (**d–e**) their increase in temperature according to application of various strengths of magnetic field (20 KA/m to 80 KA/m), (**f**) cell toxicity results measured by CCK-8 assay with different concentration of MNPs, and (**g**) fluorescent microscope images of separated HaCaT cells stained with various dyes. (Reprinted with permission from reference [161])

used for hyperthermia application after applying different field strengths. Further, the increase in surface absorption rate was also measured with an increase in an applied $B_o$ for the successful elaboration of magnetic hyperthermia application. Similar to silica, carbon nanotubes have been efficiently used for hyperthermia application with other magnetic nanoparticles like Zr, Ni, Co, etc [160]. Wei Gao et al. have designed multiple magnetic nanoparticles containing carbon nanotubes ($Zn_{0.54}Co_{0.46}Cr_{0.6}Fe_{1.4}O_4$) and have determined their cytotoxicity, magnetic properties, and heat generation ability [161]. Finally, on the application of external magnetic field (frequency, 100 kHz; intensity, 200 Oe), the heat generation reached to about 42 °C which is a feasible temperature to achieve hyperthermia application (Fig. 10.11f–g). The cytotoxicity effect of magnetic carbon nanotubes was measured on HaCaT cell line by the incubation of prepared MNPs with different concentrations and time. In Fig. 10.11g, it can be seen that cells incubated with various concentrations of magnetic carbon nanotubes for 72 h have shown Calcein-positive, Hochest 33,258-positive, and PI-negative state. This toxicity result is consistent with the quantitative CCK-8 assay in Fig. 10.11f. These outcomes have revealed that the prepared magnetic carbon nanotubes demonstrated very minute level of toxicity for HaCaT cells.

The above reports showed the promising use of MNPs as a magnetic hyperthermia agent with enhanced biocompatibility.

## 10.5 Magnetic Nanoparticles for Drug Delivery

The treatments of fatal diseases with high efficiency have been the substantial aim of the human race since the ancient times. With passage of time, scientists have effectively refined the understanding toward the functions of human organs (bones, blood vessels, various tissues, and nerves) and other related parts including living cells with their basic phenomenon. From the last two decades, there were some incredible innovations appeared in biomedical field with polymer and micro-nanoparticles related to different ways of treatments including diagnosis, sensing, drug delivery, and imaging techniques with different therapies. Among them, drug delivery mechanism has been greatly explored to cure some fatal disease with low toxic side effects and in other clinical trials. Various species of drug carriers have been reported for different treatments including polymeric nanoparticles, micelle-shaped polymeric nanoparticles, liposomes, carbon-based systems, silica nanoparticles, and magnetic nanoparticles (e.g., iron oxides) (Fig. 10.12). Here, we discuss some magnetic nanoparticles which are used for drug delivery systems for various clinical trials. Magnetic nanoparticles are used for drug delivery system because of their control mechanism with magnetic field to cure effective part of body. In the early 1970s, the concept of using magnetic micro- and nanoparticles for drug carriers was introduced to target a specific part inside the body [162].

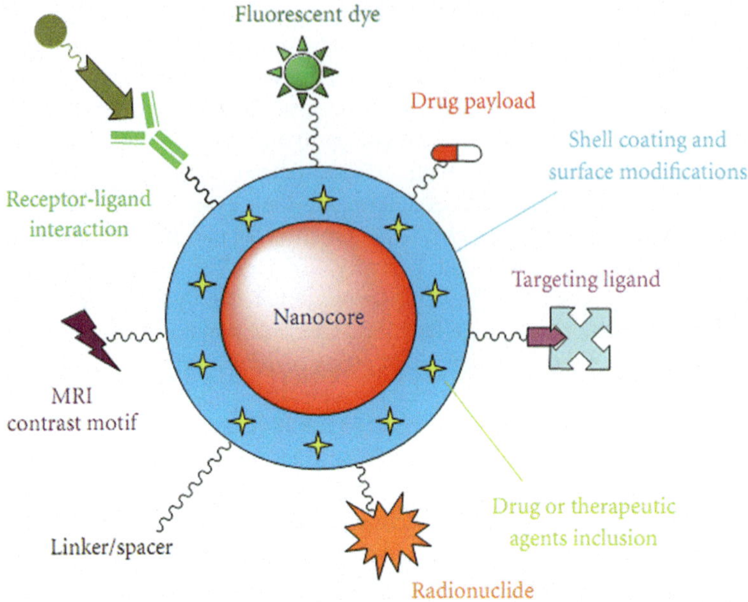


**Fig. 10.12** An overview of various drug delivery systems using nanoparticles with many targeting possibilities. (Reprinted with permission from reference [163])

### *10.5.1 Drug Delivery Mechanism*

Several drug delivery systems are being applied successfully for treatments, however still need to develop some accurate methods to overcome certain challenges that need to negotiate with development of advanced technology as well as for achieving targeted delivery of drugs. Recently, enormous progress has been devoted to hand over natural active compounds or therapeutic agents in the field of drug delivery system for various targeted treatments [164, 165]. Recently, drug delivery system based on nanotechnology studied widely that will make ease to progressive system of drug delivery. In nanomedicine drug delivery system, nanoscale structures act as a carrier of various substances like nanorobots [166], sensory [167], delivery of drugs [168], natural polysaccharides [169, 170], antibodies [171], tunable surfactants [172], cell membrane [173], and peptides [174] to target regions and improve specificity of organisms through chemical functionalization. The system for controlled drug release through employment of different nano drug carriers is shown in Fig. 10.13. There are two main paths to deliver drugs through nanostructures: first is passive and second is self-delivery. In the foregoing, hydrophobic effect is being used to incorporate the drugs in structural cavity. When nanomaterials are employed toward specific sites, the proposed drug amount is allowed due to small content of drugs which is surrounded by hydrophobic contexture [175]. Moreover, the purposive drug is directly associated with nanocarrier for simple delivery. In this way, the drug releasing time is critical as the drug will dissociate from the MNPs before

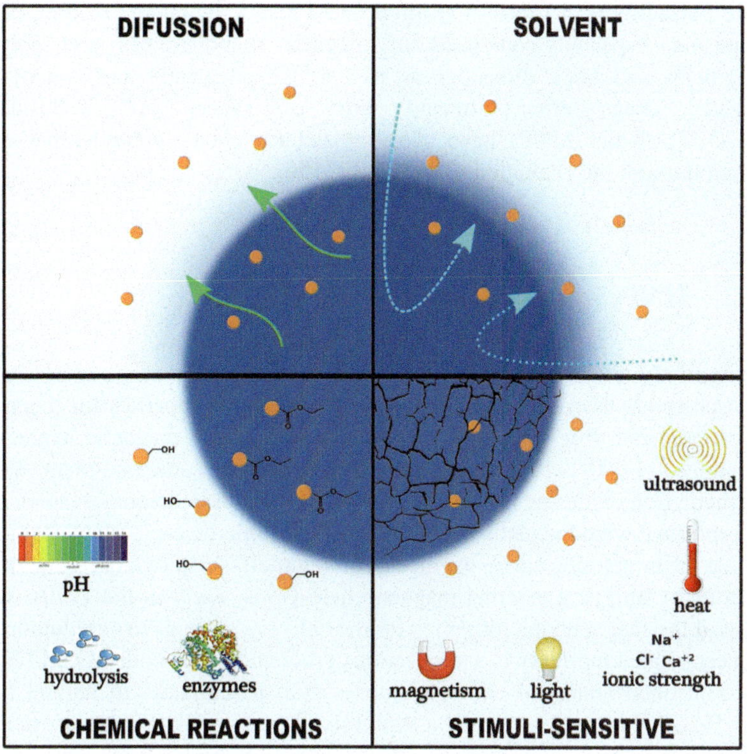


**Fig. 10.13** Systems for drug release by employment of various nanocarriers. (Reprinted with permission from reference [177])

reaching to the desired target; however, its efficiency and bioactivity will be minimized if it is dissociated at right time from the nanocarrier [175]. Targeting of drugs is further classified into two aspects, active and passive, by usage of nanoformulations. In the mechanism of active targeting, moieties, such as peptides and antibodies, are associated with delivery system to hold them toward receptors of the target site. The passive targeting mechanism is a little bit different from active targeting because the desired drug carrier circulated in the bloodstream and propelled to the target site through affinity affected by properties like temperature, shape, pH, and molecular site. The receptors are primary targets on cell membranes, antigens on the surfaces of cell, and lipid elements of cell membrane [176]. Currently, drug deliveries based on nanoformulations are mostly directed toward the cure and treatments of cancer.

Hence, stimuli-responsive carriers at nanoscale exhibit capability to regulate the drug release through external actions such as pH [178], heat [179–181], ultrasound [182], light [183], magnetism [184, 185], and ionic strength [186], which help to enhance the targeting ability and control of dosage. For example, by using external magnetic field, superparamagnetic iron oxide nanoparticles are conjugated with lipids [187] or polymeric nanocarrier [188] to activate a controlled drug release system. Furthermore, Ulbrich et al. proposed a recent attainment of drug delivery mechanisms, especially on basis of magnetic nanoparticles and polymeric nanoparticles and mark directions as well to the covalently and noncovalently conjugated drugs effects toward cure of cancer [189]. Besides it, Au/$Fe_3O_4$@polymer MNPs have been utilized for chemo-photothermal therapy application under the excitation of NIR light [190].

## 10.5.2 MNPs in Drug Delivery

Meaningful clinical investigations have been studied for the usage of MNPs especially iron oxide-based nanoparticles due to its unique properties for theranostics such as magnetic drug targeting, molecular imaging, and magnetic hyperthermia applications [117, 191–193]. In this section, we illustrate some important examples of magnetic iron oxide and iron-based nanoparticles in drug delivery system. Due to high response toward magnetic field of magnetic nanoparticles, it establishes them to penetrate in target site either by means of internally employment of permanent magnet or by applying external magnetic field [194]. Early in the 1970s, Widder et al. used the magnetic microspheres for the delivery of drugs toward tumor [195].

Cancer-related mortality is a very serious problem in this modern world because the uncontrolled abnormal cell accumulates and disseminates to normal healthy tissues [126]. Gao et al. proposed a synthesis of 8 nm of $Fe_3O_4$ nanoparticles and cisplatin which are dual loaded with magnetic liposome for generous ROS (Fig. 10.14) [196]. The apoptosis of mitochondria and DNA damage is caused by intracellular ROS, restoring the system in cisplatin-resistant A549 (A549/R) cells. In this work, novel magnetic liposome was designed, consisting of drug, targeting ligands, and iron oxide MNPs. The generation of $Fe^{2+}/Fe^{3+}$ from $Fe_3O_4$ MNPs in the acid liposome causes to cisplatin and catalyzed anticancer drug artemisinin (ART) to form toxic ROS via Fenton reaction, which improved the cisplatin anticancer effect with much less side effects.

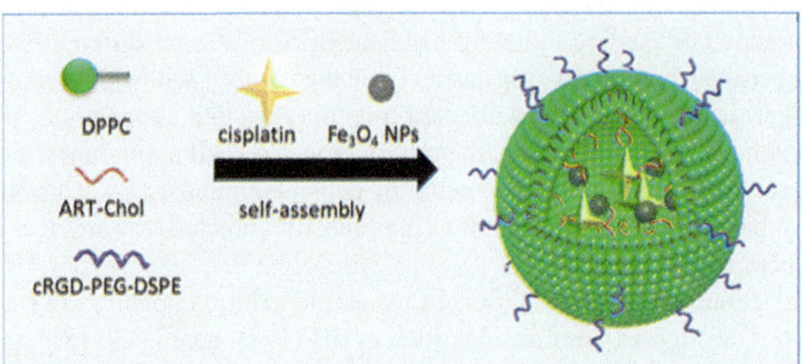


**Fig. 10.14** Graphical synthesis mechanism of the magnetic liposomes cRGD-AFePt@NPs. (Reprinted with permission from reference [196])

Further the loading of cargos carrying $Fe_3O_4$ and cisplatin was estimated by respective formula.

$$\text{DLC (wt\%)} = \frac{\text{weight of cargos (cisplatin or Fe O NPs)}}{\text{weight of cRGD AFePt@NPs}}$$

In vitro examination of A549/R cells exhibits that cRGD-AFe@NPs showed a 15.17-fold lower $IC_{50}$ value of cisplatin ($IC_{50} = 32.47$ μM). Furthermore, increased intracellular ROS formation and apoptosis of cell are showed by flow cytometry tests.

Various diseases can be cured with targeted drug delivery in better way by means of magnetic nanoparticles. According to the World Health Organization, cardiovascular diseases caused the death of 17.7 million people in 2015, a toll that comprises ~31% of mortality throughout the world [197]. Thrombolysis is the type of blood clot which may lead to different mortalities and disabilities as outcomes. Prilepskii et al. proposed a novel magnetic nanocomposite with thrombolytic drug synthesized by heparin-mediated cross-linking of urokinase (MNPs@uPA) [198].

This novel magnetic thrombolytic drug nanocomposite was synthesized by extension of urokinase-type plasminogen trigger to MNPs with heparin as the binding agent (Fig. 10.15a). This nanocomposite MNPs@uPA exhibits superparamagnetic behavior as shown in Fig. 10.15b. Subsequently, examine the nanocomposite on thrombolytic activity in clot model, found that in the presence of

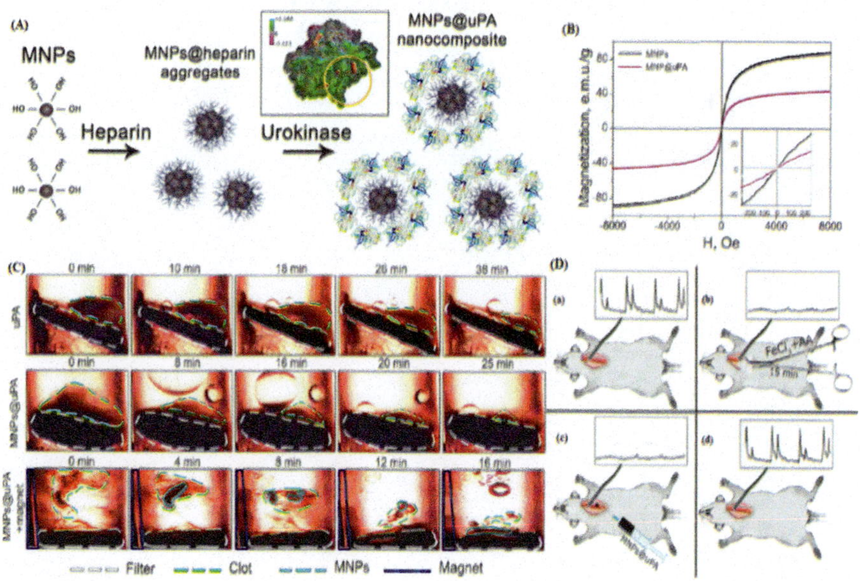


**Fig. 10.15** (**a**) Synthesis process of MNPs@uPA. (**b**) Magnetic property (M-H) curve of MNPs@uPA and of pure magnetite. (**c**) Dissolution of the model clot exposed to nanoparticles. (**d**) Thrombolytic activity of MNPs@uPA in vivo. (Reprinted with permission from reference [198])

magnetic field the rate of clot reductions being carried out in 12 min with (MNPs@uPA) and take mean value of 5 measurements (Fig. 10.15c). Furthermore, the scheme (Fig. 10.15d) investigation of this novel thrombolytic drug composite is also examined in vitro. Firstly, in rat carotid artery clot was formed through imposing of cotton piece with ascorbic acid and $FeCl_3$. After injection of MNPs@uPA within 0.5–10 min, the blood flow resumed, and the gradual decline of clot continued. This new composition is FDA approved, and biocompatible components also exhibit thrombolytic efficiency by at least 30%.

As an improvement of hydrophobic drugs in vivo achievable by nanoparticle formulation through circumventing solubility effect and handover targeted delivery, Hsiao et al. presented a mechanism for hydrophobic drug PTX attached on the surface of hydrophobic iron oxide NPs and then coated with chitosan-PEG copolymer. Later, chlorotoxin (CTX) was attached onto the MNPs as a targeting agent. This work has significance to overcome the hurdles of hydrophobic drugs to the specific area, and IONPs easily manipulate with the external magnetic field which is another advantage of this work. The hydrophobic CTX drug contains 36-amino acid peptide and particularly enchains to metalloproteinase-2 (MMP-2) and was used for proposed human glioblastoma (GBM) cell. The synthesis mechanism is given in Fig. 10.16a. The prepared PTX-NPs are very stable and did not show any

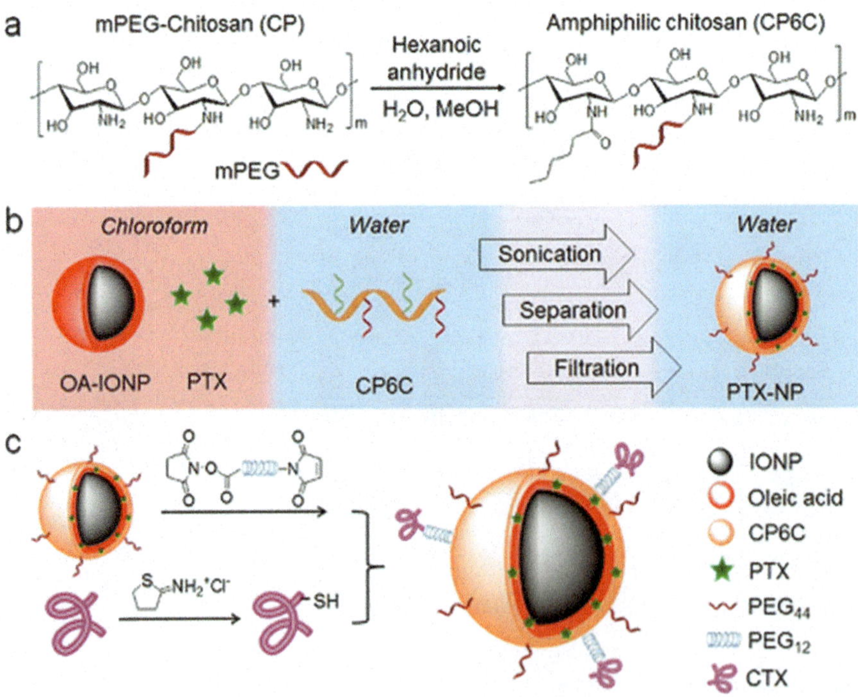


**Fig. 10.16** Schematic description of preparation of CP6C copolymer-coated, PTX-loaded, and CTX-conjugated IONPs. (Reprinted with permission from reference [199])

aggregation in the hydrodynamic size measurements. On the other hand, most of the liposomes are to be left in serum at 37 °C for few hours [200–202]. Further, investigate the cellular uptake of NPs with the help of confocal laser scanning microscope (CLSM). While, nonspecific uptake of PTX-NP by cell was investigated which shows that biniding of CTX-PTX-NP accepted by cells which shows that binding of CTX refined the cellular uptake. Next, cell viability and the therapeutic performance were investigated on U-118 MG cells using different groups of synthesized MNPs. Accordingly, improved cellular drug growth-mediated mechanism by CTX in CTX-PTX-NP enhanced the killing of tumor cell. Further, CP6C copolymer-coated iron oxide nanoparticles are promising platform for hydrophobic drug delivery system and can be replaced with different ligands for imaging and targeting applications.

As with the passage of time, nanoparticles have gained much more attention toward delivery of vaccine to tumor cells. Furthermore, vaccine delivery system based on nanoformulation has great ability to enhance the vaccine immunogenicity.

Conventionally, the nanoparticle-adjuvant vaccine articulation is classified into two types. In the first type, nanoparticles acted as delivery system by delivery of antigen to immune system either by delivery of antigen to target site or through congesting by immune cell [204]. The remaining type is based on nanoparticles used as immune potentiator to improve antigen operation. Zhao et al. reported the mechanism of vaccine delivery based on coupling of Ovalbumin Endofit with iron oxide nanoparticles ($Fe_3O_4$-OVA) to inhibit the tumor growth [203]. In vitro cellular uptake of $Fe_3O_4$-OVA is being studied on treatment of CT26 cells for 24 h. Accordingly, tumor growth was remarkably inhibited through incubation of $Fe_3O_4$-OVA NPs against single $Fe_3O_4$ and soluble OVA which had comparable PBS control group (Fig. 10.17e). Additionally, the in vivo cytotoxicity of $Fe_3O_4$-OVA nanovaccine was studied, and intratumoral injection results showed that this nanovaccine has better compatibility after investigating organ toxicity (Fig. 10.17c). Results revealed that $Fe_3O_4$-OVA NPs have strong effect in immunity reaction and tumor ablation as well as can be employed as a general system for cancer vaccines.

## 10.6 Magnetic Resonance Imaging (MRI)

Worldwide scientists have been paying attention to improve the cancer diagnosis techniques at early stage by combining different scientific disciplines. Many molecular imaging techniques have been available for early diagnosis of serious injuries and different types of cancers. However, magnetic resonance imaging gained significant interest in clinical usage and for medical research purpose owing to its special features such as noninvasive, selectivity, and sensitivity as compared to other imaging modalities. This molecular imaging technique successfully maps severe injuries, abnormalities, heart problems, and cancer tumor [205, 206].

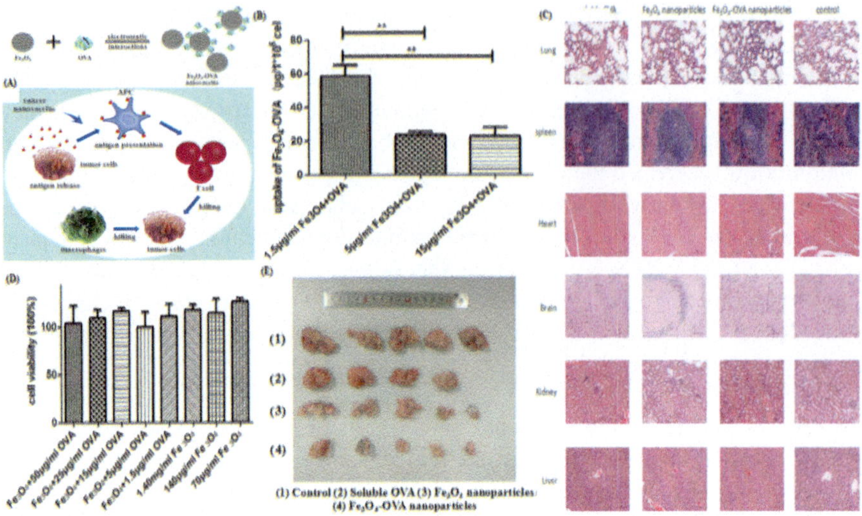


**Fig. 10.17** (**a**) Schematic and working principle of $Fe_3O_4$-OVA. (**b**) The concentration of MNPs inside DCs cells. (**c**) Histological examination of different major organs treated with different groups of NPs. (**d**) Cell toxicity experiment. (**e**) Photograph of tumors after treatment. (Reprinted with permission from reference [203])

## 10.6.1 MRI Contract Agents

Initially, MRI was designed for imaging purpose because it does not contain any harmful radiations. However, low sensitivity was the major drawback of MRI when compared with another diagnosis technique such as nuclear imaging. Later, this shortcoming is resolved by the development of magnetic contrast agents (MCAs), and then MRI attained a significant position in clinical diagnosis which is a promising tool for biomedical research [207]. MCAs increase sensitivity and/or specificity because of the shortened relaxation time of protons. The relaxivity values of $r_1$ and $r_2$, which are generally measured in $mM^{-1}\ s^{-1}$, demonstrate the increase in $1/T_1$ and $1/T_2$ per concentration [M] of contrast agent, respectively. The expression is given below.

$$\frac{1}{T_i} = \frac{1}{T_{i,0}} + r_i[M]\ (i = 1, 2) \tag{10.8}$$

The efficiency of CAs in MRI is measured in expression of the ratio between the transverse and longitudinal relaxivities ($r_2/r_1$). CAs are strongly depending on this $r_2/r_1$ ratio because by using this, we can decide whether the CAs can be employed as a positive ($T_1$) or negative ($T_2$) agent.

Furthermore, from the clinical applications point of view, clinical contrast agents can be categorized into two types: (1) positive ones (blood pool agents, hepatobiliary agents, and extracellular agents) and (2) negative ones (blood pool imaging and passive targeting) [208]. Magnetic resonance probes are further categorized into two types: the first one is paramagnetic material-based CAs such as gadolinium ($Gd^{3+}$) or manganese ($Mn^{2+}$) which are usually used as $T_1$ MR imaging, and the second type is based on superparamagnetic CAs such as pure iron oxide NPs and iron oxide-based NPs which are generally used for $T_2$ MR imaging. Paramagnetic agents are coupled with some chelate to control the toxicity of free ions. Sometimes polymer or coating substances are used to improve the biocompatibility and aggregation problems of the iron oxide NPs. Superparamagnetic agents have a considerably better magnetic moment as compared to paramagnetic agents [209, 210]. Mostly, these superparamagnetic iron oxide agents are used for $T_2$. However, a new class, ultrasmall (5 nm) iron oxide NP, has been introduced to increase the $T_1$ MR imaging [211, 212].

### *10.6.2 The Longitudinal Relaxation ($T_1$) Agents*

Positive contrast agents decrease the $T_1$ and $r_2/r_1$ and generate bright image. Generally, gadolinium (III)-based complexes, which have seven unpaired electrons on the surface, are used to create $T_1$-weighted MRI. These ions enhance the relaxation rate of the adjacent and exchangeable water molecules and as a result increase the intensity in $T_1$-weighted images. The longitudinal relaxation reveals a thermal loss from the spin system to its surroundings (lattice). The intricated paramagnetic ion is embedded by three kinds of water molecule. Firstly, water molecules are attached directly to the paramagnetic ion at inner sphere. These kinds of molecules can exchange with the surrounding bulk water and generate the strongest effect on the overall relaxivity because of direct link with the paramagnetic ion. Hydrogen bonded to the organic chelate linking the paramagnetic ion at second sphere or middle sphere of the water molecule.

They experience fast exchange with the nearby bulk water, and their relaxivity is less affected. Finally, the outer sphere or bulk water including the water from surrounding the complex as the contrast agent diffuses through the tissue (Fig. 10.18).

The magnetization of paramagnetic materials is directly based on the number of ions and electrons of unpaired ions. The configuration of paramagnetic metal ions with magnetic moment is given in Table 10.1.

Some other transition metals also have unpaired ions, but they do not fulfill the requirement of Larmor frequency rule. However, ions of iron, manganese, and gadolinium perfectly meet the above requirements. The major difficulty is to decrease the toxicity effect of paramagnetic heavy metal ions. Several gadolinium (III)-based contrast agents have been available commercially, and the most common are (1) Magnevist (Bayer HealthCare Pharmaceuticals), (2) ProHance (Bracco Diagnostics), and (3) Omniscan (GE Healthcare). These complexes are all administered intravenously (0.1 mmol/Kg dose) and are pass out from the kidneys within 24 h. They enhanced image efficiency in the brain, lesions, abnormalities, associated tissues, and spine, but in pathogenic tissues, such as certain tumors, current contrast agents are not perfect in visualization.

Magnetic nanoparticles based on gadolinium and manganese have a quite stronger paramagnetic property in comparison with iron oxide nanoparticles. For

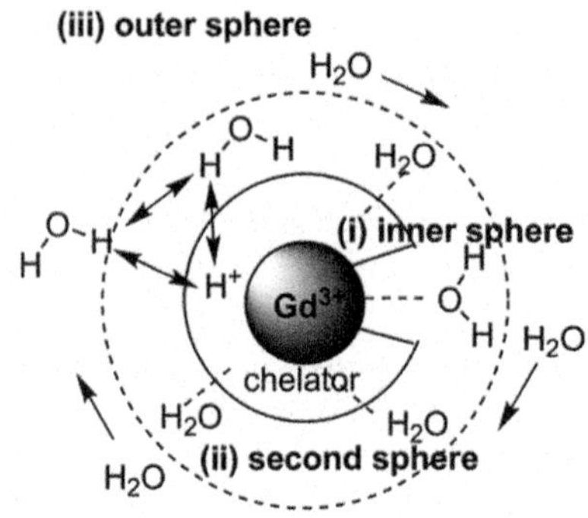


**Fig. 10.18** The three types of water molecule which undergo a change in relaxivity when in proximity to a paramagnetic ion. (Reprinted with permission from reference [213])

**Table 10.1** List of some important paramagnetic metal ions with their configuration and moment. (Reprinted with permission from reference [214])

| Ion | Configuration | Magnetic moment |
|---|---|---|
| $^{24}Cr^{3+}$ | ↑ ↑ ↑ — — | 3.9 |
| $^{25}Mn^{2+}$ | ↑ ↑ ↑ ↑ ↑ | 5.9 |
| $^{26}Fe^{3+}$ | ↑ ↑ ↑ ↑ ↑ | 5.9 |
| $^{29}Cu^{2+}$ | ↑↓ ↑↓ ↑↓ ↑↓ ↑ | 1.7 |
| $^{63}Eu^{3+}$ | ↑↓ ↑ ↑ ↑ ↑ ↑ ↑ | 3.4 |
| $^{64}Gd^{3+}$ | ↑ ↑ ↑ ↑ ↑ ↑ ↑ | 7.9 |
| $^{66}Dy^{3+}$ | ↑↓ ↑↓ ↑ ↑ ↑ ↑ ↑ | 10.6 |

example, in case of PEG (polyethylene glycol)-coated $Gd_2O_3$, the $r_2/r_1$ is around 1, while in commercial iron oxide nanoparticles (Feridex) as negative contrast agents, the $r_2/r_1$ value is about 10. Further, some $Gd_2O_3$ nanoparticles are more sensitive than clinical imaging CAs such as Gd-DOTA complex [215].

Apart from the $Gd^{3+}$-based contrast agents for $T_1$ MR imaging, recently $Mn^{2+}$-based nanoparticles have obtained a considerable attraction. Manganese is one kind of paramagnetic contrast agent which showed better efficiency in cardiac and hepatic MRI. Particularly, it is used to observe the mapping of functional brain regions and anatomical structures. In manganese case, cellular toxicity issues are still hurdle to be a perfect contrast agent [216]. Also, the unpaired electron of $Mn^{2+}$ decreases its stability, and therefore it is very exciting to prepare the stable and homogeneous CAs based on manganese. The effective performance of manganese as a MCA is strongly based on the affected site-specific delivery and low dose to avoid the negative signal, maintaining the detectability by MRI.

Another paramagnetic contrast agent is Dy. Dye-based CAs are also extensively utilized as $T_1$ contrast agents due to high value of intrinsic paramagnetism of lanthanide ion. Hydrophilic $Dy^{3+}$ agent coated by amphiphilic polymers linked with liposome enhances the visualization ability in cellular epitopes even at very low concentrations. Moreover, at 7 T, unilamellar liposome loaded with Dy-HPDO3A shows $T_2$ relaxation enhancement as compared to superparamagnetic iron oxide nanoparticles [65, 217].

Importantly, in the recent years, the Food and Drug Administration (FDA) authority has added a warning notice on gadolinium CA injection because of increasing the risk of nephrogenic fibrosing dermopathy (NFD) and nephrogenic systemic fibrosis (NSF) for patients who have renal failure problem. These patients are encouraged to find the alternative CAs instead of $Gd^{3+}$. This kind of disease can cause fibrosis of the joints, internal organs, eyes, and skin which if stern enough may lead to the patient's death. The abovementioned diseases are associated with Omniscan contrast agent, a nonionic linear agent, which has Kd $= 1.4 \times 10^{-17}$ M and shows poor kinetic stability for $Gd^{3+}$ and is easily released into the blood. The contrast agents with poor kinetic stability are easy to discharge in the kidneys due to lower pH value as compared to surrounding tissues. Recently, new research work has carried out Gd(III)-DTPA chelates with dendrimer [218, 219].

### *10.6.3 The Transverse Relaxation (T2) Agents*

Iron oxide nanoparticles present the superparamagnetic behavior which is significant for $T_2$ dark (negative) contrast on MRI in comparison with paramagnetic ions. The aptitude of SPIONs to achieve as MR imaging CAs is aggressively examined in the last 20 years. MNPs with controlled shape, size, and magnetic features are intensively studied as MRI contrast agents. However, several other organic or inorganic layers have been established to increase their stability, functionalization, targeting, and toxicity. Most SPION-based clinical-approved MRI contrast agents are coated with dextran citrate-stabilized particles (VSOP-C184) or carbohydrate [220, 221]. Some polymers, metals, and silica are also in development phase or in different stages of the preclinical research. The versatile coating layers provided nanoparticles as platform for additional fictionalization by conjugation with multi-domain targeting entities such as peptides and antibodies. The magnetic moment of SPIONPs is much higher than paramagnetic and consequently needs a low concentration for the contrast agent purposes. In the last two decades, several reports have demonstrated the detailed investigation on the iron oxide, iron oxide cores with different sizes, and their applications in theranostics [222]. The controls over size, shape, and magnetic, intrinsic, and extrinsic properties of the nanoparticles are major factors to develop excellent $T_2$ contrast agents. It is obvious, MNPs showed contrast enhancment under applied alternative magnetic fields compare with water [223].

It is observed that there was no dipole moment in the absence of magnetic field, and by applying an external field, the MNPs induced magnetic dipole moment in the surrounding water molecules. And it results in the shortening of the spin-spin relaxation time of the proton. Various concentrations of MNPs are uptaken by different tissues with variable $T_2$ values and create distinct MR images. Healthy cells contain an effective reticuloendothelial system (RES) as compared to the cancerous cells. Therefore the relaxation time of the tumor cells will not have reformed by the CAs. These phenomena make them apparent from the adjacent healthy cells [224].

MNPs of iron oxide are confirmed to be a better substitution for the gadolinium-based complexes like Gd-DTPA as CAs owing to their high relaxivity, biocompatibility, and low toxicity [225]. Therefore, little dose of SPIONPs was required into the human body than paramagnetic material-based CAs. Currently, commercialized available gadolinium required 10-7 mol/g to obtain high-quality image which is very high, and typical receptors usually need low concentrations of $10^{-9}$ to $10^{-13}$ mol/g. Therefore, it is very important to resolve the sensitivity issue. Many MNPs have been under investigation as CAs, but magnetite is a well-examined material with very low toxicity and rapidly cleared from the organism [226]. Wittenberg performed a very refined and detailed research on different sizes, their effects on magnetization, and relaxivities of MRI as shown in Fig. 10.19. When the particle size is increasing, the $T_2$-weighted contrast is enhanced significantly. In order to increase the biocompatibility, inorganic oxides (silica and carbon), organic surfactants (dodecylamine and sodium oleate), organic polymers (chitosan, dextran, polyethylene glycol, polyaniline, and polysorbate), inorganic metals (gold), as well as bioactive structures and molecules (ligands/receptors, liposomes, and peptides) have been introduced on the surface of magnetic nanoparticles [227].

The aggregation and embolization are the important issues related to MNPs that start to restrict blood flow. In addition, most of the MNPs entered in the liver and produced cytotoxicity which is another unwanted side effect. Small particles with biodegradable coating substance provide a platform to overcome these circumstances. Previous studies have shown that 15.0 nm nanoparticles tend to be cleared by the liver, whereas ultrasmall particles (5.0 nm) can be excreted quickly by the kidney [228].

## 10.7 Conclusion and Prospective

Nanotechnology holds an immense potential in energy, agriculture, and environmental and biomedical applications. Nanomaterials offer a significant opportunity to ease the daily life from traditional to modern. Engineered magnetic nanoparticles present multiple solutions for the diagnosis and treatment of many dangerous diseases due to their unique characteristics. The magnetic characteristics of single-domain iron oxide nanoparticles are strongly based on the saturation magnetization, nanoparticle size and aggregation, Néel relaxation time, and anisotropy constant.

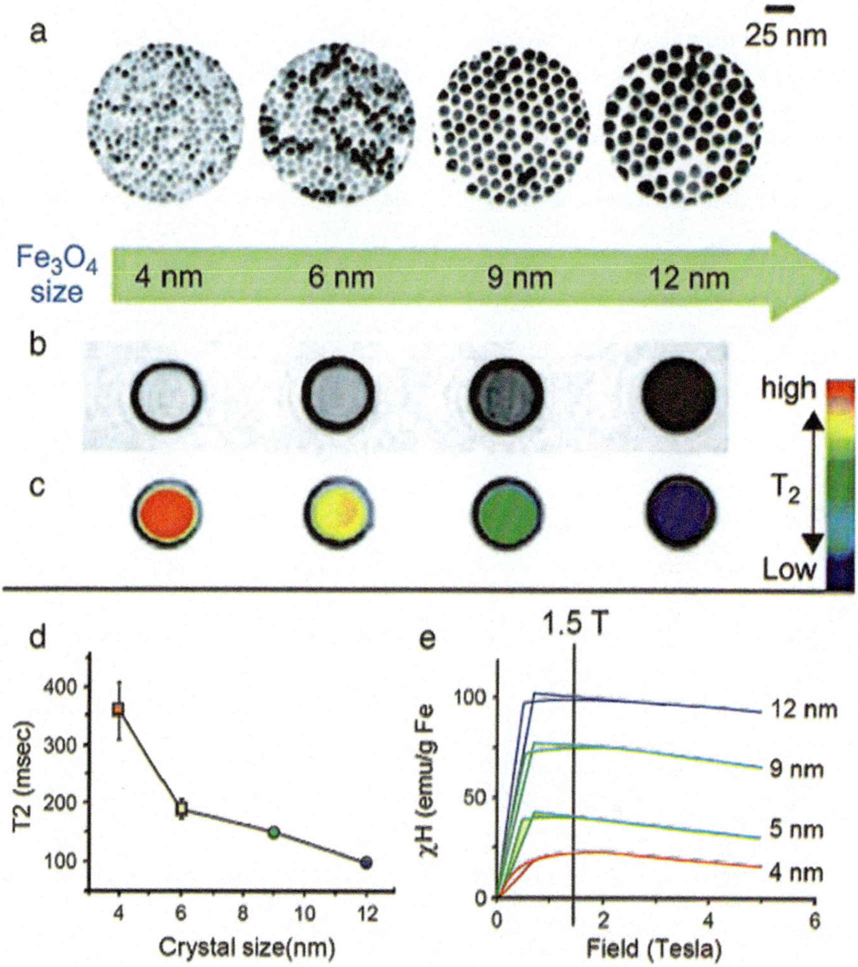


**Fig. 10.19** (**a**) TEM images of nanoparticles of $Fe_3O_4$ nanoparticle of different sizes, (**b**, **c**) size-based $T_2$ characteristics and (**d**) $T_2$ values with respect to size, and (**e**) magnetization characteristics according to size [229]

MNPs have showed hyperthermia effect by applying an alternative field, that is, activation of MNP brings targeted therapeutic heating of tumor. MNPs actively participated as a therapeutic agent, but there is still a need to consider the size-dependent temperature change. Also, mostly studies explained about the intratumor killing effect, therefore, intravenous targeting and produce hyperthermia inside cancer cell will be remarkable edition in the field on cancer theranostics. Delivery systems using magnetic nanoparticles are promising drug transporter. However, there is still a need to develop some accurate methods to improve the delivery efficiency that is needed to have proper surface functionalities, achieving a maximum targeted delivery of drugs. These properties of the magnetic nanoparticles play a significant character where these features are essential not only in MR imaging and relaxivities properties of $T_1$- and $T_2$-weighted contrast agents but also are potential applicability in medical areas. Finally, the improvement of synthesizing magnetic nanoparticle probes is still in its infancy, but it has contributed a lot in developing MRI. The use of nanoparticles as contrast agents, advancing toward clinical implementation, to improve the relaxivity coefficient, long-term cytotoxicology, storage, and kinetic stability, should continue to be taken into consideration.

**Acknowledgments** The authors would like to thank the continuous support by National Key R&D Program of China (2018YFC0910601), Natural Science Foundation of China (U1432114 to Aiguo Wu and 81950410638 and 81650410654 to M. Zubair Iqbal), Zhejiang Province Financial Supporting (2017C03042, LY18H180011), and the Science & Technology Bureau of Ningbo City (2015B11002, 2017C110022). Furthermore, the authors also acknowledge Shanghai Synchrotron Radiation Facility at Line BL15U (No. h15 sr0021) used for X-ray fluorescence imaging and National Synchrotron Radiation Laboratory in Hefei used for soft X-ray imaging (No. 2016-HLS-PT-002193).

## References

1. Baker JR, Quintana A, Piehler L, Banazak-Holl M, Tomalia D, Raczka E. The synthesis and testing of anti-Cancer therapeutic Nanodevices. Biomed Microdevices. 2001;3:61–9.
2. Savage N, Thomas TA, Duncan JS. Nanotechnology applications and implications research supported by the US Environmental Protection Agency STAR grants program. J Environ Monit. 2007;9:1046–54.
3. Martin CR. Welcome to nanomedicine. Nanomedicine. 2006;1:5.
4. Sonawane GH, Patil SP, Sonawane SH. Chapter 1 - Nanocomposites and its applications. In: Mohan Bhagyaraj S, Oluwafemi OS, Kalarikkal N, Thomas S, editors. Applications of nanomaterials. Cambridge, UK: Woodhead Publishing; 2018. p. 1–22.
5. Rahman M, Rebrov E. Microreactors for gold nanoparticles synthesis: from faraday to flow. PRO. 2014;2:466.
6. Schoonman J. Nanostructured materials in solid state ionics. Solid State Ion. 2000;135:5–19.
7. Rabia Riasat NG, Riasat Z, Aslam I, Sakeena M. Effects of nanoparticles on gastrointestinal disorders and therapy. J Clin Toxicol. 2014;6:10.
8. Iqbal MZ, Wang F, Zhao H, Rafique MY, Wang J, Li Q. Structural and electrochemical properties of SnO nanoflowers as an anode material for lithium ion batteries. Scr Mater. 2012;67:665–8.
9. Iqbal MZ, Wang F, Feng T, Zhao H, Rafique MY, Rafi ud D, Farooq MH, Javed Q u a J, Khan DF. Facile synthesis of self-assembled SnO nano-square sheets and hydrogen absorption characteristics. Mater Res Bull. 2012;47:3902–7.
10. Salata OV. Applications of nanoparticles in biology and medicine. J Nanobiotechnol. 2004;2:3–3.
11. Yang L, Zhang X, Ye M, Jiang J, Yang R, Fu T, Chen Y, Wang K, Liu C, Tan W. Aptamer-conjugated nanomaterials and their applications. Adv Drug Deliv Rev. 2011;63:1361–70.
12. Reeves DB, Weaver JB. Approaches for modeling magnetic nanoparticle dynamics. Crit Rev Biomed Eng. 2014;42:85–93.
13. Gun'ko Y. Magnetic nanomaterials and their applications. Nano. 2014;4:505.

14. Al Lehyani SHA, Hassan R, Alharbi AA, Alomayri T, Alamri H. Magnetic hyperthermia using cobalt ferrite nanoparticles: the influence of particle size. Research Article. Int J Adv Tech. 2017;8:6.
15. Pankhurst QA, Connolly J, Jones SK, Dobson J. Applications of magnetic nanoparticles in biomedicine. J Phys D Appl Phys. 2003;36:R167.
16. Malekigorji M, Curtis ADM, Hoskins C. The use of iron oxide nanoparticles for pancreatic cancer therapy. J Nanomed Res. 2014;1:12.
17. Lee H, Shin T-H, Cheon J, Weissleder R. Recent developments in magnetic diagnostic systems. Chem Rev. 2015;115:10690–724.
18. Ito A, Shinkai M, Honda H, Kobayashi T. Medical application of functionalized magnetic nanoparticles. J Biosci Bioeng. 2005;100:1–11.
19. Masih D, Frank S, Joachim L, Nathalie R, Biplab S, Werner K, Heiko W. Nanoscale size effect on surface spin canting in iron oxide nanoparticles synthesized by the microemulsion method. J Phys D Appl Phys. 2012;45:195001.
20. Kai W, Liang T, Diqing S, Jian-Ping W. Magnetic dynamics of ferrofluids: mathematical models and experimental investigations. J Phys D Appl Phys. 2017;50:085005.
21. Kim T, Shima M. Reduced magnetization in magnetic oxide nanoparticles. J Appl Phys. 2007;101:09M516.
22. Dutta P, Pal S, Seehra MS, Shah N, Huffman GP. Size dependence of magnetic parameters and surface disorder in magnetite nanoparticles. J Appl Phys. 2009;105:07B501.
23. Demortière A, Panissod P, Pichon BP, Pourroy G, Guillon D, Donnio B, Bégin-Colin S. Size-dependent properties of magnetic iron oxide nanocrystals. Nanoscale. 2011;3:225–32.
24. Teja AS, Koh P-Y. Synthesis, properties, and applications of magnetic iron oxide nanoparticles. Prog Cryst Growth Charact Mater. 2009;55:22–45.
25. Lu A-H, Salabas EL, Schüth F. Magnetic nanoparticles: synthesis, protection, functionalization, and application. Angew Chem Int Ed. 2007;46:1222–44.
26. Mohammadzadeh A, Sadri M, Seyed Afghahi SS, Alizadeh Y, Najafian S, Hosseini H. In vitro biocompatibility of low and medium molecular weight chitosan–coated Fe3O4 nanoparticles. Nanomed Res J. 2017;2:250–9.
27. Atila Dinçer C, Yildiz N, Karakeçili A, Aydoğan N, Çalimli A. Synthesis and characterization of Fe3O4-MPTMS-PLGA nanocomposites for anticancer drug loading and release studies. Artif Cells Nanomed Biotechnol. 2017;45:1408–14.
28. Jeun M, Lee S, Kang JK, Tomitaka A, Kang KW, Kim YI, Takemura Y, Chung K-W, Kwak J, Bae S. Physical limits of pure superparamagnetic Fe3O4 nanoparticles for a local hyperthermia agent in nanomedicine. Appl Phys Lett. 2012;100:092406.
29. Ma P, Luo Q, Chen J, Gan Y, Du J, Ding S, Xi Z, Yang X. Intraperitoneal injection of magnetic $Fe_3O_4$-nanoparticle induces hepatic and renal tissue injury via oxidative stress in mice. Int J Nanomedicine. 2012;7:4809–18.
30. Patil JV, Mali SS, Kamble AS, Hong CK, Kim JH, Patil PS. Electrospinning: a versatile technique for making of 1D growth of nanostructured nanofibers and its applications: An experimental approach. Appl Surf Sci. 2017;423:641–74.
31. Khan I, Saeed K, Khan I. Nanoparticles: properties, applications and toxicities. Arab J Chem. 2017; https://doi.org/10.1016/j.arabjc.2017.05.011.
32. Goel S, Chen F, Cai W. Synthesis and biomedical applications of copper sulfide nanoparticles: from sensors to theranostics. Small (Weinheim an der Bergstrasse, Germany). 2014;10:631–45.
33. Annu, Ali A, Ahmed S. Green synthesis of metal, metal oxide nanoparticles, and their various applications. In: Martínez LMT, Kharissova OV, Kharisov BI, editors. Handbook of ecomaterials. Cham: Springer; 2018. p. 1–45.
34. Grill L, Dyer M, Lafferentz L, Persson M, Peters MV, Hecht S. Nano-architectures by covalent assembly of molecular building blocks. Nat Nanotechnol. 2007;2:687.
35. Ikkala O, ten Brinke G. Functional materials based on self-assembly of polymeric Supramolecules. Science. 2002;295:2407–9.

36. Rawat RS. Dense plasma focus-from alternative fusion source to versatile high energy density plasma source for plasma nanotechnology. J Phys Conf Ser. 2015;591:25.
37. Sun S, Murray CB, Weller D, Folks L, Moser A. Monodisperse FePt nanoparticles and ferromagnetic FePt nanocrystal superlattices. Science. 2000;287:1989–92.
38. Shevchenko EV, Talapin DV, Rogach AL, Kornowski A, Haase M, Weller H. Colloidal synthesis and self-assembly of CoPt3 nanocrystals [J. Am. Chem. Soc. 2002, 124, 11480−11485]. J Am Chem Soc. 2002(124):13958.
39. Park J, An K, Hwang Y, Park J-G, Noh H-J, Kim J-Y, Park J-H, Hwang N-M, Hyeon T. Ultra-large-scale syntheses of monodisperse nanocrystals. Nat Mater. 2004;3:891.
40. Armijo LM, Brandt YI, Mathew D, Yadav S, Maestas S, Rivera AC, Cook NC, Withers NJ, Smolyakov GA, Adolphi NL, Monson TC, Huber DL, Smyth HDC, Osiński M. Iron oxide nanocrystals for magnetic hyperthermia applications. Nano. 2012;2:134.
41. Grasset F, Labhsetwar N, Li D, Park DC, Saito N, Haneda H, Cador O, Roisnel T, Mornet S, Duguet E, Portier J, Etourneau J. Synthesis and magnetic characterization of zinc ferrite nanoparticles with different environments: powder, colloidal solution, and zinc ferrite−silica Core−Shell nanoparticles. Langmuir. 2002;18:8209–16.
42. Park S-J, Kim S, Lee S, Khim ZG, Char K, Hyeon T. Synthesis and magnetic studies of uniform Iron Nanorods and Nanospheres. J Am Chem Soc. 2000;122:8581–2.
43. Puntes VF, Krishnan KM, Alivisatos AP. Colloidal nanocrystal shape and size control: the case of cobalt. Science. 2001;291:2115–7.
44. Benjamin JS. Dispersion strengthened superalloys by mechanical alloying. Metall Trans. 1970;1:2943–51.
45. Lukashev RV, Alekova AF, Korchagina SK, Chibirova FK. Mechanical processing of γ-Fe2O3. Inorg Mater. 2015;51:134–7.
46. Arbain R, Othman M, Palaniandy S. Preparation of iron oxide nanoparticles by mechanical milling. Miner Eng. 2011;24:1–9.
47. Chen C-N, Chen Y-L, Tseng WJ. Surfactant-assisted de-agglomeration of graphite nanoparticles by wet ball mixing. J Mater Process Technol. 2007;190:61–4.
48. Jiang Y, Liu J, Suri PK, Kennedy G, Thadhani NN, Flannigan DJ, Wang J-P. Preparation of an α″-Fe16N2 magnet via a ball milling and shock compaction approach. Adv Eng Mat. 2016;18:1009–16.
49. Chakka VM, Altuncevahir B, Jin ZQ, Li Y, Liu JP. Magnetic nanoparticles produced by surfactant-assisted ball milling. J Appl Phys. 2006;99:08E912.
50. Yiping W, Yang L, Chuanbing R, Liu JP. Sm–co hard magnetic nanoparticles prepared by surfactant-assisted ball milling. Nanotechnology. 2007;18:465701.
51. Kim EH, Lee HS, Kwak BK, Kim B-K. Synthesis of ferrofluid with magnetic nanoparticles by sonochemical method for MRI contrast agent. J Magn Magn Mater. 2005;289:328–30.
52. De Matteis L, Custardoy L, Fernández-Pacheco R, Magén C, de la Fuente JM, Marquina C, Ibarra MR. Ultrathin MgO coating of superparamagnetic magnetite nanoparticles by combined coprecipitation and sol–gel synthesis. Chem Mater. 2012;24:451–6.
53. Xie C-Y, Meng S-X, Xue L-H, Bai R-X, Yang X, Wang Y, Qiu Z-P, Binks BP, Guo T, Meng T. Light and magnetic dual-responsive Pickering emulsion micro-reactors. Langmuir. 2017;33:14139–48.
54. Salazar-Alvarez G, Muhammed M, Zagorodni AA. Novel flow injection synthesis of iron oxide nanoparticles with narrow size distribution. Chem Eng Sci. 2006;61:4625–33.
55. Basak S, Chen D-R, Biswas P. Electrospray of ionic precursor solutions to synthesize iron oxide nanoparticles: modified scaling law. Chem Eng Sci. 2007;62:1263–8.
56. Rasekh M, Ahmad Z, Cross R, Hernández-Gil J, Wilton-Ely JD, Miller PW. Facile preparation of drug-loaded tristearin encapsulated superparamagnetic iron oxide nanoparticles using coaxial electrospray processing. Mol Pharm. 2017;14:2010–23.
57. Zhang Y, Yang Y, Duan H, Lü C. Mussel-inspired catechol-formaldehyde resin coated Fe3O4 Core-Shell magnetic Nanospheres: An effective catalyst support for highly active palladium nanoparticles. ACS Appl Mater Interfaces. 2018;10:44535.

58. Unni M, Uhl AM, Savliwala S, Savitzky BH, Dhavalikar R, Garraud N, Arnold DP, Kourkoutis LF, Andrew JS, Rinaldi C. Thermal decomposition synthesis of iron oxide nanoparticles with diminished magnetic dead layer by controlled addition of oxygen. ACS Nano. 2017;11:2284–303.
59. De Palma R, Peeters S, Van Bael MJ, Van den Rul H, Bonroy K, Laureyn W, Mullens J, Borghs G, Maes G. Silane ligand exchange to make hydrophobic superparamagnetic nanoparticles water-dispersible. Chem Mater. 2007;19:1821–31.
60. Shevchenko EV, Talapin DV, Rogach AL, Kornowski A, Haase M, Weller H. Colloidal synthesis and self-assembly of CoPt3 nanocrystals. J Am Chem Soc. 2002;124:11480–5.
61. Chen D, Zhang Y, Chen B, Kang Z. Coupling effect of microwave and mechanical forces during the synthesis of ferrite nanoparticles by microwave-assisted ball milling. Ind Eng Chem Res. 2013;52:14179–84.
62. Iida H, Takayanagi K, Nakanishi T, Osaka T. Synthesis of Fe3O4 nanoparticles with various sizes and magnetic properties by controlled hydrolysis. J Colloid Interface Sci. 2007;314:274–80.
63. Shen Z, Chen T, Ma X, Ren W, Zhou Z, Zhu G, Zhang A, Liu Y, Song J, Li Z, Ruan H, Fan W, Lin L, Munasinghe J, Chen X, Wu A. Multifunctional Theranostic nanoparticles based on exceedingly small magnetic Iron oxide nanoparticles for T1-weighted magnetic resonance imaging and chemotherapy. ACS Nano. 2017;11:10992–1004.
64. Barrow M, Taylor A, García Carrión J, Mandal P, Park BK, Poptani H, Murray P, Rosseinsky MJ, Adams DJ. Co-precipitation of DEAE-dextran coated SPIONs: how synthesis conditions affect particle properties, stem cell labelling and MR contrast. Contrast Media Mol Imaging. 2016;11:362–70.
65. Sato T, Iijima T, Seki M, Inagaki N. Magnetic properties of ultrafine ferrite particles. J Magn Magn Mater. 1987;65:252–6.
66. Berkowitz A, Schuele W, Flanders P. Influence of crystallite size on the magnetic properties of acicular γ-Fe2O3 particles. J Appl Phys. 1968;39:1261–3.
67. Morales M, Andres-Verges M, Veintemillas-Verdaguer S, Montero M, Serna C. Structural effects on the magnetic properties of γ-Fe2O3 nanoparticles. J Magn Magn Mater. 1999;203:146–8.
68. Coey JMD. Noncollinear spin arrangement in ultrafine ferrimagnetic crystallites. Phys Rev Lett. 1971;27:1140.
69. Sun S, Zeng H. Size-controlled synthesis of magnetite nanoparticles. J Am Chem Soc. 2002;124:8204–5.
70. Xie J, Xu C, Kohler N, Hou Y, Sun S. Controlled PEGylation of monodisperse Fe3O4 nanoparticles for reduced non-specific uptake by macrophage cells. Adv Mater. 2007;19:3163–6.
71. Sun S, Zeng H, Robinson DB, Raoux S, Rice PM, Wang SX, Li G. Monodisperse mfe 2o4 (m= fe, co, mn) nanoparticles. J Am Chem Soc. 2004;126:273–9.
72. Peng X, Wickham J, Alivisatos A. Kinetics of II-VI and III-V colloidal semiconductor nanocrystal growth: "focusing" of size distributions. J Am Chem Soc. 1998;120:5343–4.
73. O'Brien S, Brus L, Murray CB. Synthesis of monodisperse nanoparticles of barium titanate: toward a generalized strategy of oxide nanoparticle synthesis. J Am Chem Soc. 2001;123:12085–6.
74. Redl FX, Black CT, Papaefthymiou GC, Sandstrom RL, Yin M, Zeng H, Murray CB, O'Brien SP. Magnetic, electronic, and structural characterization of nonstoichiometric iron oxides at the nanoscale. J Am Chem Soc. 2004;126:14583–99.
75. Rockenberger J, Scher EC, Alivisatos AP. A new nonhydrolytic single-precursor approach to surfactant-capped nanocrystals of transition metal oxides. J Am Chem Soc. 1999;121:11595–6.
76. Samia AC, Hyzer K, Schlueter JA, Qin C-J, Jiang JS, Bader SD, Lin X-M. Ligand effect on the growth and the digestion of co nanocrystals. J Am Chem Soc. 2005;127:4126–7.

77. Li Y, Afzaal M, O'Brien P. The synthesis of amine-capped magnetic (Fe, Mn, Co, Ni) oxide nanocrystals and their surface modification for aqueous dispersibility. J Mater Chem. 2006;16:2175–80.
78. Jana NR, Chen Y, Peng X. Size-and shape-controlled magnetic (Cr, Mn, Fe, Co, Ni) oxide nanocrystals via a simple and general approach. Chem Mater. 2004;16:3931–5.
79. Zeng H, Rice PM, Wang SX, Sun S. Shape-controlled synthesis and shape-induced texture of MnFe2O4 nanoparticles. J Am Chem Soc. 2004;126:11458–9.
80. Maity D, Ding J, Xue J-M. Synthesis of magnetite nanoparticles by thermal decomposition: time, temperature, surfactant and solvent effects. Funct Mater Lett. 2008;1:189–93.
81. Kahlweit M. Ostwald ripening of precipitates. Adv Colloid Interf Sci. 1975;5:1–35.
82. LaMer VK, Dinegar RH. Theory, production and mechanism of formation of monodispersed hydrosols. J Am Chem Soc. 1950;72:4847–54.
83. Groult H, Poupard N, Herranz F, Conforto E, Bridiau N, Sannier F, Bordenave S, Piot J-M, Ruiz-Cabello J, Fruitier-Arnaudin I. Family of bioactive heparin-coated iron oxide nanoparticles with positive contrast in magnetic resonance imaging for specific biomedical applications. Biomacromolecules. 2017;18:3156–67.
84. Mahmoudi M, Sahraian MA, Shokrgozar MA, Laurent S. Superparamagnetic iron oxide nanoparticles: promises for diagnosis and treatment of multiple sclerosis. ACS Chem Neurosci. 2011;2:118–40.
85. Huang K-W, Chieh J-J, Yeh C-K, Liao S-H, Lee Y-Y, Hsiao P-Y, Wei W-C, Yang H-C, Horng H-E. Ultrasound-induced magnetic imaging of tumors targeted by biofunctional magnetic nanoparticles. ACS Nano. 2017;11:3030–7.
86. Wang Y-XJ, Hussain SM, Krestin GP. Superparamagnetic iron oxide contrast agents: physicochemical characteristics and applications in MR imaging. Eur Radiol. 2001;11:2319–31.
87. Lu M, Cohen MH, Rieves D, Pazdur R. FDA report: ferumoxytol for intravenous iron therapy in adult patients with chronic kidney disease. Am J Hematol. 2010;85:315–9.
88. Hufschmid R, Arami H, Ferguson RM, Gonzales M, Teeman E, Brush LN, Browning ND, Krishnan KM. Synthesis of phase-pure and monodisperse iron oxide nanoparticles by thermal decomposition. Nanoscale. 2015;7:11142–54.
89. Iqbal MZ, Ma X, Chen T, Zhang L e, Ren W, Xiang L, Wu A. Silica-coated super-paramagnetic iron oxide nanoparticles (SPIONPs): a new type contrast agent of T 1 magnetic resonance imaging (MRI). J Mater Chem B. 2015;3:5172–81.
90. Song Q, Zhang ZJ. Shape control and associated magnetic properties of spinel cobalt ferrite nanocrystals. J Am Chem Soc. 2004;126:6164–8.
91. Wetz F, Soulantica K, Falqui A, Respaud M, Snoeck E, Chaudret B. Hybrid co–au nanorods: controlling au nucleation and location. Angew Chem Int Ed. 2007;46:7079–81.
92. Puntes VF, Zanchet D, Erdonmez CK, Alivisatos AP. Synthesis of hcp-co nanodisks. J Am Chem Soc. 2002;124:12874–80.
93. Mao B, Kang Z, Wang E, Lian S, Gao L, Tian C, Wang C. Synthesis of magnetite octahedrons from iron powders through a mild hydrothermal method. Mater Res Bull. 2006;41:2226–31.
94. Zhu H, Yang D, Zhu L. Hydrothermal growth and characterization of magnetite (Fe3O4) thin films. Surf Coat Technol. 2007;201:5870–4.
95. Giri S, Samanta S, Maji S, Ganguli S, Bhaumik A. Magnetic properties of α-Fe2O3 nanoparticle synthesized by a new hydrothermal method. J Magn Magn Mater. 2005;285:296–302.
96. Sobal NS, Hilgendorff M, Moehwald H, Giersig M, Spasova M, Radetic T, Farle M. Synthesis and structure of colloidal bimetallic nanocrystals: the non-alloying system ag/co. Nano Lett. 2002;2:621–4.
97. Chen D, Xu R. Hydrothermal synthesis and characterization of nanocrystalline Fe3O4 powders. Mater Res Bull. 1998;33:1015–21.
98. Xuan S, Wang F, Wang Y-XJ, Jimmy CY, Leung KC-F. Facile synthesis of size-controllable monodispersed ferrite nanospheres. J Mater Chem. 2010;20:5086–94.
99. Wu W, Wu Z, Yu T, Jiang C, Kim W-S. Recent progress on magnetic iron oxide nanoparticles: synthesis, surface functional strategies and biomedical applications. Sci Technol Adv Mater. 2015;16:023501.

100. Xuan S, Wang Y-XJ, Yu JC, Cham-Fai Leung K. Tuning the grain size and particle size of superparamagnetic Fe3O4 microparticles. Chem Mater. 2009;21:5079–87.
101. Kim J, Tran VT, Oh S, Kim C-S, Hong JC, Kim S, Joo Y-S, Mun S, Kim M-H, Jung J-W. Scalable Solvothermal synthesis of superparamagnetic Fe3O4 nanoclusters for bio-separation and Theragnostic probes. ACS Appl Mater Interfaces. 2018;10:41935.
102. Jalajerdi R, Gholamian F, Shafie H, Moraveji A, Ghanbari D. Thermal and magnetic characteristics of cellulose acetate-Fe3O4. J Nanostruct. 2011;1:105–9.
103. Ghanbari D, Salavati-Niasari M. Hydrothermal synthesis of different morphologies of MgFe 2 O 4 and magnetic cellulose acetate nanocomposite. Korean J Chem Eng. 2015;32:903–10.
104. Ghanbari D, Salavati-Niasari M, Sabet M. Preparation of flower-like magnesium hydroxide nanostructure and its influence on the thermal stability of poly vinyl acetate and poly vinyl alcohol. Compos Part B. 2013;45:550–5.
105. Hedayati K, Goodarzi M, Ghanbari D. Hydrothermal synthesis of Fe3O4 nanoparticles and flame resistance magnetic poly styrene nanocomposite. J Nanostruct. 2017;7:32–9.
106. Li J, Pei Q, Wang R, Zhou Y, Zhang Z, Cao Q, Wang D, Mi W, Du Y. Enhanced photocatalytic performance through magnetic field boosting carrier transport. ACS Nano. 2018;12:3351–9.
107. Kim J, Tran VT, Oh S, Kim C-S, Hong JC, Kim S, Joo Y-S, Mun S, Kim M-H, Jung J-W, Lee J, Kang YS, Koo J-W, Lee J. Scalable Solvothermal synthesis of superparamagnetic Fe3O4 nanoclusters for bioseparation and Theragnostic probes. ACS Appl Mater Interfaces. 2018;10:41935–46.
108. West JL, Halas NJ. Applications of nanotechnology to biotechnology: commentary. Curr Opin Biotechnol. 2000;11:215–7.
109. Davis S. Biomedical applications of nanotechnology—implications for drug targeting and gene therapy. Trends Biotechnol. 1997;15:217–24.
110. Hussein AK. Applications of nanotechnology in renewable energies—a comprehensive overview and understanding. Renew Sust Energ Rev. 2015;42:460–76.
111. Li X, Zhang F, Zhao D. Lab on upconversion nanoparticles: optical properties and applications engineering via designed nanostructure. Chem Soc Rev. 2015;44:1346–78.
112. Pfeiffer C, Rehbock C, Hühn D, Carrillo-Carrion C, de Aberasturi DJ, Merk V, Barcikowski S, Parak WJ. Interaction of colloidal nanoparticles with their local environment: the (ionic) nanoenvironment around nanoparticles is different from bulk and determines the physico-chemical properties of the nanoparticles. J R Soc Interface. 2014;11:20130931.
113. Kelly KL, Coronado E, Zhao LL, Schatz GC. The optical properties of metal nanoparticles: the influence of size, shape, and dielectric environment. J Phys Chem B. 2003;107(3):668–77.
114. Gupta AK, Gupta M. Synthesis and surface engineering of iron oxide nanoparticles for biomedical applications. Biomaterials. 2005;26:3995–4021.
115. Xu C, Sun S. Monodisperse magnetic nanoparticles for biomedical applications. Polym Int. 2007;56:821–6.
116. Villanueva A, Cañete M, Roca AG, Calero M, Veintemillas-Verdaguer S, Serna CJ, del Puerto Morales M, Miranda R. The influence of surface functionalization on the enhanced internalization of magnetic nanoparticles in cancer cells. Nanotechnology. 2009;20:115103.
117. Arruebo M, Fernández-Pacheco R, Ibarra MR, Santamaría J. Magnetic nanoparticles for drug delivery. Nano Today. 2007;2:22–32.
118. Hugander A, Robins HI, Martin P, Schmitt C. Temperature distribution during radiant heat whole-body hyperthermia: experimental studies in the dog. Int J Hyperth. 1987;3:199–208.
119. Fortin J-P, Wilhelm C, Servais J, Ménager C, Bacri J-C, Gazeau F. Size-sorted anionic iron oxide nanomagnets as colloidal mediators for magnetic hyperthermia. J Am Chem Soc. 2007;129:2628–35.
120. Chung S, Hoffmann A, Bader S, Liu C, Kay B, Makowski L, Chen L. Biological sensors based on Brownian relaxation of magnetic nanoparticles. Appl Phys Lett. 2004;85:2971–3.
121. Kötitz R, Weitschies W, Trahms L, Brewer W, Semmler W. Determination of the binding reaction between avidin and biotin by relaxation measurements of magnetic nanoparticles. J Magn Magn Mater. 1999;194:62–8.

122. Soukup D, Moise S, Céspedes E, Dobson J, Telling ND. In situ measurement of magnetization relaxation of internalized nanoparticles in live cells. ACS Nano. 2015;9:231–40.
123. Dieckhoff J, Eberbeck D, Schilling M, Ludwig F. Magnetic-field dependence of Brownian and Néel relaxation times. J Appl Phys. 2016;119:043903.
124. Sharma P, Allison JP. Immune checkpoint targeting in cancer therapy: toward combination strategies with curative potential. Cell. 2015;161:205–14.
125. Wicki A, Witzigmann D, Balasubramanian V, Huwyler J. Nanomedicine in cancer therapy: challenges, opportunities, and clinical applications. J Control Release. 2015;200:138–57.
126. Qin S-Y, Zhang A-Q, Cheng S-X, Rong L, Zhang X-Z. Drug self-delivery systems for cancer therapy. Biomaterials. 2017;112:234–47.
127. Li S, Li C, Jin S, Liu J, Xue X, Eltahan AS, Sun J, Tan J, Dong J, Liang X-J. Overcoming resistance to cisplatin by inhibition of glutathione S-transferases (GSTs) with ethacraplatin micelles in vitro and in vivo. Biomaterials. 2017;144:119–29.
128. Espinosa A, Di Corato R, Kolosnjaj-Tabi J, Flaud P, Pellegrino T, Wilhelm C. Duality of iron oxide nanoparticles in cancer therapy: amplification of heating efficiency by magnetic hyperthermia and photothermal bimodal treatment. ACS Nano. 2016;10:2436–46.
129. Zhang Z, Wang J, Nie X, Wen T, Ji Y, Wu X, Zhao Y, Chen C. Near infrared laser-induced targeted cancer therapy using thermoresponsive polymer encapsulated gold nanorods. J Am Chem Soc. 2014;136:7317–26.
130. Singh A, Sahoo SK. Magnetic nanoparticles: a novel platform for cancer theranostics. Drug Discov Today. 2014;19:474–81.
131. Ling D, Lee N, Hyeon T. Chemical synthesis and assembly of uniformly sized iron oxide nanoparticles for medical applications. Acc Chem Res. 2015;48:1276–85.
132. Shi D, Sadat M, Dunn AW, Mast DB. Photo-fluorescent and magnetic properties of iron oxide nanoparticles for biomedical applications. Nanoscale. 2015;7:8209–32.
133. Zhou Q, Zhang B, Han D, Chen R, Qiu F, Wu J, Jiang H. Photo-responsive reversible assembly of gold nanoparticles coated with pillar [5] arenes. Chem Commun (Camb). 2015;51:3124–6.
134. Fantechi E, Innocenti C, Zanardelli M, Fittipaldi M, Falvo E, Carbo M, Shullani V, Di Cesare Mannelli L, Ghelardini C, Ferretti AM. A smart platform for hyperthermia application in cancer treatment: cobalt-doped ferrite nanoparticles mineralized in human ferritin cages. ACS Nano. 2014;8:4705–19.
135. Deatsch AE, Evans BA. Heating efficiency in magnetic nanoparticle hyperthermia. J Magn Magn Mater. 2014;354:163–72.
136. Pankhurst Q, Thanh N, Jones S, Dobson J. Progress in applications of magnetic nanoparticles in biomedicine. J Phys D Appl Phys. 2009;42:224001.
137. Hedayatnasab Z, Abnisa F, Daud WMAW. Review on magnetic nanoparticles for magnetic nanofluid hyperthermia application. Mater Des. 2017;123:174–96.
138. Conde J, Doria G, Baptista P. Noble metal nanoparticles applications in cancer. J Drug Deliv. 2012;2012:1.
139. Rosensweig RE. Heating magnetic fluid with alternating magnetic field. J Magn Magn Mater. 2002;252:370–4.
140. Garitaonandia JS, Insausti M, Goikolea E, Suzuki M, Cashion JD, Kawamura N, Ohsawa H, Gil de Muro I, Suzuki K, Plazaola F. Chemically induced permanent magnetism in Au, Ag, and Cu nanoparticles: iocalization of the magnetism by element selective techniques. Nano Lett. 2008;8:661–7.
141. Brezovich IA. Low frequency hyperthermia: capacitive and ferromagnetic thermoseed methods. Med Phys Mono. 1988;16:82–111.
142. Dennis C, Jackson A, Borchers J, Hoopes P, Strawbridge R, Foreman A, Van Lierop J, Grüttner C, Ivkov R. Nearly complete regression of tumors via collective behavior of magnetic nanoparticles in hyperthermia. Nanotechnology. 2009;20:395103.
143. Chatterjee J, Bettge M, Haik Y, Chen CJ. Synthesis and characterization of polymer encapsulated Cu–Ni magnetic nanoparticles for hyperthermia applications. J Magn Magn Mater. 2005;293:303–9.

144. Das R, Rinaldi-Montes N, Alonso J, Amghouz Z, Garaio E, García J, Gorria P, Blanco J, Phan M, Srikanth H. Boosted hyperthermia therapy by combined AC magnetic and photothermal exposures in Ag/Fe3O4 nanoflowers. ACS Appl Mater Interfaces. 2016;8:25162–9.
145. Jiang Q, Zheng S, Hong R, Deng S, Guo L, Hu R, Gao B, Huang M, Cheng L, Liu G. Folic acid-conjugated Fe3O4 magnetic nanoparticles for hyperthermia and MRI in vitro and in vivo. Appl Surf Sci. 2014;307:224–33.
146. Parchur AK, Sharma G, Jagtap JM, Gogineni VR, LaViolette PS, Flister MJ, White SB, Joshi A. Vascular interventional radiology-guided Photothermal therapy of colorectal Cancer liver metastasis with Theranostic gold Nanorods. ACS Nano. 2018;12:6597–611.
147. Han X, Huang J, Jing X, Yang D, Lin H, Wang Z, Li P, Chen Y. Oxygen-deficient Black Titania for synergistic/enhanced Sonodynamic and Photoinduced Cancer therapy at near infrared-II biowindow. ACS Nano. 2018;12:4545–55.
148. Saeed M, Iqbal MZ, Ren W, Xia Y, Liu C, Khan WS, Wu A. Controllable synthesis of Fe 3 O 4 nanoflowers: enhanced imaging guided cancer therapy and comparison of photothermal efficiency with black-TiO 2. J Mater Chem B. 2018;6:3800–10.
149. Gangopadhyay P, Gallet S, Franz E, Persoons A, Verbiest T. Novel superparamagnetic core (shell) nanoparticles for magnetic targeted drug delivery and hyperthermia treatment. IEEE Trans Magn. 2005;41:4194–6.
150. Martinez-Boubeta C, Simeonidis K, Serantes D, Conde-Leborán I, Kazakis I, Stefanou G, Peña L, Galceran R, Balcells L, Monty C. Adjustable hyperthermia response of self-assembled ferromagnetic Fe-MgO Core–Shell nanoparticles by tuning dipole–dipole interactions. Adv Funct Mater. 2012;22:3737–44.
151. Tian Q, Hu J, Zhu Y, Zou R, Chen Z, Yang S, Li R, Su Q, Han Y, Liu X. Sub-10 nm $Fe_3O_4$@ $Cu_{2-x}$ S Core–Shell nanoparticles for dual-modal imaging and Photothermal therapy. J Am Chem Soc. 2013;135:8571–7.
152. Tang J, Zhou H, Liu J, Liu J, Li W, Wang Y, Hu F, Huo Q, Li J, Liu Y. Dual-mode imaging-guided synergistic chemo-and magnetohyperthermia therapy in a versatile nanoplatform to eliminate cancer stem cells. ACS Appl Mater Interfaces. 2017;9:23497–507.
153. Di Corato R, Béalle G, Kolosnjaj-Tabi J, Espinosa A, Clement O, Silva AK, Menager C, Wilhelm C. Combining magnetic hyperthermia and photodynamic therapy for tumor ablation with photoresponsive magnetic liposomes. ACS Nano. 2015;9:2904–16.
154. Berry CC, Curtis AS. Functionalisation of magnetic nanoparticles for applications in biomedicine. J Phys D Appl Phys. 2003;36:R198.
155. Guibert C, Dupuis V, Peyre V, Fresnais J. Hyperthermia of magnetic nanoparticles: experimental study of the role of aggregation. J Phys Chem C. 2015;119:28148–54.
156. Georgiadou V, Tangoulis V, Arvanitidis I, Kalogirou O, Dendrinou-Samara C. Unveiling the physicochemical features of CoFe2O4 nanoparticles synthesized via a variant hydrothermal method: NMR relaxometric properties. J Phys Chem C. 2015;119:8336–48.
157. Kotoulas A, Dendrinou-Samara C, Sarafidis C, Kehagias T, Arvanitidis J, Vourlias G, Angelakeris M, Kalogirou O. Carbon-encapsulated cobalt nanoparticles: synthesis, properties, and magnetic particle hyperthermia efficiency. J Nanopart Res. 2017;19:399.
158. Neuberger T, Schöpf B, Hofmann H, Hofmann M, Von Rechenberg B. Superparamagnetic nanoparticles for biomedical applications: possibilities and limitations of a new drug delivery system. J Magn Magn Mater. 2005;293:483–96.
159. Chen X, Klingeler R d, Kath M, El Gendy AA, Cendrowski K, Kalenczuk RJ, Borowiak-Palen E. Magnetic silica nanotubes: synthesis, drug release, and feasibility for magnetic hyperthermia. ACS Appl Mater Interfaces. 2012;4:2303–9.
160. Beg S, Rizwan M, Sheikh AM, Hasnain MS, Anwer K, Kohli K. Advancement in carbon nanotubes: basics, biomedical applications and toxicity. J Pharm Pharmacol. 2011;63:141–63.
161. Zuo X, Wu C, Zhang W, Gao W. Magnetic carbon nanotubes for self-regulating temperature hyperthermia. RSC Adv. 2018;8:11997–2003.
162. Widder KJ, Senyei AE, Scarpelli DG. Magnetic microspheres: a model system for site specific drug delivery in vivo. Proc Soc Exp Biol Med. 1978;158:141–6.

163. Zhe Liu FK, Gätjens J. Advanced nanomaterials in multimodal imaging: design, functionalization, and biomedical applications. J Nanomater. 2010;2010:15.
164. Obeid MA, Al Qaraghuli MM, Alsaadi M, Alzahrani AR, Niwasabutra K, Ferro VA. Delivering natural products and biotherapeutics to improve drug efficacy. Ther Deliv. 2017;8:947–56.
165. Miele E, Spinelli GP, Miele E, Di Fabrizio E, Ferretti E, Tomao S, Gulino A. Nanoparticle-based delivery of small interfering RNA: challenges for cancer therapy. Int J Nanomedicine. 2012;7:3637.
166. Saadeh Y, Vyas D. Nanorobotic applications in medicine: current proposals and designs. Am J Robot Surg. 2014;1:4–11.
167. Holzinger M, Le Goff A, Cosnier S. Nanomaterials for biosensing applications: a review. Front Chem. 2014;2:63.
168. De Jong WH, Borm PJ. Drug delivery and nanoparticles: applications and hazards. Int J Nanomedicine. 2008;3:133.
169. Almalik A, Benabdelkamel H, Masood A, Alanazi IO, Alradwan I, Majrashi MA, Alfadda AA, Alghamdi WM, Alrabiah H, Tirelli N. Hyaluronic acid coated chitosan nanoparticles reduced the immunogenicity of the formed protein Corona. Sci Rep. 2017;7:10542.
170. Martens TF, Remaut K, Deschout H, Engbersen JF, Hennink WE, Van Steenbergen MJ, Demeester J, De Smedt SC, Braeckmans K. Coating nanocarriers with hyaluronic acid facilitates intravitreal drug delivery for retinal gene therapy. J Control Release. 2015;202:83–92.
171. Kolhar P, Anselmo AC, Gupta V, Pant K, Prabhakarpandian B, Ruoslahti E, Mitragotri S. Using shape effects to target antibody-coated nanoparticles to lung and brain endothelium. Proc Natl Acad Sci. 2013;110:10753–8.
172. Müller J, Bauer KN, Prozeller D, Simon J, Mailänder V, Wurm FR, Winzen S, Landfester K. Coating nanoparticles with tunable surfactants facilitates control over the protein corona. Biomaterials. 2017;115:1–8.
173. Gao W, Zhang L. Coating nanoparticles with cell membranes for targeted drug delivery. J Drug Target. 2015;23:619–26.
174. Gao H, Yang Z, Zhang S, Cao S, Shen S, Pang Z, Jiang X. Ligand modified nanoparticles increases cell uptake, alters endocytosis and elevates glioma distribution and internalization. Sci Rep. 2013;3:2534.
175. Lu H, Wang J, Wang T, Zhong J, Bao Y, Hao H. Recent progress on nanostructures for drug delivery applications. J Nanomater. 2016;2016:20.
176. Kumari A, Kumar V, Yadav S. Nanotechnology: a tool to enhance therapeutic values of natural plant products. Trends Med Res. 2012;7:34–42.
177. Patra JK, Das G, Fraceto LF, Campos EVR, del Pilar Rodriguez-Torres M, Acosta-Torres LS, Diaz-Torres LA, Grillo R, Swamy MK, Sharma S. Nano based drug delivery systems: recent developments and future prospects. J Nanobiotechnol. 2018;16:71.
178. Xu L, Qiu L, Sheng Y, Sun Y, Deng L, Li X, Bradley M, Zhang R. Biodegradable pH-responsive hydrogels for controlled dual-drug release. J Mater Chem B. 2018;6:510–7.
179. Al-Ahmady Z, Kostarelos K. Chemical components for the design of temperature-responsive vesicles as cancer therapeutics. Chem Rev. 2016;116:3883–918.
180. Zhang Z, Zhang D, Wei L, Wang X, Xu Y, Li H-W, Ma M, Chen B, Xiao L. Temperature responsive fluorescent polymer nanoparticles (TRFNPs) for cellular imaging and controlled releasing of drug to living cells. Colloids Surf B: Biointerfaces. 2017;159:905–12.
181. Bai Y, Xie F-Y, Tian W. Controlled self-assembly of Thermo-responsive amphiphilic H-shaped polymer for adjustable drug release. Chin J Polym Sci. 2018;36:406–16.
182. Anirudhan T, Nair AS. Temperature and ultrasound sensitive gatekeepers for the controlled release of chemotherapeutic drugs from mesoporous silica nanoparticles. J Mater Chem B. 2018;6:428–39.
183. Mathiyazhakan M, Wiraja C, Xu C. A concise review of gold nanoparticles-based photo-responsive liposomes for controlled drug delivery. Nano-Micro Lett. 2018;10:10.

184. Hervault A, Thanh NTK. Magnetic nanoparticle-based therapeutic agents for thermo-chemotherapy treatment of cancer. Nanoscale. 2014;6:11553–73.
185. Guo Y, Zhang Y, Ma J, Li Q, Li Y, Zhou X, Zhao D, Song H, Chen Q, Zhu X. Light/magnetic hyperthermia triggered drug released from multi-functional thermo-sensitive magnetoliposomes for precise cancer synergetic theranostics. J Control Release. 2018;272:145–58.
186. Ma G, Lin W, Yuan Z, Wu J, Qian H, Xu L, Chen S. Development of ionic strength/pH/enzyme triple-responsive zwitterionic hydrogel of the mixed L-glutamic acid and L-lysine polypeptide for site-specific drug delivery. J Mater Chem B. 2017;5:935–43.
187. Alonso J, Khurshid H, Devkota J, Nemati Z, Khadka NK, Srikanth H, Pan J, Phan M-H. Superparamagnetic nanoparticles encapsulated in lipid vesicles for advanced magnetic hyperthermia and biodetection. J Appl Phys. 2016;119:083904.
188. Grillo R, Gallo J, Stroppa DG, Carbó-Argibay E, Lima R, Fraceto LF, Bañobre-López M. Sub-micrometer magnetic nanocomposites: insights into the effect of magnetic nanoparticles interactions on the optimization of SAR and MRI performance. ACS Appl Mater Interfaces. 2016;8:25777–87.
189. Ulbrich K, Hola K, Subr V, Bakandritsos A, Tucek J, Zboril R. Targeted drug delivery with polymers and magnetic nanoparticles: covalent and noncovalent approaches, release control, and clinical studies. Chem Rev. 2016;116:5338–431.
190. Chen C-W, Syu W-J, Huang T-C, Lee Y-C, Hsiao J-K, Huang K-Y, Yu H-P, Liao M-Y, Lai P-S. Encapsulation of au/Fe 3 O 4 nanoparticles into a polymer nanoarchitecture with combined near infrared-triggered chemo-photothermal therapy based on intracellular secondary protein understanding. J Mater Chem B. 2017;5:5774–82.
191. Wu W, Jiang CZ, Roy VA. Designed synthesis and surface engineering strategies of magnetic iron oxide nanoparticles for biomedical applications. Nanoscale. 2016;8:19421–74.
192. Wahajuddin SA. Superparamagnetic iron oxide nanoparticles: magnetic nanoplatforms as drug carriers. Int J Nanomedicine. 2012;7:3445.
193. Prijic S, Sersa G. Magnetic nanoparticles as targeted delivery systems in oncology. Radiol Oncol. 2011;45:1–16.
194. El-Boubbou K. Magnetic iron oxide nanoparticles as drug carriers: clinical relevance. Nanomedicine. 2018;13:953–71.
195. Laurent S, Saei AA, Behzadi S, Panahifar A, Mahmoudi M. Superparamagnetic iron oxide nanoparticles for delivery of therapeutic agents: opportunities and challenges. Expert Opin Drug Deliv. 2014;11:1449–70.
196. Gao Z, Li Y, You C, Sun K, An P, Sun C, Wang M, Zhu X, Sun B. Iron oxide nanocarrier-mediated combination therapy of cisplatin and artemisinin for combating drug resistance through highly increased toxic reactive oxygen species generation. ACS Appl Bio Mater. 2018;1:270–80.
197. Kosmas C, Muñoz Estrella A, Sourlas A, Silverio D, Hilario E, Montan P, Guzman E. Inclisiran: a new promising agent in the management of hypercholesterolemia. Diseases. 2018;6:63.
198. Prilepskii AY, Fakhardo AF, Drozdov AS, Vinogradov VV, Dudanov IP, Shtil AA, Bel'tyukov PP, Shibeko AM, Koltsova EM, Nechipurenko DY. Urokinase-conjugated magnetite nanoparticles as a promising drug delivery system for targeted thrombolysis: synthesis and preclinical evaluation. ACS Appl Mater Interfaces. 2018;10:36764–75.
199. Hsiao M-H, Mu Q, Stephen ZR, Fang C, Zhang M. Hexanoyl-chitosan-PEG copolymer coated iron oxide nanoparticles for hydrophobic drug delivery. ACS Macro Lett. 2015;4:403–7.
200. Clary L, Verderone G, Santaella C, Vierling P. Membrane permeability and stability of liposomes made from highly fluorinated double-chain phosphocholines derived from diaminopropanol, serine or ethanolamine. Biochim Biophys Acta Biomembr. 1997;1328:55–64.
201. Gabizon A, Dagan A, Goren D, Barenholz Y, Fuks Z. Liposomes as in vivo carriers of adriamycin: reduced cardiac uptake and preserved antitumor activity in mice. Cancer Res. 1982;42:4734–9.

202. Immordino ML, Brusa P, Arpicco S, Stella B, Dosio F, Cattel L. Preparation, characterization, cytotoxicity and pharmacokinetics of liposomes containing docetaxel. J Control Release. 2003;91:417–29.
203. Zhao Y, Zhao X, Cheng Y, Guo X, Yuan W. Iron oxide nanoparticles-based vaccine delivery for Cancer treatment. Mol Pharm. 1791-1799;2018:15.
204. Mody KT, Popat A, Mahony D, Cavallaro AS, Yu C, Mitter N. Mesoporous silica nanoparticles as antigen carriers and adjuvants for vaccine delivery. Nanoscale. 2013;5:5167–79.
205. Gillies RJ, Bhujwalla ZM, Evelhoch J, Garwood M, Neeman M, Robinson SP, Sotak CH, Van Der Sanden B. Applications of magnetic resonance in model systems: tumor biology and physiology. Neoplasia (New York, NY). 2000;2:139–51.
206. Furman-Haran E, Schechtman E, Kelcz F, Kirshenbaum K, Degani H. Magnetic resonance imaging reveals functional diversity of the vasculature in benign and malignant breast lesions. Cancer. 2005;104:708–18.
207. Bowtell R. Colourful future for MRI. Nature. 2008;453:993.
208. Mornet S, Vasseur S, Grasset F, Veverka P, Goglio G, Demourgues A, Portier J, Pollert E, Duguet E. Magnetic nanoparticle design for medical applications. Prog Solid State Chem. 2006;34:237–47.
209. Hoehn M, Himmelreich U, Kruttwig K, Wiedermann D. Molecular and cellular MR imaging: potentials and challenges for neurological applications. J Magn Reson Imaging. 2008;27:941–54.
210. Alford R, Ogawa M, Choyke PL, Kobayashi H. Molecular probes for the in vivo imaging of cancer. Mol Bio Syst. 2009;5:1279–91.
211. Longmire M, Choyke PL, Kobayashi H. Dendrimer-based contrast agents for molecular imaging. Curr Top Med Chem. 2008;8:1180–6.
212. Bouzigues C, Gacoin T, Alexandrou A. Biological applications of rare-earth based nanoparticles. ACS Nano. 2011;5:8488–505.
213. Hanaoka K. Development of responsive lanthanide-based magnetic resonance imaging and luminescent probes for biological applications. Chem Pharm Bull. 2010;58:1283–94.
214. Na HB, Song IC, Hyeon T. Inorganic nanoparticles for MRI contrast agents. Adv Mater. 2009;21:2133–48.
215. Ivanuša T, Beravs K, Medič J, Serša I, Serša G, Jevtič V, Demsar F, Mikac U. Dynamic contrast enhanced MRI of mouse fibrosarcoma using small-molecular and novel macromolecular contrast agents. Phys Med. 2007;23:85–90.
216. Terreno E, Castelli DD, Viale A, Aime S. Challenges for molecular magnetic resonance imaging. Chem Rev. 2010;110:3019–42.
217. Ripoll J, Ntziachristos V, Cannet C, Babin AL, Kneuer R, Gremlich H-U, Beckmann N. Investigating pharmacology in vivo using magnetic resonance and optical imaging. Drugs RD. 2008;9:277–306.
218. Reilly RF. Risk for nephrogenic systemic fibrosis with gadoteridol (pro Hance) in patients who are on long-term hemodialysis. Clin J Am Soc Nephrol CJASN. 2008;3:747–51.
219. Reiter T, Ritter O, Prince MR, Nordbeck P, Wanner C, Nagel E, Bauer WR. Minimizing risk of nephrogenic systemic fibrosis in cardiovascular magnetic resonance. J Cardiovasc Magn Reson. 2012;14:31.
220. McCarthy JR, Weissleder R. Multifunctional magnetic nanoparticles for targeted imaging and therapy. Adv Drug Deliv Rev. 2008;60:1241–51.
221. Lawaczeck R, Menzel M, Pietsch H. Superparamagnetic iron oxide particles: contrast media for magnetic resonance imaging. Appl Organomet Chem. 2004;18:506–13.
222. Hu S-H, Gao X. Nanocomposites with spatially separated functionalities for combined imaging and Magnetolytic therapy. J Am Chem Soc. 2010;132:7234–7.
223. Jun Y-w, Choi J-s, Cheon J. Heterostructured magnetic nanoparticles: their versatility and high performance capabilities. Chem Commun. 2007:1203–14.

224. Fulton DA, O'Halloran M, Parker D, Senanayake K, Botta M, Aime S. Efficient relaxivity enhancement in dendritic gadolinium complexes: effective motional coupling in medium molecular weight conjugates. Chem Commun. 2005:474–6.
225. Haris M, Yadav SK, Rizwan A, Singh A, Wang E, Hariharan H, Reddy R, Marincola FM. Molecular magnetic resonance imaging in cancer. J Transl Med. 2015;13:313.
226. Singh N, Jenkins GJS, Asadi R, Doak SH. Potential toxicity of superparamagnetic iron oxide nanoparticles (SPION). Nano Rev. 2010;1 https://doi.org/10.3402/nano.v3401i3400.5358.
227. Veiseh O, Gunn JW, Zhang M. Design and fabrication of magnetic nanoparticles for targeted drug delivery and imaging. Adv Drug Deliv Rev. 2010;62:284–304.
228. Lin MM, Kim DK, Haj AJE, Dobson J. Development of Superparamagnetic Iron Oxide Nanoparticles (SPIONS) for translation to clinical applications. IEEE Trans Nano Biosci. 2008;7:298–305.
229. Nathan J, Wittenberg CLH. Using nanoparticles to push the limits of detection. Wiley Interdiscip Rev Nanomed Nanobiotechnol. 2009;1:237–54.

The manufacturer's authorised representative in the EU is Springer
Nature Customer Service Centre GmbH, Europaplatz 3, 69115 Heidelberg,
Germany. If you have any concerns regarding our products, please
contact ProductSafety@springernature.com

Printed and bound by CPI Group (UK) Ltd, Croydon, CR0 4YY
28/07/2026
02179235-0002